DISTANT WORLDS

DISTANT

WORLDS

MILESTONES IN PLANETARY EXPLORATION

PETER BOND

Copernicus Books
An Imprint of Springer Science+Business Media

In Association with
Praxis Publishing Ltd

Published in the United States by Copernicus Books, an imprint of Springer Science+Business Media.

Copernicus Books
Springer Science+Business Media
233 Spring Street
New York, NY 10013
www.springer.com

Library of Congress Control Number:
2006931779

Manufactured in China.
Printed on acid-free paper.

9 8 7 6 5 4 3 2 1

ISBN-10: 0-387-40212-8   e-ISBN-10: 0-387-68367-4
ISBN-13: 978-0-387-40212-3   e-ISBN-13: 978-0-387-68367-6

To Edna,
in memory of a kitchen conversation many years ago

CONTENTS

PREFACE

Until about 500 years ago, the Earth was believed to lie at the center of the Universe, with the Sun and five planets revolving around it. The planets themselves were merely points of light that drifted across the stellar constellations. Then came the invention of the telescope that enabled human eyes to see the planets as colorful disks, each with its own unique characteristics and quirks.

Fifty years ago, the population of the Solar System had swollen to include nine planets, 31 satellites and thousands of comets and asteroids. However, many mysteries remained. As recently as the early 1960s, scientists were still arguing about the existence of canals and vegetation on Mars, or the presence of oceans on Venus.

Today, the number of planets has risen to 10, the tally of satellites has passed 150 and the number of identified small objects is climbing rapidly as increasingly sensitive searches discover thousands of Sun-grazing comets and huge ice balls in the dark regions beyond Neptune.

During the first age of exploration, courageous navigators sailed the seven seas in search of new lands that would bring them fame and fortune. Now, with the exception of the ocean floors, there are few places on Earth that have not been explored. Nevertheless, our thirst for knowledge and desire to understand the unknown—characteristics that make our species unique—remain undiminished.

We are fortunate to be alive during the second great age of discovery, when modern technology is enabling us to construct automated spacecraft and robots that can take our place as explorers and ambassadors, venturing forth into the vast, hostile ocean of space to seek out and study new worlds.

For half a century, robotic spacecraft have been venturing vast distances to examine at close quarters all of the planets in our Solar System, with the exception of Pluto and its recently discovered, larger cousin in the far reaches of the Sun's realm. For the first time, human eyes have been able to see towering cliffs, dust devils, erupting volcanoes, dry river beds and ice formations on dozens of distant worlds, most of them totally alien to our experience here on Earth.

This book recounts the faltering, but inexorable, search for scientific truth and knowledge that, over thousands of years has enabled us to explore beyond the bounds of Earth and understand our place in the Universe.

Inevitably, much of the epic, long-running saga of Solar System exploration and discovery is devoted to the key missions that have unlocked the secrets of these strange worlds. A vast stream of data from a half dozen manned expeditions to the Moon and hundreds of robotic spacecraft has enabled scientists to assemble, piece by piece, a realistic picture of our planetary system.

However, as I have attempted to show in each chapter, long journeys often begin with a few faltering steps. This story of exploration would not be complete without reference to the research and insights of the early pioneers, people such as Aristarchus, Copernicus, Galileo, Kepler and Newton.

Many years ago, my imagination was captured by books that described the family of distant, alien worlds that circle our Sun. I have been hooked on the planets ever since. It is my hope that readers of this book will be similarly fascinated and inspired.

First considered 20 years ago, this book has eventually reached fruition with the support and encouragement of publisher Clive Horwood. My sincere thanks also go to John Mason for finding the time to read each chapter and make invaluable suggestions for improvement, and to Alex Whyte for his careful editing of the draft text.

Much of the information it contains has been provided by the public information officers and other employees of the major space agencies and companies, assembled over many years. Numerous other scientific sources—many now available on the Internet—are listed in the final pages. I am also very grateful to everyone who helped me to obtain the spectacular images that illuminate this story of outreach and discovery.

Finally, I would like to thank my wife, Edna, whose encouragement and patient forbearance enabled me to achieve a long-lasting ambition.

Peter Bond
Cranleigh, Surrey
September 2005

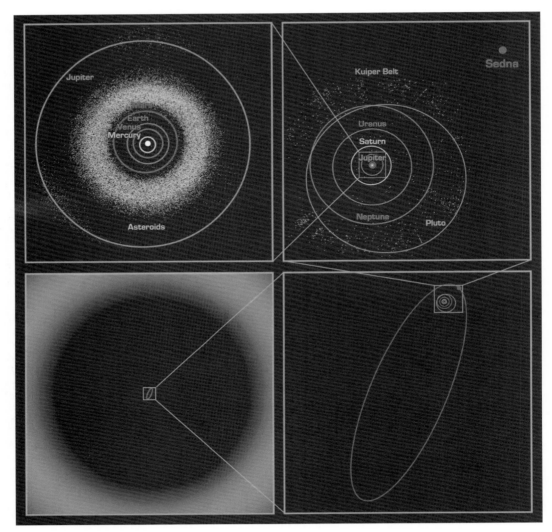

The Solar System—The scale of the Solar System as we know it today. The first panel (top left) shows the orbits of the inner planets and the asteroid belt between Mars and Jupiter. The second panel (top right) shows the outer planets and the Edgeworth–Kuiper Belt. The third panel (lower right) shows the orbit and current location of Sedna, which travels further from the Sun than any known object in the Solar System. The final box shows that even Sedna's orbit lies well inside the inner edge of the Oort cloud (shown in blue). (NASA/JPL/R. Hurt, SSC–Caltech)

1 DISCOVERING DISTANT WORLDS

If I can see further, it is because I stand on the shoulders of giants.

SIR ISAAC NEWTON

October 7, 1959, marked a new era in the history of humanity. On that day, 29 fuzzy black-and-white images trickled back to Earth from a Soviet spacecraft that had successfully looped around the Moon, venturing some 400,000 km (250,000 mi) from Earth. After staring at the wonders of the night sky for millions of years, technology had enabled human eyes to see an uncharted, extraterrestrial terrain—the far side of the Moon.

Since this first, faltering step, humanity has embarked on a wonderful episode of exploration that has enabled the inhabitants of our little blue world to unveil the secrets of seven planets and hundreds of smaller alien worlds that circle the Sun.

Hundreds of robotic ambassadors have been despatched across the Solar System to establish contact with every corner of the Sun's realm. Their electronic eyes have revealed alien landscapes that are far stranger than anything imagined by our forebears, and helped us to appreciate the fertile oasis in space that is our cradle and our home. Four spacecraft have even overcome the domineering gravitational grasp of our nearest star, carrying messages from Earth as they venture forth on never-ending voyages to distant star systems.

Since the pioneering breakthrough of Luna 3, we have discovered a menagerie of worlds unimagined only half a century ago. The most exotic of these include:

- Mercury—the little "winged messenger" flies around the Sun in just 88 days and, despite a midday temperature of 440°C, it may harbor water ice within its polar craters;
- Venus—a suffocating oven blanketed by sulfuric acid clouds and circled by super-rotating winds;
- Moon—the two-faced satellite that is a product of a cataclysmic planetary collision and innumerable asteroid impacts;
- Mars—an arid, icy world where liquid water once flowed and, perhaps, primitive life evolved;
- Jupiter—the gaseous king of the planets, home of a 300-year-old storm and ruler of more than 60 satellites;
- Io—a violent world of never-ending volcanic eruptions;
- Europa—a smooth ice world hiding a briny ocean;
- Saturn—the lightweight lord of the rings;
- Titan—a smog-shrouded giant where liquid methane takes the place of water;
- Uranus—a toppled giant where summers last for 21 years;
- Neptune—an icy giant dominated by huge storms and fierce, hurricane-force winds;
- Triton—where nitrogen ice is smudged by trails from alien geysers.
- Pluto and Charon—a double planet born in a swam of icy objects on the edge of the Solar System.

In many ways these are the stars of this story of discovery, but, in the words of Isaac Newton, the epic saga of robotic planetary exploration would not have been possible without the ability of modern scientists and engineers to "stand on the shoulders of giants."

The Wandering Stars

Until the advent of the Space Age, our view of the Universe around us had been severely restricted by the limitations of our vantage point and the difficulty in bridging the vast distances that separate Earth from its so-called neighbors. In the night sky, only our familiar Moon displayed a visible disk, its perfection spoiled by mysterious dark markings.

Seven star-like interlopers drifted among the fixed constellations, changing brightness and sometimes reversing direction. By the sixth century BC it was realized that the brilliant "evening star," known to the ancient Greeks as Hesperus, and the "morning star," known as Phosphorus, were one and the same. The same was true of the two most elusive wandering stars or "planets," Lucifer and Hermes, which always lingered close to the horizon at sunrise or sunset. This brought the number of starlike planets down to five. (A further three planets were subsequently discovered: Uranus in 1781, Neptune in 1846, and Pluto in 1930.)

Simply by studying the sky with the naked eye, it was possible to draw certain conclusions.

Since the Sun and Moon were the two brightest objects, dominating either the day or night sky, they were considered to be more important from a theological or astrological point of view.

It also seemed clear that they were closer to Earth than the planets because they displayed visible disks, and because they traveled across the sky more quickly. At times the Moon could be seen to move in front of a star or planet, thus hiding it from view or occluding it. On rare occasions the Sun could be eclipsed by the new Moon passing in front of it. In these ways the relative distances of the Sun, Moon and planets were established by the time of Aristotle in the fourth century BC.

The nearest body had to be the Moon, followed by the rapidly moving Venus and Mercury, which were clearly closely associated with the Sun. Further out were Mars and Jupiter, with Saturn as the outpost of the Solar System.

The absolute scale of the Universe was a major problem. To ancient civilizations the Earth seemed huge compared with all of the celestial objects, and since it was also the home of intelligent life, especially humans, it was assumed that Earth held a pre-eminent position. It also seemed obvious that, with the exception of mountains and valleys, the Earth was flat, while the Sun and Moon were round and the stars and planets were mere points of light.

The first civilizations of the eastern Mediterranean had knowledge of a very limited area, which suggested to them that the Earth was rectangular, though elongated in an east–west direction to allow for the Mediterranean Sea. The early Greek philosophers preferred some sort of flat disk or cylinder, with its rim marked by an expanse of water known as the "Ocean River." What supported this disk or cylinder caused a great deal of debate throughout the ancient world.

The Hindus, for example, believed that the Earth rested on four pillars that were based on the backs of elephants, which in turn stood on the back of a gigantic turtle swimming in a huge ocean. Beyond the ocean they did not go, merely saying that it was wicked to inquire further. Greek myths told of the giant Atlas who rebelled against Zeus and was punished by having to carry the Earth on his shoulders for the rest of time. By the sixth century BC, the Greek philosopher Thales was suggesting that the Earth floated in water.

From this time onward, ideas concerning the Earth began to change radically. Around 550 BC another Greek, Anaximander, proposed that the Earth was curved in a north–south direction, though it took another 100 years to progress to the idea of a spherical Earth. One piece of evidence was the fact that as ships sailed toward the horizon they did not simply appear smaller and smaller, which would be the case on a flat Earth, but they disappeared prow first, with the sails being visible longest. Furthermore, this happened no matter in which direction the ship was sailing. Also, if the Earth was flat, the same stars ought to be visible from anywhere on its surface, but travelers

Sizes of the Planets—The approximate sizes of the planets relative to the Sun. Outward from the Sun (left to right), they are Mercury, Venus, Earth, Mars, Jupiter, Saturn, Uranus, Neptune, and Pluto. Jupiter's diameter is about 11 times that of the Earth, and the Sun's diameter is about 10 times that of Jupiter. Pluto's diameter is slightly less than one-fifth that of Earth. The distances of the planets are not shown to scale. (Lunar and Planetary Laboratory)

reported that different stars appeared above the horizon if they journeyed north or south.

Another convincing piece of evidence involved lunar eclipses, when the full Moon became darker, often turning a deep orange. The Greeks finally realized that this temporary darkening could be explained by the Sun shining on the Earth and creating a shadow through which the Moon passed. Since the shadow always appeared curved in outline, Anaximander's cylinder was not an adequate explanation.

The first to realize that the Earth is a sphere was Philolaus, a follower of Pythagoras, in about 450 BC. Anaximander had already proposed the idea that the Earth was surrounded by empty space and kept its place "because of its equal distance from everything." These two advances in knowledge were tremendously important. The idea of "down" became relative to the observer, since all falling objects move toward the center of a spherical Earth, whether in Britain or Australia, and there is no question of people falling off the edge of the world.

If the Earth was spherical, how large was it? By 250 BC the known world stretched nearly 10,000 km (6,000 mi)

from the Atlantic Ocean to the borders of India, yet there was still no sign of the surface doubling back on itself, so it was obviously much larger than this. The actual size of the Earth was eventually calculated by another Greek, Eratosthenes, toward the end of the third century BC. He realized that the Sun was overhead at noon on June 21 (midsummer's day) at the Tropic of Cancer, but a little lower in the sky further north. After comparing the length of the shadow cast at two widely separated places—Syene (modern Aswan) and Alexandria, which lies on the coast of Egypt—he was able to use basic geometry to show that the circumference of the Earth must be about 40,000 km (25,000 mi).

Eratosthenes also knew how to calculate the diameter of the Earth from this and obtained the figure of about 12,500 km (8,000 mi), which is remarkably close to the true figure. Unfortunately, later astronomers revised his figures downward, and so when Columbus set sail for China he believed the voyage would be quite short, with no space available for an intervening continent in the form of America.

Once the size of the Earth had been determined, the next step was to discover the sizes and distances of the Sun and

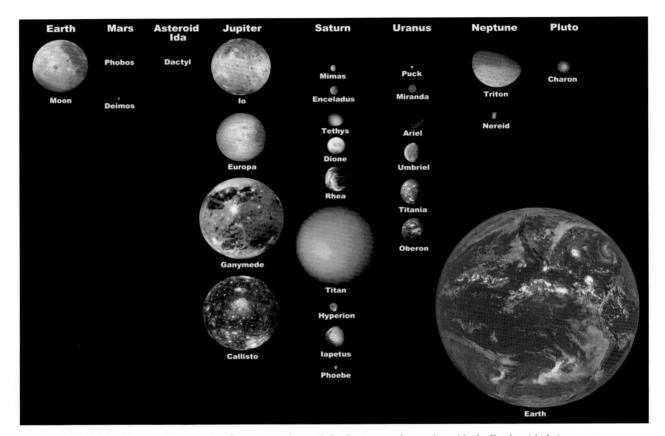

The Major Moons—The most significant moons in our Solar System are shown alongside the Earth, with their correct relative sizes and true color. Two of them (Ganymede and Titan) are larger than the planet Mercury, and eight are larger than Pluto. Earth's Moon is the fifth largest, with a diameter of 3,476 km (2,160 mi). Most are thought to have formed from a disk of debris left over from the formation of their home planet. However Triton and many of the smallest satellites are thought to be captured objects. Earth's Moon is thought to have formed from the debris ejected when a Mars-sized object collided with the young Earth. (NASA)

the Moon. In the fifth century BC the Greek philosopher Anaxagoras startled the court of Pericles by stating that the Sun was a "great hot stone" the size of Greece, a claim that led to his exile for blasphemy. However, studies of lunar eclipses soon indicated that this was a considerable underestimate rather than an overestimate.

At about the same time that Eratosthenes was working in Egypt, the Greek astronomer Aristarchus was making the first serious attempt to calculate the Moon's distance. By comparing the diameter of the Moon with that of Earth's shadow during a lunar eclipse, he decided that the Sun was 20 times further from Earth than the Moon and seven times

Earth's diameter. These were staggering figures at the time, for they began to suggest that perhaps the Earth was not as important in the Universe as had previously been supposed.

About 150 BC, Aristarchus' methods were refined by Hipparchus, who stated that the Moon's distance was 30 times Earth's diameter. Using Eratosthenes' calculation for the diameter of the Earth, this is very close to the actual figure. So by the mid-second century BC it was known that the Moon orbited the Earth at a distance of about 400,000 km (250,000 mi), and that the Sun was perhaps 8 million km (5 million mi) from the Earth. This meant that the celestial sphere that contained all the fixed stars had to be even further away. Man's horizons were beginning to expand considerably, while at the same time the once dominant Earth was shrinking in importance.

Unfortunately the five planets were posing problems that no one seemed able to solve satisfactorily. From the earliest times it was recognized that they varied in brightness, and it seemed logical to assume that they were brightest when they were nearest to the Earth (the astronomical term is "in opposition"), though this is not actually true in the cases of Venus and Mercury because they are lost in the Sun's glare when at their nearest to us. Such times were specially noted since the planets could be expected to affect human destiny most strongly when they were very bright and close.

After centuries of careful observation these occasions could be accurately predicted, but the planets' intermediate positions defied prediction. The main difficulty arose over the peculiar apparent motions of the planets across the sky. The stars were supposed to be attached to a crystal sphere that revolved daily around a central Earth, and the Sun, Moon, and each planet were also thought to be attached to seven smaller, rotating spheres.

The general movement of the stars was from east to west, while the "superior" planets (Mars, Jupiter, and Saturn) slowly shifted eastward against the background of the different Zodiac constellations. However, at times these planets seemed to come to a halt, then move in a reverse or retrograde direction for weeks or even months before resuming their eastward shift. The overall shape of their paths was a flattened circle or "loop," which was very

difficult to explain, especially if the Earth was at the center of the Universe. (One of the few astronomers to question the established beliefs was Aristarchus, but his view that Earth and the planets orbited the Sun was too advanced for the time.)

The Greeks made numerous attempts to explain this phenomenon. The first step was taken in the fourth century BC by Heraclides who realized that the sky only appeared to move slowly from east to west because the Earth is spinning in the opposite direction. However, he retained the Earth in its central position despite correctly placing Venus and Mercury in orbits around the Sun.

Ironically, the excellence of the Greeks as geometers proved a major obstacle to progress because they saw the circle as the perfect figure and, therefore, insisted that the Sun, Moon, and planets should all have circular orbits.

As observations became more accurate, the explanations put forward became increasingly complex and unrealistic. In the fourth century BC Eudoxus resorted to combinations of 30 circles to account for the apparent motions. A little later Aristotle proposed 55 circles. Even this proved inadequate, so that by the time of Hipparchus many more little circles (known as epicycles) had been added to the main circular orbits.

This system was modified by Claudius Ptolemy of Alexandria in about 140 AD so that it gave a reasonably accurate model for explaining the planetary motions, and his version became the universally accepted explanation for well over 1,000 years. Although observations continued during the following centuries, notably by the Arab and Chinese astronomers, no one dared to challenge the pre-eminence of Ptolemy. Until, that is, the science of astronomy was re-awakened by a Polish priest and astronomer named Nicolaus Copernicus.

The Renaissance

Although he was not a great observer, Copernicus gained a reputation as an original thinker. He led the study of

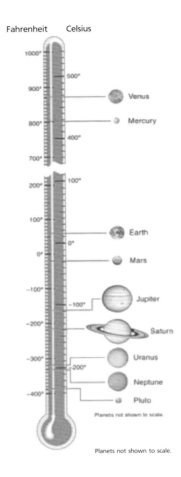

Planets not shown to scale.

The Largest Known Kuiper Belt Objects—Beyond the orbit of Neptune lie Pluto, its moon Charon, and millions of icy Kuiper Belt Objects. On July 29, 2005, astronomers announced that they had discovered a "tenth planet" that is larger than Pluto. Many more objects larger than Pluto may be found with increasingly powerful ground-based and orbital instruments. (NASA, ESA and A. Feild/ STScI)

Solar System Temperatures—In general, the surface temperature of the planets decreases with their distance from the Sun. Venus is an exception because its dense carbon dioxide atmosphere acts as a greenhouse. Mercury's night-side temperature is more than 500°C colder than the day-side temperature shown above. Temperatures for the giant outer planets are shown for an altitude in the atmosphere where pressure is equal to that at sea level on Earth. Earth lies in the center of the so-called habitable zone, where water can exist as a liquid and conditions are favourable to life. (Lunar and Planetary Institute) classical Greek astronomy and found support for his dissatisfaction with Ptolemy among the writings of ancient philosophers and scientists such as Philolaus and Aristarchus. Finally, he decided to rearrange the Solar System in a book called *De Revolutionibus Orbium Celestium* (On the Revolutions of the Celestial Spheres). Published just before his death in 1543, the book concluded that all the planets orbited the Sun, *not* the Earth as Ptolemy and most classical astronomers had insisted. The Earth was thus relegated to one of six planets, revolving on its axis once every 24 hours and orbited by only one body, the Moon.

Strangely, although the Copernican system became seen as a possible alternative to the Ptolemaic system, there was no violent reaction or revolution in scientific thought. Even Copernicus was obliged to resort to epicycles to explain the detailed planetary movements. The impasse lasted for more than half a century.

The next breakthrough was made in 1609 by a brilliant young German mathematician named Johannes Kepler. By a strange twist of irony, Kepler worked as an assistant to one of the leading opponents of the Copernican order, the Dane, Tycho Brahe. Able to benefit from the excellent observations recorded by Brahe, Kepler spent years laboriously calculating all the alternatives to circles and epicycles that he could think of in order to explain the looping motion of Mars.*

By this method of trial and error, he eventually realized that the planetary orbits were not combinations of circles, but simple, elongated circles known as ellipses. Planets would move faster when they were close to the Sun, and slow down when they were more remote. As a result, the outer planets would take longer to complete one orbit than their inner cousins.

Kepler's laws were eventually to transform orbital calculations. Unfortunately, although the orbital periods of the planets were well known by this time, there were no trustworthy figures for planetary distances. Nevertheless, an accurate picture of their spacing was available for the first time. Saturn, the most remote of the known planets, turned out to be nearly 10 times further from the Sun than Earth was. Since the actual distances in kilometers or miles were unknown, the standard unit of measurement became the "astronomical unit," which represented the Earth's mean distance from the Sun.

Kepler died in poverty in 1630, with the battle between the new ideas and the establishment still raging. By then, his laws and the Copernican theory in general had received visual confirmation through the pioneering telescopic observations of the Italian, Galileo Galilei. In the period 1609–1610, Galileo became the first human to record mountains and craters on the Moon, and to see the phases of Venus—proof that the planet moved around the Sun. Through his primitive instruments, the planets could be seen as spheres for the first time, rather than mere points of light.

Most significant of all was his discovery of four star-like objects in orbit around Jupiter. By watching their motions from day to day, he was able to calculate the time each moon took to complete one circuit. There could hardly be more convincing proof that the Earth was not at the center of the Universe and that everything did not revolve around our world. Unfortunately, the leaders of the Roman Catholic Church could not accept such evidence, obstinately continuing to support an Earth-centered Universe.

Finally, in 1633 Galileo was brought before the Inquisition and forced to recant under threat of torture. According to legend, he still managed to mutter, "Nevertheless it (the Earth) moves!" At least he did not undergo the ordeal of another Copernican follower, Giordano Bruno, who was burned to death for his beliefs. However, as more people acquired telescopes and saw the undeniable evidence for themselves, it became clear that the Church would have to modify its views.

As a result of the work of these pioneers, the Universe was better understood than ever before, and most of the irregular motions of the planets could be accounted for. The next step was to find an explanation for Kepler's laws.

---

* He was also the first to coin the term "satellite" for planetary moons, and wrote one of the first science fiction novels, the *Somnium*, which described an imaginary trip to the Moon.

Although Galileo conducted gravitational experiments—almost certainly *not* including the dropping of two objects of different weights from the top of the Leaning Tower of Pisa—he did not realize the full significance of his discoveries. This was left to an Englishman, Isaac Newton, who was born in 1642, the year that Galileo died.

Unlike many such stories in history, that of the apple falling from the tree acting as the source of inspiration for Newton's discovery of universal gravitation may contain some substance. Certainly, by 1684 Newton was able to explain planetary motions. His first premise was that all objects attract each other, and that this gravitational attraction is proportional to their mass. Clearly, since the Sun has nearly all of the mass in the Solar System, it should pull all of the other bodies into it. Newton explained that this did not happen because their orbital velocities are just sufficient to counteract the Sun's gravity. The result is that the planets fall toward the Sun in such a way that the curve

of their fall takes them completely around it. (The same explanation, of course, applies to artificial satellites.)

Newton's second fundamental law was that gravitational attraction decreases with distance. For example, if planet A is twice as far from the Sun as planet B, then the gravitational force exerted by the Sun on planet A is one-quarter of that exerted on planet B. In practical terms, this means that a satellite in low Earth orbit must travel at 8 km/s (5 mi/s), whereas the Moon only has to circle the Earth at 1 km/s (0.6 mi/s) in order to avoid crashing into our planet. Similarly, the planets further from the Sun need only move relatively slowly around their orbits compared with those in the inner Solar System. Newton's laws also explained why a planet speeded up as it approached perihelion (closest point to the Sun) and slowed down near aphelion (furthest point from the Sun).

From this time on, the Solar System was very well understood. The main difficulties involved minor variations

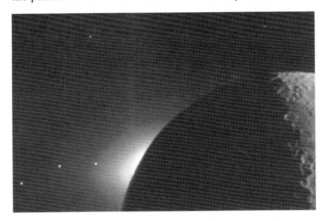

The Ecliptic—The plane of the ecliptic is illustrated in this image taken by the Clementine spacecraft, which shows (from right to left) the Moon lit by Earthshine, the Sun's corona rising over the Moon's dark limb, and the planets Saturn, Mars, and Mercury. Most of the planets orbit near this plane, since they were formed from a spinning, flattened disk of dust and gas. The main exception is Pluto, whose orbit is tilted at 17 degrees to the ecliptic. Many comets and asteroids follow paths that are even more steeply inclined. (NASA/SDIO)

The Formation of the Planets—The planets formed about 4.5 billion years ago from a huge nebula—cloud of gas and dust—that surrounded the young Sun. Within a few million years, colliding particles in the nebula snowballed until sizeable boulders appeared. Further collisions caused these to grow into the planets we see today. Some of these, further from the Sun, were able to pull in huge atmospheres of hydrogen and helium. Those in the warmer, inner regions were made of rock rather than ices and light gases. The remnants formed comets and asteroids. (NASA/JPL–Caltech/ T. Pyle, SSC)

in orbits caused by gravitational interactions between the planets, particularly those involving massive Jupiter. Careful study of such irregularities in the orbit of Uranus even enabled the position of an unknown planet, Neptune, to be successfully calculated (see Chapter 10)—although there are those who consider the discovery to be pure chance.

Unfortunately, the inner planet, Mercury, refused to follow the orbit predicted by Newton's laws, even when the perturbations caused by other planets were taken into account. This led the French astronomer Le Verrier, who had so successfully forecast the location of Neptune in 1846, to predict the presence of another new planet orbiting between Mercury and the Sun. He was so sure that it would be found that he named it Vulcan, but no such planet was ever discovered, despite several false alarms.

The explanation of Mercury's strange behaviour had to wait until 1915, when Albert Einstein produced his revolutionary *General Theory of Relativity*. He knew that mutual perturbations by the planets resulted in a gradual change in the direction of the axis joining the positions of aphelion and perihelion—the two extremes of an elliptical orbit. For Mercury this precession of perihelion should be 532 seconds of arc per century, but measurements showed that it is actually 574 arc seconds per century, indicating an extra factor not included in Newton's theory.

Einstein's theory stated that the mass of a body increases with an increase in velocity. In relation to the planets, this means that their masses increase as they accelerate toward perihelion, which in turn causes a slight increase in the curvature of the orbit. The extra shift of about 42 arc seconds per century for the orbit of Mercury was exactly predicted by Einstein's theory, as were the similar but smaller shifts of the perihelia of Venus and the Earth. The reasons their shifts are smaller are that their orbits are not as eccentric as that of Mercury, and their orbital velocities are lower.

Scaling the Solar System

As mentioned above, Kepler's third law enabled the scale of the Solar System to be calculated for the first time in terms of astronomical units, but the actual distances of the planets remained doubtful for some time, since no one could find an accurate method of measuring even one planetary distance.

In the second half of the seventeenth century, astronomers turned to a basic method first used by the ancient Greeks. This involved a principle called parallax, the apparent shift in position of an object when viewed from two different locations. To illustrate this, hold one finger upright in front of your face and close first one eye and then the other. The finger seems to shift position against the background, although it is, of course, stationary. When the finger is moved closer, the shift appears larger, and vice versa.

Astronomers realized that, if a shift in a planet's position could be detected from two widely separated observatories due to parallax effects, then its distance could be calculated. This method was first used by a French astronomer, Jean Richer, working in Cayenne (French Guiana) together with Giovanni Domenico Cassini and Jean Picard in Paris. They made simultaneous parallax observations of Mars during its opposition in 1671, using the newly invented pendulum clocks to ensure that the measurements were made at precisely the same moment. A by-product of this experiment was the discovery that a pendulum swung more slowly at Cayenne than at Paris, showing that gravity is slightly weaker at the equator. Newton later used this result to show that the Earth is flattened at the poles.

Cassini's calculations led to a value of about 140 million km (87 million mi) for the astronomical unit, a reasonably accurate result which meant that the actual distances of the Sun and planets were approximately known for the first time.

During the eighteenth century a great deal of time and money was spent in attempting to increase the accuracy of these figures. One method was to observe transits of Venus across the face of the Sun from all over the world. These events are rare because Venus's orbit is slightly inclined to that of Earth, but two transits took place very close together in 1761 and 1769. On the latter occasion, the British explorer Captain Cook was one of 150 observers scattered across the globe. Their task was to note the exact time that Venus made apparent contact with the Sun's disk, the time it appeared to separate from the disk, and the duration of the transit. Unfortunately, the results proved disappointing, partly because of a "black drop effect" which made it impossible to detect when the Sun and the planet came into contact.

More successful was the world-wide effort in 1931 to determine the parallax, and therefore the distance, of the near-Earth asteroid Eros. Highly accurate measurements were possible since Eros has no atmosphere and appears as a mere point of light in even the largest telescopes. The mean Earth–Sun distance or astronomical unit was found to be 149.6 million km (93 million mi).

More recently still, incredibly accurate figures for the distances of the planets have been obtained by bouncing radio waves off their surfaces. Since the velocity of these microwaves is known and the time taken between emission and reception can be measured to a fraction of a second, the distance can be readily calculated.

Once the distance of an object is accurately known, its diameter can be determined from its apparent size, as seen in a telescope. Unfortunately, this is very difficult for the smaller or more distant members of the Solar System, particularly if their albedo, or surface reflectivity, is uncertain. For example, the diameters of some members of the Edgeworth–Kuiper Belt, beyond the orbit of Pluto, have "shrunk" as new observations indicate that their surface brightness is greater than previously believed.

Another method, involving the occultation of a star by a planet or other object, is especially valuable in relation to bodies that are normally difficult to observe. The planet's diameter is calculated from the length of time during which it hides the star from view. Unfortunately, for a small distant body such as Pluto these events are very rare. (The

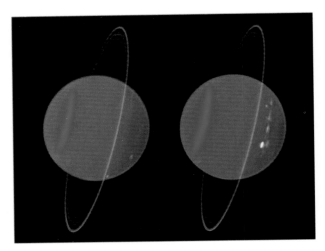

Stormy Uranus—These images of both hemispheres of Uranus were taken at near-infrared wavelengths with the 10-m (33-ft) Keck telescope in Hawaii in July 2004. With the aid of an adaptive optics system that corrects most of the atmospheric effects that blur viewing, they reveal a major increase in storm activity since the Voyager 2 flyby in 1986. More than 30 cloud features are visible, exceeding the total number observed up to the year 2000. The planet's northern hemisphere (right) is coming out of its long period of winter darkness. The rings are shown in red. (Lawrence Sromovsky, UW–Madison Space Science and Engineering Center)

blinking in and out of a star also led to the discovery of the rings of Neptune.)

Radar has also been a valuable astronomical tool since the 1960s, when the first views of the hidden surface of Venus were obtained by analyzing echoes from our mysterious planetary neighbor. However, some of the most sensational results have involved asteroids that cross the Earth's orbit.

By bouncing radio waves off the surfaces of objects, Steve Ostro, a pioneering research scientist at NASA's Jet Propulsion Laboratory, has been able to reveal the sizes and shapes of hundreds of asteroids. In place of the points of light seen in the best optical telescopes, his radar experiments have revealed exotic shapes, such as the dog bone configuration of asteroid Kleopatra and the elongated

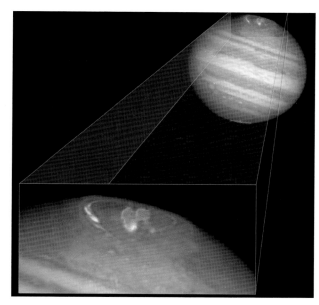

An X-ray Mystery—Observations at different wavelengths enable astronomers to learn more about the physical processes taking place on planets. This image combines X-ray data from Chandra (magenta), ultraviolet data from the Hubble Space Telescope (blue), and a Hubble visible light image of Jupiter. As a result, astronomers were able to discover a fixed, pulsating hot spot of X-rays in the upper atmosphere, near Jupiter's north magnetic pole. The cause of this feature is unclear. (X-ray: NASA/SWRI/R. Gladstone et al.; UV: NASA/HST/J. Clarke et al.; Optical:NASA/HST/R. Beebe et al.)

shape of Geographos. Others have been found to have unusual motions, such as the slow wobbling of Toutatis. The observations have even identified some asteroids as metallic, some as unconsolidated heaps of rubble, and some as pairs in orbit around each other.

Ground-based observations have also advanced in leaps and bounds during the last few decades. Today, amateurs using telescopes equipped with charge-coupled devices and image-processing software can produce pictures that far surpass those generated by the world's largest professional instruments in the 1960s.

Meanwhile, telescopes in the 8- and 10-m (26- and 33-ft) class, such as the Kecks in Hawaii and the Very Large Telescope in Chile, have been equipped with adaptive optics that overcome the turbulence of Earth's atmosphere. Although much of their time is allocated to observations of distant galaxies and quasars, astronomers are now able to obtain stunning views of storms on Uranus and Neptune in place of the fuzzy images of a few decades ago.

However, the ultimate Solar System observatory is not perched on a remote mountain top, but in a 600-km (380-mi) orbit around the Earth. Since 1990, the Hubble Space Telescope has been used to study every planet apart from Mercury (which is too near the glare of the Sun for safe observation). Highlights of its 15 years of Solar System exploration include images of shrinking ice caps and dust storms during the changing Martian seasons; remarkable views of a planet-girdling storm on Saturn; glowing polar auroras on Jupiter and Saturn; shifting storms on wind-swept Neptune and topsy-turvy Uranus; and the first maps of Pluto and Charon.

Observations at "invisible" wavelengths have also become commonplace. Ground-based telescopes have been able to take advantage of "windows" in the atmosphere that allow them to observe in the infrared—particularly valuable when attempting to unravel the temperature or composition of a planet. Radio telescopes can detect signals involving interactions between magnetic fields and charged particles, such as the remarkable tube of high-voltage plasma that links Jupiter and its volcanic moon Io.

Up above, a number of Earth-orbiting observatories (most notably Hubble and the International Ultraviolet Explorer) have been able to study the planets and their companions in infrared and ultraviolet light. Recently, spacecraft such as Rosat, Chandra, and XMM-Newton have given astronomers X-ray vision, revealing unexpected high-energy emissions from comets and planets, including various auroral light shows and hot spots on the gas giants.

The Space Age

Remarkable though they are, modern telescopes and Earth-orbiting observatories inevitably have their limitations, the most obvious being the vast distances that separate them from their targets. Even the Hubble Space Telescope's sharpest photo of Mars, taken during the favorable opposition of August 2003, could only reveal small craters and other surface markings a few tens of kilometers (about a dozen miles) across. The only solution is to despatch automated spacecraft to survey distant worlds from close range.

Since the pioneering success of the Soviet Luna 3 in 1959, more than 200 robotic ambassadors from Earth have ventured forth to explore remote members of the Sun's realm. At first, it was a struggle for the primitive rockets to even boost the spacecraft to escape velocity, and many would-be probes were lost during launch or failed to reach their targets.

Considering its relative proximity, it was not surprising that the first objective was the Moon. Over the six-year period from 1959 to 1965, two dozen spacecraft flew past Earth's sole natural satellite or crashed into its cratered surface. The lunar program gained tremendous impetus on May 25, 1961, when President Kennedy committed his country to a 25-billion-dollar program that would challenge the existing Soviet supremacy in space.

In his address to Congress, the President declared: "I believe that this nation should commit itself to achieving the goal, before this decade is out, of landing a man on the Moon and returning him safely to Earth. No single space project in this period will be more impressive to mankind, or more important in the long-range exploration of space; and none will be so difficult or expensive to accomplish."

Approval was not universal. An informal coalition of scientists decried the decision, arguing that the "First Man on the Moon" program would swallow up huge sums and provide a poorer return than more modest investment in unmanned scientific missions—a debate that has raged ever since.

Nevertheless, the political drive for international prestige and superpower supremacy became the dominating factor over the next decade. Ironically, as a by-product of the fevered preparations to send astronauts to another world, the science programs of the United States and the Soviet Union became major beneficiaries of the huge concentration of national resources on human spaceflight.

An armada of robotic spacecraft set sail for the Moon in order to prepare the way for their more fragile human successors. First came the American Rangers and the early Soviet Lunas, which sent back preliminary data during kamikaze dives onto the rugged landscape. By the mid-1960s, technology had evolved sufficiently for orbiters to map the entire surface in a search for suitable sites that astronauts could explore. Meanwhile, sophisticated soft-landers provided ground truth, sending back information on surface roughness, soil strength, and composition.

Fortunately, improvements in rocketry and spacecraft design also brought the planets within reach. After a series of Soviet failures, NASA finally made the breakthrough on December 14, 1962, when Mariner 2 completed the first successful flyby of Venus. Among the data that streamed back to Earth at 8.3 bits/s was the first confirmation that Venus was a hell-hole with a surface temperature of at least 425°C.

However, progress was slow, and five years passed before Mariner 5 made a much closer fly by of cloud-shrouded Venus. Not until 1970 did the Soviets succeed in delivering an operational capsule to that planet's sizzling surface, signaling that its atmospheric pressure was a crushing 90 times greater than on Earth. Having perfected the survival technique, the Soviets returned to Venus at every opportunity during the 1970s and early 1980s. Their remarkably robust vehicles survived long enough to send back panoramic views of a ruddy, rock-strewn terrain that resembled rugged terrestrial volcanic landscapes.

In contrast to their successes at Venus, the Soviets suffered many failures when they attempted to repeat the feat at Earth's other neighbor, Mars. Of 18 attempts over 36 years, none could be said to have completely fulfilled its objectives.

Curiously, Mars has proved to be the most challenging

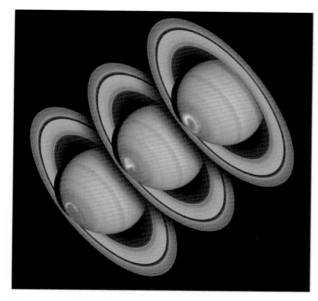

Shimmering Lights—These images of the aurora above Saturn's south pole were taken by the Hubble Space Telescope on three days in January 2005. Each image combines views of the ultraviolet auroral emissions with visible wavelength images of the planet and rings. The images were obtained during a joint campaign with the Cassini spacecraft to measure the interaction of the solar wind with the giant planet. The strong brightening of the aurora on January 26 (top right) corresponded with the arrival of a large disturbance in the solar wind. (NASA/STScI/Z. Levay and J. Clarke)

of all planetary targets. NASA, too, has suffered its fair share of disappointments and failures, interspersed between some wonderful technological and scientific triumphs. Of particular interest is the way in which the scientific consensus has changed over the years as new images and data from Mars have winged their way back to Earth. In the early 1960s, there was still a sizeable community that considered it possible for Mars to support some primitive form of vegetation. This optimistic viewpoint was severely compromised by the first Mariners, which returned pictures of an ancient, cratered landscape not too different from that of the arid, lifeless Moon.

The pendulum of opinion experienced a major swing in the opposite direction when Mariner 9 outlasted the dust storm of 1971 to reveal a world blessed with country-sized volcanoes, a rift valley system large enough to cross continents, and numerous dry channels that testified to a warmer, wetter past. The possibility of rainfall and melt-water feeding rivers that flowed into ancient oceans inevitably led to speculation that Martian organisms could have evolved in such relatively benign conditions. Unfortunately, these hopes were dashed in 1976, when experiments on board the Viking landers failed to find any compelling evidence for the existence of such life forms.

The first golden age of planetary exploration came to an end in the 1980s. After almost 100 Solar System missions in the 1960s and over 50 such missions in the 1970s, the number of flights plummeted to just 16 during the fourth decade of the Space Age. There were many reasons for this. On both sides of the East–West divide, budgets for space exploration nosedived. With national priorities once again directed toward programs involving development of human-rated space shuttles and space stations, unmanned missions had to settle for a small slice of a much more modest cake.

The precarious situation in the US planetary community was highlighted by NASA's decision not to take advantage of the 1-in-76-year opportunity to investigate Halley's comet during its return to the inner Solar System. While the European Space Agency, Japan, and the Soviet Union worked together on a multinational investigation of this primordial planetary building block, the United States had to settle for observations by Earth-orbiting astronauts and a handful of other non-dedicated spacecraft, including one in orbit around Venus.

It is also true to say that all of the "easy" missions had been flown, and the most obvious targets had been met by the 1980s. After the first flybys of the Moon, Venus, and Mars during the 1960s, the level of difficulty had escalated to include orbiters and landers. Once Mariner 10 completed three flybys of Mercury in 1974–1975, the initial reconnaissance of the inner Solar System was complete.

Scientists' attention inevitably turned toward the huge worlds that lingered mysteriously in the dark outer reaches of the Solar System, beyond the asteroid belt. First to penetrate the millions of rocky objects that inhabit the space between Mars and Jupiter were two American Pioneers, each carrying a plaque to inform curious aliens of its terrestrial origin.

The first golden age culminated in 1977 in most spectacular style with the launch of two spacecraft on a grand tour of the gas giants. Initially targeted only at Jupiter and Saturn, the nuclear-powered Voyagers just kept on going, with Voyager 2 eventually providing human eyes with their first close-up glimpses of the Solar System's frigid outer worlds, Uranus and Neptune, along with their menagerie of icy moons and coal-black rings. Today, Earth-based antennas can still detect their feeble signals as they probe the boundary with interstellar space, more than 14 billion km (8.8 billion mi) from the Sun.

The successes of the Voyagers, the Vikings, the Mariners, and the Veneras provided the basis of today's revived planetary exploration programs. The first step was the utilization of radar to pierce the cloud layers that blanket Venus and had frustrated astronomers for centuries. After the pioneering efforts of the Soviet Veneras 15 and 16, the US Magellan orbiter completed the mapping of Earth's half sister.

Meanwhile, the Voyagers' tantalizing glimpses of Jupiter and its planet-sized satellites, followed by scintillating images of Saturn and its smog-shrouded satellite, Titan, inspired scientists to seek further missions that would provide deeper insights into the workings of these complex systems.

The first of the new generation of planetary explorers was NASA's Galileo, which was deployed from the Shuttle's cargo bay in October 1989. After almost six years of bouncing around the inner Solar System in an effort to pick up speed through gravity assists from Earth and Venus, Galileo finally arrived at Jupiter in December 1995. After releasing an instrumented probe into the planet's dense cloud decks, Galileo made history by becoming the first spacecraft to enter orbit around a gas giant. Despite the handicap of a main antenna that failed to open, scientists and engineers managed to retrieve hundreds of images and other data that revolutionized

Cassini–Huygens—The spacecraft sent to explore the Solar System today are much larger and more sophisticated than their forebears, often involving international collaboration. Cassini–Huygens weighed in at more than 5.6 tonnes. NASA's Cassini orbiter will study the Saturn system for at least four years, while ESA's Huygens probe (gold dish) landed on the surface of Titan in January 2005. The Cassini orbiter has 12 instruments and the Huygens probe had six. The instruments often have multiple functions, enabling 27 different science investigations. Power is provided by three Radioisotope Thermoelectric Generators. (NASA–JPL)

our knowledge of the king of the planets and its entourage.

An even more ambitious mission set off for Saturn in 1997. Once again, the sheer size of the Cassini spacecraft necessitated a roundabout route to its target, with the orbiter finally arriving at the ringed planet in July 2004.

Since then, it has already begun to realize its promise to unveil the secrets of the multiple rings, the tiny shepherd satellites, and the wind-swept atmosphere. Intriguing Titan is also a primary objective, with a European-built probe making a daring parachute descent onto its icy surface, and instruments on the orbiter attempting to penetrate the all-embracing orange haze.

Faster, Better, Cheaper

Unfortunately, while these "Battlestar Galactica" style ships fulfill scientists' dreams of multi-instrumented missions that can simultaneously investigate every conceivable aspect of a planet's environment, space agencies can only afford to fly a few of these every decade. Scientists may have to wait many years for a rare opportunity to carry out their research.

The possibility of an expensive failure has also driven agencies away from this one-off approach, in which a single malfunction of a spacecraft can be catastrophic. One catalyst for this policy shift was NASA's billion-dollar Mars Observer, which disappeared without trace on arrival at the Red Planet in 1993.

Coinciding with this embarrassing and expensive loss was the installation of new administrator, Dan Goldin, who was charged to increase efficiency and reliability at the agency. Impressed with the success of a Department of Defense/NASA collaboration on the cut-price Clementine lunar orbiter, Goldin introduced a new mantra, "faster, better, cheaper," in an effort to meet demands for more frequent missions that could be flown, despite limited funding.

Goldin's approach included the Discovery program, which required innovative planetary missions to be developed for less than $245 million, with spacecraft development capped at $150 million. In addition, the missions, which would fly every 12–18 months, would have to use a rocket no larger than a Delta 2, and aim for launch within three years of approval for development. A similar Mars program was introduced, in which modest, targeted missions would be flown during each launch window.

Universities, government laboratories, and aerospace companies responded enthusiastically to NASA requests for proposals, leaving agency officials spoiled for choice during the annual selection process. The program got off to a flying start with the successes of Mars Pathfinder, with its small automated rover, and Mars Global Surveyor, which achieved many of the Mars Observer's goals at a fifth of the cost. These were followed by Lunar Prospector, which mapped the Moon and found evidence for the existence of water ice at the lunar poles, and NEAR, which became the first spacecraft to orbit and land on an asteroid.

At the same time, NASA introduced the New Millennium program, whose primary function was to test new technologies. The first planetary mission in the program, known as Deep Space 1, carried 12 innovative, advanced technologies, most notably an ion propulsion system that was 10 times more fuel-efficient than traditional chemical rocket engines, and a Remote Agent artificial intelligence program that was intended to operate and control the spacecraft with the minimum of human assistance. This was to be followed by Deep Space 2, which would fire two penetrators into Mars, and Deep Space 4/Champollion, which would drop a lander onto comet Tempel 1.

The wisdom of this quick-fire, piecemeal approach was driven home in 1996 when one of the largest planetary exploration spacecraft ever launched ended up in the Pacific Ocean. The Russian Mars-96 spacecraft, which included an orbiter, two small surface stations, and two penetrators, proved to be the swansong of the once powerful Soviet program to explore the Solar System. Since then, Russian investigators have been restricted to playing minor roles in other nations' endeavors.

A few years later, some equally high-profile failures forced NASA to reconsider its policy. The first warning sign came in 1997, when Mars Global Surveyor flirted with disaster after one of its solar panels failed to latch properly and nearly snapped during the commencement of aero-braking in the Red Planet's upper atmosphere. The spacecraft had to be carefully nursed for one and a half years before it finally reached its operational orbit.

In December 1998, the NEAR spacecraft was almost

lost when its engine misfired. Fortunately, the rendezvous with asteroid Eros was merely delayed, and the mission eventually became a resounding success. A few months later, Deep Space 1 made the closest ever flyby of an asteroid; but as its camera was pointing the wrong way, and only a few long-range images were returned to Earth. Once again, the flaws were corrected, enabling the spacecraft to complete an historic encounter with Comet Borrelly in September 2001.

Then, in late 1999, two Mars missions were lost in quick succession. First, Mars Climate Orbiter disappeared as it prepared to brake into orbit, then contact was lost with Mars Polar Lander (along with the Deep Space 2 penetrators). Particularly embarrassing was the revelation by the Climate Orbiter Board of Inquiry that NASA and prime contractor Lockheed Martin had been using different

units of measurement. As a result, the spacecraft had hit the atmosphere at too low an angle and burned up.

While the shock waves reverberated around NASA, discussions focused around whether the agency's cost-cutting efforts had gone too far. A Mars lander scheduled for launch in 2001 was canceled, and the agency considered pruning one of two Mars Exploration Rovers penciled in for 2003. Eventually, managers agreed to take a more pragmatic approach, with mission success taking a higher priority, even if it meant increased expenditure in the preparatory stage. Even so, the rover teams complained that they were being asked to sandwich development of the $400-million vehicles into 34 months, about one third of the time normally allocated for such a complex mission.*

Into the Twenty-First Century

Mars moved even further up the ladder of importance in January 2004, when President George W. Bush unveiled his administration's new vision of Solar System exploration. According to the new approach, NASA would be required to retire the Shuttle by 2010, redirecting its resources into returning humans to the Moon and preparing for a human expedition to Mars by 2035.

Exactly how this revisionary approach impacts America's future Solar System program remains to be seen, although robotic missions to the Moon and Mars will obviously be given high priority. At the same time, there is some concern among scientists about how this change in priorities and major redistribution of funds will affect future projects that are not related to human spaceflight. Meanwhile, Europe, Russia, and Japan have been considering for some years how they might participate in an

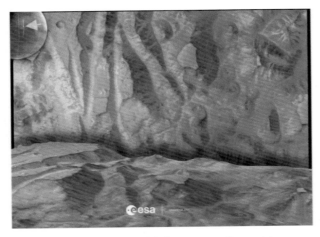

Martian Canyons in Stereo—Modern spacecraft are giving us an entirely new view of our planetary neighbors. This composite view of Martian canyons and mesas was taken by the High Resolution Stereo Camera on board ESA's Mars Express orbiter on January 14, 2004. It shows one section of a 1,700-km (1,060-mi) long and 65-km (40-mi) wide swath across the Valles Marineris. It was the first 3D color image of this size to show the surface of Mars in high resolution. The perspective view (bottom) was computer-generated from the original data. The globe (top left) shows the location on the Martian surface. (ESA/DLR/FU Berlin—G. Neukum)

* Another Discovery mission, known as Genesis, almost ended in disaster when its parachute failed to open during its return to Earth in September 2004. Its fragile solar wind collectors shattered into a multitude of fragments.

international endeavor to send astronauts beyond low-Earth orbit. These efforts are continuing, notably under ESA's Aurora program, which plans to send a rover to search for life on Mars around 2011.

One of the most notable trends of recent years has been the willingness of Europe and Japan to initiate ambitious robotic missions of planetary exploration. ESA's Mars Express orbiter has already sent back stunning stereo images of the Red Planet and conducted the first radar search for subsurface water. Another resounding success was Europe's Huygens probe, which touched down on Saturn's moon Titan amid a plethora of icy pebbles scattered across a dry river bed carved by liquid methane.

Although Japan's first Mars orbiter, Nozomi, was irreparably damaged by a solar flare, and its Lunar-A mission to fire penetrators into the Moon has been grounded by technical problems, the eastern space power continues to pursue a program that will send future craft to the Moon and Venus. Meanwhile, the exploration community is about to be enlarged with the entrance of India and China. As high-profile demonstrations of their growing technological prowess and economic advancement, these future Asian powerhouses intend to pursue programs of extraterrestrial exploration, beginning with the launches of two Moon orbiters in 2007.

At the same time, even Russia is expressing its desire to launch a home-grown mission to the Martian moon Phobos in 2009. The mission's objectives are to collect soil samples from the small satellite and return them to Earth for analysis, as well as to study the atmosphere and surface of the Red Planet.

As this brief summary demonstrates, mankind's desire to reach for the stars and explore to the furthest outposts of the Solar System seems certain to continue for decades to come. If this search for knowledge leads to a better understanding of our own fragile world, then so much the better. What other surprises and revelations await remains to be seen. The discovery of life on Mars or Europa perhaps—the long-awaited revelation that we are not alone in the vast Universe?

One day, these pioneering endeavours will enable future generations to leave Earth behind and establish a new cradle of civilization on some alien world. Meanwhile, we can only marvel at the thousands of philosophers, scientists, engineers, and dreamers whose efforts over more than two millennia have enabled us to understand and explore our small corner of the Universe.

At the same time, it is salutary to remember that our planet—the "pale blue dot" described by Carl Sagan—is a mere speck in the vast ocean of space. If the Earth's distance from the Sun could be reduced to a mere 15 cm (6 in), then Pluto, on the outer edge of the Solar System, would be nearly 6 m (20 ft) from the Sun. On this scale Earth would be almost invisible (about 0.1 mm) and even mighty Jupiter would only be the size of a grain of sand. Spaceship Earth, for so long thought to be the center of the Universe, is actually just one small world out of millions that populate the Milky Way galaxy.

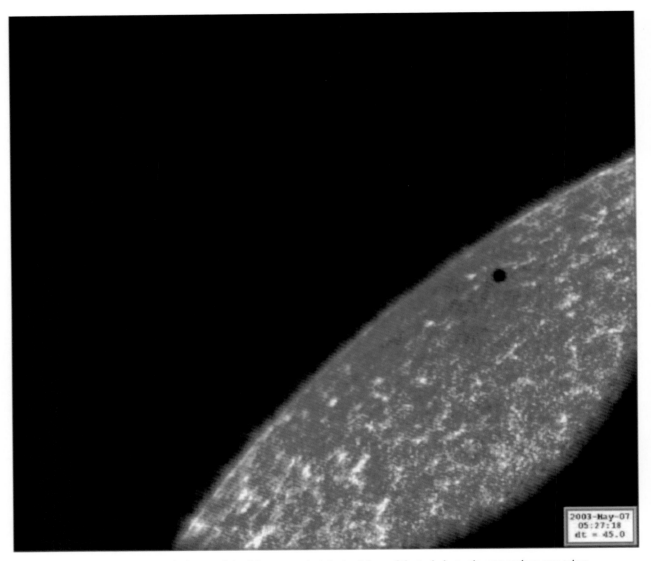

2003-May-07
05:27:18
dt = 45.0

2003 Mercury Transit—An image of tiny Mercury against the backdrop of the turbulent solar atmosphere was taken by the TRACE satellite on May 7, 2003. The sharp outline of Mercury seen during transits clearly shows that the planet has no appreciable atmosphere. (NASA)

2 MERCURY: THE IRON PLANET

The Winged Messenger

Mercury is one of Earth's cosmic neighbors, regularly approaching to within 80 million km (50 million mi)—closer than any planet other than Venus and Mars. At such times, it becomes a tiny beacon that rivals Sirius, the brightest star in the sky. Yet, although the existence of this elusive little planet has been recognized since prehistoric times, it remains one of the most mysterious inhabitants of our Solar System.

Part of the explanation for this lack of knowledge lies in its proximity to the Sun. We never see the planet more than 28 degrees from the Sun—a little more than the width of a fully spread hand at arm's length. At such times, it peeps above the horizon for about two hours after sunset or before sunrise—although for much of this short window of opportunity it is invisible in the solar glare. Since Mercury's orbit is highly elliptical, some less favorable elongations occur at a mere 18 degrees.

Mercury also changes position so quickly that it is only visible for a few weeks before being lost once more in the brilliant solar glare. Not surprisingly, the Sun's faithful companion confused ancient skywatchers, who believed that separate, twin-like planets periodically appeared alongside the Sun in the morning or evening.

The morning star that climbed above the eastern horizon shortly before sunrise was known to the Greeks as Apollo, after the Sun god. Its counterpart that appeared in the west at the end of the day was named after Hermes, the winged messenger of the gods. Not until around 550 BC was it realized that these two "wandering stars," so similar in behavior and appearance, yet never visible on the same day, just had to be one and the same. So it was that Hermes (now known to us by the name of Mercury, his Roman counterpart) came to be recognized as one of the five naked-eye planets recorded in pre-telescopic days.

With its insistence on lingering in the twilight, it was clear that Mercury must lie closer to the Sun than any of the other planets. We now know that it approaches to within 46 million km (29 million mi) of the star's scorching surface at perihelion, although at its furthest point (aphelion) it is about 69.8 million km (44 million mi) away—half as far again. This unusual orbit means that an observer on Mercury would see the dazzling Sun grow from twice to three times its apparent size as seen from Earth over a period of just six weeks.

As its name implies, Mercury is the fastest moving of all the planets. By traveling at an average speed of 48 km/s (30 mi/s), the little world manages to overcome the enormous gravitational pull of its gigantic neighbor and avoids falling into the Sun. Like an athlete on the inside track, it is able to overtake all of its more distant cousins, winging its way around the Sun in just 88 Earth days. Speedy Mercury finishes four circuits of the Sun before the more leisurely Earth completes one.

However, Mercury's eccentric (elongated) orbit means that its speed through space varies greatly, ranging from a relatively modest 39 km/s (24 mi/s) at aphelion to 56 km/s (35 mi/s) during its sunward sweep. Not only is Mercury's orbit far from circular, but it is tilted by 7 degrees to the ecliptic plane, so that it dips far above and below the paths of almost all the other planets. Indeed, only Pluto follows a more stretched and steeply inclined orbit.

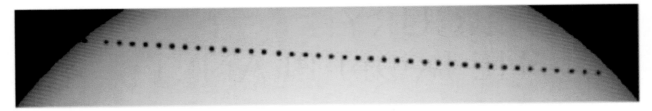

Transit of Mercury—Although it orbits closer to the Sun than any other planet, transits of Mercury across the face of the Sun are rare, occurring only a dozen times or so per century. This picture is a time sequence of images taken by the ESA–NASA SOHO spacecraft during the transit of May 7, 2003. The entire journey across the face of the Sun lasted about 5½ hours. (ESA–NASA)

Despite Mercury's frequent passages between the Sun and the Earth, the different orbital inclinations of the two planets mean that we rarely see it pass across the fiery face of our star. Indeed, Nature's planetary timetable means that such transits can only take place in May or November. The next is scheduled for November 8, 2006. On these special occasions, Mercury appears as a small black disk, so tiny that it cannot be seen with the naked eye.

Nineteenth-century observers also noticed that Mercury's orbit drifts in space, so that the point of closest approach to the Sun noticeably shifts as time goes by. Although similar orbital shifts (known as precession) occur on a much smaller scale for the other planets, the advance of Mercury's orbit was too large to be explained by Newtonian physics.

One explanation, put forward in 1859 by the Director of the Paris Observatory, Urbain Jean Joseph Le Verrier, was that some unknown object must be pulling little Mercury off its path. (Neptune had been discovered only 13 years earlier as the result of a similar calculation.) His theory seemed to be vindicated soon after when a doctor and amateur observer named Edmond Lescarbault reported seeing an 'intra-Mercurian' planet pass across the face of the Sun.

Confident that his hidden attractor had been discovered, the jubilant Le Verrier named the new planet Vulcan. He calculated that its average distance from the Sun was 21 million km (13 million mi), so that it completed one orbit every 19 days and 17 hours. Sadly, his prediction that Vulcan would transit the Sun once more on March 22, 1877, failed to materialize, and, although some astronomers still believe that "Vulcanoid" asteroids may exist inside Mercury's orbit, we now know that Le Verrier's planet does not exist.

The explanation for Mercury's errant orbit had to wait until 1915, when Albert Einstein published his *General Theory of Relativity*. One of its predictions was that a large, massive body—in this case, the Sun—will significantly curve the space around it. In the case of Mercury, such warping of space will be most marked when the planet is at perihelion, so causing the orbit to drift like an errant hula-hoop.

Mysterious Mercury

Even in the dark, unpolluted skies enjoyed by ancient observers, Mercury was difficult to observe. These problems did not go away with the invention of the telescope in the early seventeenth century, but the new instrument did make it possible for Italian observer Giovanni Zupi to confirm Galileo's suspicion that Mercury displays a succession of different illuminated phases, similar to those of the Moon. This is because, like Venus, Mercury is an "inferior" planet: it orbits between the Earth and the Sun.

When Mercury is on the far side of the Sun, in the position known as superior conjunction, it appears as a tiny "full" disk, though it can never be properly viewed because

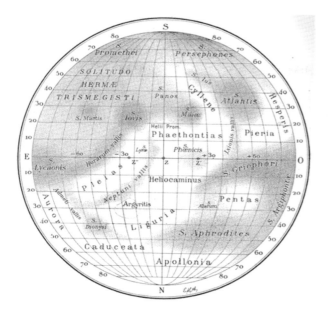

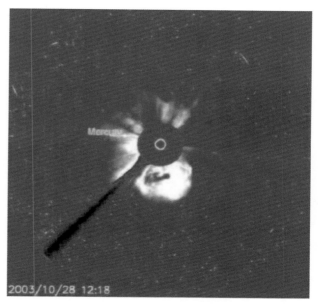

Antoniadi's Chart—Greek astronomer Eugène Antoniadi produced this map of Mercury, based on telescopic observations made during daylight hours between 1924 and 1929. Note that north is at the bottom—matching the inverted view obtained with the 84-cm (33-in) refractor at the Observatory of Meudon. Although most of the shaded regions bear little resemblance to actual features, the names have been adopted where practical on modern maps. However, the grid coordinates are very different from those used today. (Royal Astronomical Society)

Solar Impact—Mercury imaged by the SOHO spacecraft during a huge coronal mass ejection (CME) from the Sun on October 28, 2003. The planet's sparse atmosphere allows charged particles from the solar wind and spectacular eruptions such as CMEs to impact the planet's surface. Ground-based observations indicate that atomic oxygen, sodium, and potassium, as well as atomic hydrogen and helium, are mixed in with the dusty surface and then slowly released, so creating a tenuous atmosphere. The sodium and potassium are probably sprayed off the surface by the impact of high energy solar wind particles. (NASA–ESA)

of the intervening glare of the Sun. As Mercury approaches the Earth, one would expect more surface detail to become visible. Unfortunately, as Mercury appears to move further from its dazzling neighbor and becomes brighter to the naked eye, the illuminated area also diminishes. At greatest elongation, the planet resembles a quarter Moon, then, as it becomes ever nearer to the Earth, it shrinks to a slim crescent. Finally, during its closest approach (inferior conjunction), Mercury is invisible—unless it makes a rare transit across the solar disk.

Ironically, although it orbits closer to the Sun than any other planet, Mercury only crosses the face of the Sun on rare occasions—only a dozen times or so per century. This is because of the fairly steep inclination of the planet's orbit to the plane of the ecliptic, so Earth, Mercury and the Sun rarely line up when the innermost planet is at inferior conjunction.

Transits of Mercury can only occur during May or November. The May events take place when Mercury is near aphelion, while the autumn transits occur when it is near perihelion. The interval between successive crossings varies from 3 to 13 years, most recently with transits in 2003 and 2006, to be followed by another pair in 2016 and 2019.

Another problem in trying to discover the secrets of Mercury is its modest size. Indeed, until the late 1970s it was thought to be the smallest of the nine planets. We now know that Mercury is not only smaller than all of its planetary cousins other than Pluto (which some argue is not really a planet at all), but it is also dwarfed by Ganymede and Titan, two moons of Jupiter and Saturn. With a diameter of 4,878 km (3,030 mi), Mercury would fit nicely into the Atlantic Ocean between Europe and North America.

The result is that, even at its closest, Mercury appears smaller in a telescope than any of the planets apart from Pluto and the remote giants Uranus and Neptune. Features less than about 300 km (190 mi) across cannot be resolved in even the largest telescopes.

Not surprisingly, the best maps of Mercury until 1974 merely showed vague variations in shading, though markings along the terminator—the boundary between night and day—suggested a mountainous or cratered surface. Some observers suggested that Mercury might resemble the Moon, since studies of the way the surface reflected sunlight suggested a similar type of fragmented material. Others disagreed, arguing that its greater distance from the asteroid belt would have minimized the impact rate, producing a surface that exhibited fewer craters than either Mars or the Moon.

Although some observers noted whitish features that they described as mists or thin clouds, there was no direct evidence of an atmosphere: for example, no blurring of the planet's edge had been noticed during transits. Furthermore, with a mass only 5% that of the Earth and an escape velocity of 4.25 km/s (2.6 mi/s) it seemed clear that Mercury could not hold on to an atmosphere, even if it originally had one. Without any appreciable atmosphere, the sky would appear black and full of stars even in the middle of the day, and there would be no natural protection from the Sun's deadly ultraviolet radiation.

An apparent breakthrough came when Italian astronomer Giovanni Schiaparelli published a chart showing linear features and dark areas that no one had seen before. Schiaparelli's technique involved observing the planet's pale disk in full daylight during favorable apparitions between 1881 and 1889. By studying the changing patterns in his sketches, he came to the surprising conclusion that Mercury keeps one hemisphere turned permanently toward the Sun, while the other is condemned to perpetual darkness. According to Schiaparelli, Mercury's day was equal to its year—approximately 88 Earth days.

Using the 61-cm (24-in) refractor at Flagstaff, millionaire businessman and planetary astronomer Percival Lowell supported Schiaparelli's conclusions, though his credibility was not improved when he drew a map covered with dark lines and patches.

In the early twentieth century, Greek-born Eugène Antoniadi—again observing in broad daylight—used the 84-cm (33-in) refracting telescope at Meudon to produce a remarkably detailed map of Mercury. The various shades of gray showed "permanent" features to which he grandly donated Latin names. With Antoniadi also coming down on the side of an 88-day synchronized spin, the question of Mercury's rotation seemed to be settled.

If this arrangement was correct, then Mercury could justly be described as a world of two halves, with one hemisphere that is permanently exposed to the blazing Sun, in contrast to the never-ending cold and darkness of the planet's shady side. Scientists speculated that a twilight zone along the day–night terminator might offer a narrow, habitable haven.

The Twilight Zone

Mercury's proximity to the Sun means that solar radiation is up to 11 times stronger than on Earth, explaining why the planet's day side is much hotter than a domestic oven set at maximum. On the other hand, the shaded night side would be expected to be much cooler—but for how long?

Right up until the early 1960s, Mercury was portrayed as a planet with one hemisphere baked incessantly by a scorching Sun that always dominated the sky. Thermocouple measurements made at Mt Wilson Observatory by

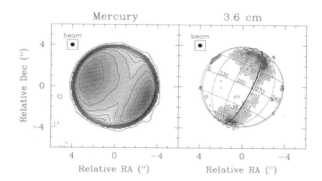

Hot Spots—Mercury's two hot spots (dark red) are visible in the left-hand image, which shows microwave energy escaping to space from a depth of about 70 cm (28 in) beneath the surface. The hot spots appear on both the day (left) and night (right) hemispheres, and coincide with regions which are exposed to maximum solar radiation. The Caloris Basin is located near the spot on the right. In the right-hand image, the cooler (blue) regions lie at the poles and in shadows cast by hills and craters along the morning terminator (bold line, Sun is to the right). (David Mitchell)

Edison Pettit and Seth Nicholson indicated a maximum surface temperature of 410°C, hot enough to melt zinc, tin, and lead. In contrast, with the presumed absence of an atmosphere to distribute heat and a rotation that always kept one hemisphere directed away from the Sun, temperatures in the region of permanent night were expected to plummet to around –150°C.

The only respite would come in a twilight zone, 320 km (200 mi) wide, that stretched along the terminator—the zone where day merged into night and temperatures would be less extreme. The likely existence of such a temperate zone between the scorching day side and the frigid nocturnal hemisphere inevitably led to excited speculation by science fiction writers who envisaged human explorers landing there, possibly to discover alien life forms.

The first small doubt arose in 1962, when studies of radio waves from Mercury suggested that the dark side was warmer than expected. However, reluctant to discard their existing theories, astronomers explained the observations away by assuming that, after all, Mercury had an atmo-

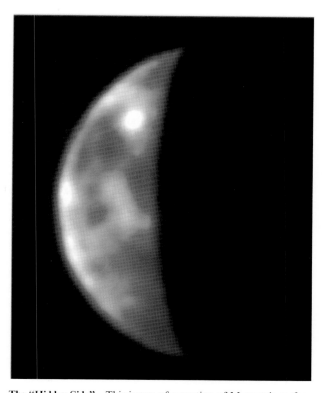

The "Hidden Side"—This image of a portion of Mercury's surface not photographed by Mariner 10 was obtained in August 1998 by Boston University astronomers using the Mt Wilson Observatory. Hundreds of thousands of pictures taken with short time exposures (1/60 second) were examined to find the 30 clearest images taken during instances of "perfect seeing" through the Earth's atmosphere. There seems to be little correlation between bright features in the visible image and radar-bright features. (Baumgardner et al., Boston University)

sphere capable of transporting heat from one side to the other.

It was not until 1965 that the debate was finally laid to rest, when Gordon Pettengill and Rolf Dyce used the world's largest radio telescope, the 305-m (1,000-ft) Arecibo dish in Puerto Rico, to bounce radar waves off the limbs of the planet. The frequency of the returned signals indicated that the planet spins once every 59 days or so—a far cry from the synchronous rotation that had been so widely accepted for more than 70 years.

The error of Schiaparelli and his successors was soon explained by Giuseppe "Bepi" Colombo, a distinguished Italian specialist in celestial mechanics, who noted that the 59-day rotation period was about two-thirds of the Mercurian year. In other words, Mercury rotates three times while circling the Sun twice—an example of what scientists call "spin–orbit coupling."

Colombo's explanation was confirmed by further radar studies, which showed that Mercury spins once every 58.6 Earth days. As a result, it presents the same face to Earth-bound observers during each alternate orbit. Presented with familiar, dusky markings at many apparitions, astronomers had tended to ignore any occasional variations and jumped to the wrong conclusion.

Astronomers now believe that Mercury once rotated much faster than it does today, but tidal drag created by the overwhelming gravity of the Sun slowed it over the aeons until its spin became locked with the orbital period. This unusual coupling arrangement leads to some strange side-effects.

One "day" for an inhabitant of Mercury (from sunrise to sunrise) would last for 176 Earth days or two Mercurian years. The slow rotation period and the lack of an atmosphere mean that Mercury has the harshest thermal environment of any planet, with a day–night variation of up to 600°C. A thermometer measuring the noon temperature on the sunlit side at perihelion would soar to 430°C, then plummet to –170°C on the dark side, reaching a low of –183°C just before sunrise.

We now know that Mercury's north–south axis is almost upright (unlike the Earth and most of the other planets), so there are no seasons as we understand them and every part of the surface, apart from deep craters at the poles, suffers from lengthy exposure to the Sun's fearsome heat.

For anyone foolhardy enough to set foot on this furnace of a world, the ferocious, fiery globe of the Sun would dominate the sky, remaining almost overhead for several weeks. The planet's slow rate of spin and rapid orbital motion in the inner part of its orbit (around perihelion) would cause the glowing orb to describe a loop in the sky as it stalls, move slightly eastward—the reverse to the usual direction—for 8 days, then resume its westward trek.

The overall effect is that there are two "hot spots," on opposite sides of Mercury, which endure even more insufferable heat than the rest of the planet. These regions, one around 0 degrees longitude and the other at 180 degrees longitude, receive the maximum possible solar radiation at the time of closest approach to the Sun.

Close range mapping by the Mariner 10 spacecraft has shown that Mercury's equivalent of Death Valley at 180 degrees coincides with the location of the Caloris ("Hot") impact basin, while the antipodal hot spot is marked by a so-called "peculiar terrain" of hills and linear features. These hellish regions receive 2.5 times more radiation overall than places at 90 degrees and 270 degrees longitude, which bathe beneath the Sun's relatively benign rays when Mercury is at its furthest point from the Sun.

Mariner 10

Most of what we know about Mercury comes courtesy of a small NASA spacecraft that skimmed over its surface on three separate occasions almost three decades ago.

Mariner 10 was the last in a series of US spacecraft that pioneered the exploration of the terrestrial planets during the 1960s and early 1970s. After successful flybys of Mars and Venus, it seemed logical for NASA to complete the set by heading for miniature Mercury. However, there was a major problem. In order to make a direct flight to Mercury, a spacecraft leaving Earth would require a very powerful, expensive launch vehicle. Fortunately, there was an alternative: use an intervening planet (i.e. Venus) to give it a gravitational boost—a "free ride" achieved without using any additional fuel, courtesy of Mother Nature.

What made Mariner 10's "billiard ball" mission possible was an obscure study on spacecraft trajectories undertaken

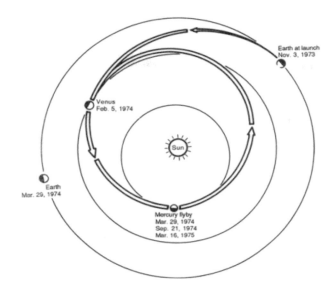

It would also require a special type of spacecraft that could survive and operate in an environment where solar radiation is 10 times more intense than at the Earth. The fragile explorer would have to carry a large, gold-coated sunshade and power-producing arrays that could tilt in order to limit the effects of solar heating. Since Mariner would be skimming past Mercury at high speed, it was also necessary to maximize the science return by developing new, high-resolution TV cameras that could zoom in on the surface from long distance.

Another, unexpected, breakthrough that would triple the scientific return from the proposed mission was provided by Giuseppe (Bepi) Colombo, the discoverer of

A Roundabout Route—Mariner 10 was the first spacecraft to use a gravity assist from one planet in order to reach another planet. Passing within 5,800 km (3,600 mi) of Venus, Mariner 10 was slowed so that it "fell" closer to the Sun, crossing Mercury's orbit to coincide with the arrival of the little planet. The spacecraft was able to rendezvous twice more with Mercury at the same point in its orbit. Unfortunately, the same surface area was illuminated during each Mercury flyby. (NASA–JPL)

by a UCLA graduate student, Michael Minovich. By October 1962, after several summers working with a room-sized mainframe computer, Minovich had discovered that the innermost planet could be reached by modifying a spacecraft's trajectory during a close flyby of Venus. Two suitable swing-by opportunities would occur in 1970 and 1973.

The main snag was that the incoming Mariner spacecraft would have to enter a very narrow window in Venusian space if it was to target the Mercurian minnow successfully. NASA described it as being equivalent to shooting a rifle bullet through a 5-cm (2-in) hole more than 160 km (100 mi) away. It was clear that the "economy class" voyage to Mercury would require pinpoint navigation.

Mariner 10—Mariner 10 was the first spacecraft to fly past Mercury. The spacecraft completed three flybys between March 29, 1974 and March 16, 1975, imaging about half of the planet's surface. Visible at the top are the steerable high-gain antenna and the large sunshade. The "eyes" at the bottom are the TV cameras. The magnetometers that discovered Mercury's magnetic field are on the long boom (left). (NASA–JPL)

Mercury's spin–orbit coupling. A distinguished specialist in celestial mechanics, Colombo realized that, after its first flyby of Mercury, Mariner 10 would enter a 176-day orbit around the Sun. This was exactly double Mercury's 88-day orbital period. With a few minor trajectory adjustments, the craft could return for further planetary surveys. The only drawback was that the surface lighting conditions would be the same for each encounter.

Mariner 10 eventually lifted off from Cape Canaveral on November 3, 1973, sweeping over the Moon's northern regions at a speed of 40,962 km/h (25,452 mph). Three months later, after a series of near-fatal technical problems and a pinpoint encounter with Venus, the spacecraft was redirected toward Mercury.

Only a few weeks before arrival, there was a minor panic when Mariner 10 lost its celestial bearings and had to be commanded from the ground to carry out a search for its main guide star, but the realignment was successful. This cleared the way for an engine burn on March 16 that shifted the Mercury overflight from the sunlit side to the dark side of the planet. It would now be possible for Mariner to return for a second look six months after its initial reconnaissance.

The next day, the non-imaging experiments were turned on in preparation for the Mercury encounter. Finally, on March 23, the long-awaited first pictures began to come in to the Mariner control center at JPL. Over the next few days, the excited scientists watched the planet grow on the TV monitors from a fuzzy crescent to a fat, featureless "first quarter" phase.

As the distance decreased, one bright spot stood out—later resolved as a 25-km (15-mi) rayed crater that was named after planetary astronomer Gerard Kuiper, who had died a few months earlier. It soon became apparent from Mercury's mottled appearance that its pockmarked surface closely resembled the Moon, with craters and impact basins blanketing every part of the tortured terrain. Scientists scrambled to assemble mosaics that revealed for the first time the true nature of the illuminated part of mysterious Mercury. By March 28, one picture was being returned every 42 seconds.

The New Mercury

Mariner 10's first flyby took place on March 29, 1974, when the spacecraft swept within 703 km (436 mi) of the planet. The trajectory was carefully chosen so that it could image a large part of the sunlit hemisphere from close range. Since the robotic explorer would also pass through Mercury's shadow, it would be able to measure the planet's interaction with the solar wind as well as temperature differences between the light and dark sides. To everyone's surprise and delight, it discovered that Mercury possessed a surprisingly potent magnetic field.

Although out of radio communication with Earth for part of the encounter, the onboard cameras merrily scanned the surface. The images were recorded on tape for later transmission back to JPL. As the spacecraft sped away it turned back to see a little more than half the planet illuminated and scientists were excited to see part of a huge circular impact structure (now known as the Caloris Basin) straddling the day–night terminator. By the time imaging was completed on April 3, more than 1,500 pictures had been returned in just 11 days—six days before closest encounter and five days after.

Meanwhile, the perils of premature announcements were made apparent on April 1 when the media discovered that the ultraviolet spectrometer experiment team had reported seeing the signature of a Mercurian moon. The puzzled TV imaging group searched in vain for the invisible intruder, and soon concluded that the "moon" was a brightish star that happened to be visible to Mariner's instruments at that particular moment. The disappointed press referred to the incident as "Mariner's April Fool."

Mariner 10's travails were not over. The tough little spacecraft was able to continue circling the Sun in order to complete two more flybys of Mercury in the same region of space as the first encounter (Mercury was near aphelion—furthest point from the Sun—on each occasion).

After surviving several minor course corrections and hardware headaches, Mariner returned, six months later,

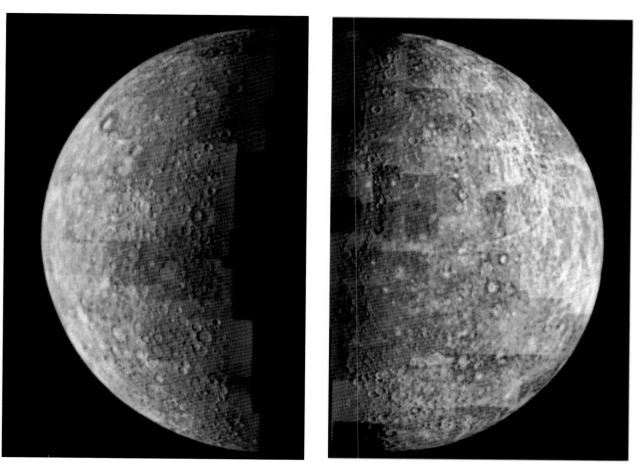

Mariner 10—Ingoing (left): Mariner 10's first close encounter with Mercury took place on March 29, 1974. This mosaic was assembled from 18 photographs taken at 14-second intervals at a distance of about 200,000 km (124,274 mi), six hours before closest approach. **Outgoing (right)**: This mosaic of 18 images was taken six hours after closest approach at a distance of 210,000 km (130,488 mi). The most notable feature, on the sunrise terminator (left), is the giant Caloris Basin. (NASA–JPL)

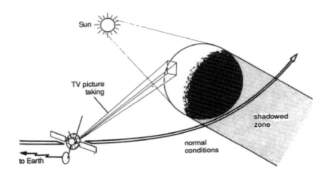

Imaging conditions for the first Mariner 10 flyby—During the first Mercury flyby on March 29, 1974, Mariner 10 imaged one illuminated hemisphere, then passed through the planet's shadow before imaging more of the dayside as it retreated. (NASA–JPL)

for its second photo opportunity. As the mechanical marvel closed on its target once again, project scientists decided to try out a new navigational technique that, if successful, would prove to be invaluable for future missions to the outer planets. Instead of simply relying on Earth-based radio measurements, they were able to fix the spacecraft's speed and location by using some 100 images to analyze the changing relative positions of Mercury and the "fixed" background stars.

September 21, 1974 marked a historic moment in Solar System exploration—the first time that a spacecraft had returned to the same world for a second look. This time the flyby took place 48,000 km (30,000 mi) above the sunlit southern hemisphere. Another 500 pictures were returned, revealing new details of the cratered south polar region.

Confirmation of Mercury's magnetic field had to wait until the third encounter on March 16, 1975, when Mariner made its closest approach of all, skimming just 327 km (204 mi) above the heavily cratered northern hemisphere. By this time, the battered spacecraft was hanging on by a thread. Not only was it virtually out of gas, but it had once again lost contact with the Earth just a few days before the final flyby, causing a scramble to re-establish communications with only hours to spare.

Unfortunately, problems with the ground-based antennas at Canberra and Goldstone meant that only one-quarter of each planned picture could be received. However, scientists were delighted to receive high-quality images of weird, jumbled terrain, almost free of impact craters, on the opposite side of Mercury to the giant Caloris Basin. Views of Caloris itself showed complex patterns of cracks and ridges on its floor.

Eight days after the treble was completed, the redoubtable spacecraft's gas tank ran dry. Its mission accomplished, the controllers in California bade farewell to the exhausted explorer as they switched off its transmitter.

Without attitude control the craft will one day fall into the Sun, but its exploits will live on in the annals of Solar System exploration. Mariner 10 will be remembered as one of the most successful space missions ever, and the spacecraft that brought Mercury from obscurity.

Mercury Unveiled

The most obvious part of the harvest from the cut-price Mariner 10 mission was the 2,400 pictures of Mercury. Unfortunately, the repeat nature of the three flybys meant that this image inventory only covered about 45% of the planet's surface. However, there was plenty for the scientists to mull over. The best pictures showed surface features as small as 140 m (450 ft) across, a spatial resolution comparable to the best views of the Moon through telescopes on Earth.

Studies of the illuminated hemisphere imaged by Mariner 10 soon made it clear that the innermost planet and Earth's Moon bore a remarkable resemblance. Not only did both bodies have the same relative brightness and reflect light in the same way, but they had also suffered major abuse over billions of years, as evidenced by countless craters and impact basins. Indeed, Mercury's terrain revealed ancient, heavily bombarded areas comparable to the lunar highlands and more recent, relatively uncratered, plains similar to the lunar maria or "seas"

The Caloris Basin—The Caloris Basin is one of the largest impact features in the Solar System, with a diameter of more than 1,300 km (800 mi). Although it was partially hidden by shadow, Mariner 10's cameras revealed an outer ring of smoothly rounded block mountains rising 1–2 km (approx. 1 mi) above the surrounding terrain, and an interior criss-crossed by ridges and fractures. Smaller craters excavated by later impacts pepper the surface of the basin. (NASA–JPL)

(though not as dark). Temperature measurements of the night and day hemispheres also suggested that the surfaces of both worlds are covered by a blanket of dust pulverized by aeons of meteorite impacts.

Craters on Mercury range in size from less than 100 m (330 ft) up to the largest impact feature, the Caloris Basin.

As on the Moon, the craters are usually bowl-shaped, sometimes with small central peaks. The larger basins tend to have terraced inner walls and flat floors, sometimes with inner mountain rings. Around them are mountains assembled from material ejected by the impact and fields of smaller secondary craters excavated by boulders blasted from the original site. Some fresh craters have bright rays leading from them, also as found on the Moon.

And yet there are subtle differences. The ejecta deposits are generally less extensive and the secondary craters are closer to their parents on Mercury than on the Moon. The craters also seem to be shallower. All of this is the result of Mercury's higher surface gravity—twice that of the Moon—which reduces the distance over which the material thrown outward by impacts can travel. Another difference is the presence of smooth plains between the heavily cratered areas, unlike the lunar highlands, which are saturated by craters of all sizes. These "intercrater plains" appear to be the remains of the original surface, thus predating the final major period of bombardment.

The "outgoing" hemisphere seen by Mariner 10, however, was very different from the heavily cratered "ingoing" hemisphere. Large areas of fairly smooth plains were visible, and it seems likely that these represent a past phase of widespread volcanic activity. (Interestingly, the Moon also has two different hemispheres, one of them—the Earth-facing side—showing similar evidence of lava flooding to that on Mercury.)

Highly fluid basaltic lava must have flooded huge areas, covering hundreds of kilometers before solidifying. No cones or vents are normally visible with such activity. One other possibility is that the lava was produced by heat energy resulting from huge impacts. Based on the number of craters—evidence of how many times the surface has been struck—these plains appear to be younger than the highlands of the "ingoing" hemisphere, though older than the rayed craters.

The most striking feature observed by Mariner 10 was the great Caloris Basin, located, as its name suggests, at one of Mercury's "hot spots." Although half of this imposing circular structure was hidden in shadow, it probably

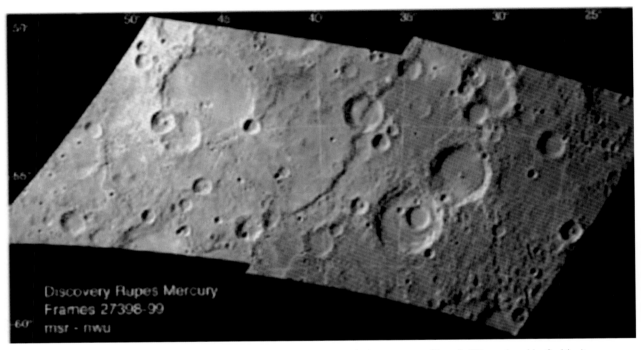

Discovery Rupes Mercury
Frames 27398-99
msr - nwu

Discovery Rupes—In the center of this image is Discovery Rupes, the longest scarp (line of cliffs produced by faulting) that crosses Mercury's intercrater plains. About 1 km (0.6 mi) high and 500 km (310 mi) long, this feature is thought to have been formed by shrinkage of the planet as it cooled over time. (Mark Robinson, Northwestern University)

measures about 1,300 km (800 mi) across, wide enough to swallow the whole of Britain. This makes it comparable in size to the Imbrium Basin on the Moon and the Hellas Basin on Mars.

The outer rim of the basin is marked by a main scarp or line of cliffs along the inner edge of a huge curved mountain range that rises up to 2 km (1 mi) above the surrounding surface. These mountains were probably thrown up by the impact that created the basin. Extending from the mountains are radial lines of hills and valleys.

The floor of the basin is covered with plains material but interlaced by fractures and ridges which presumably are the results of some kind of readjustment of the molten lava while in the process of cooling. Some lineated terrain around Caloris indicates that huge missiles shot out almost horizontally as a result of the impact, gouging long straight valleys.

The explosive creation of the Caloris Basin may have shaken the planet so much that it created an area of "peculiar" hummocky terrain on the opposite side of Mercury. This unique area of jumbled mounds and hollows seems to have arisen when enormous seismic (shock) waves created by the impact of an asteroid-size body in Caloris traveled around the planet and met on the far side.

Another unusual feature that is not found on the Moon is the so-called lobate scarps. These abrupt cliffs, which reach up to 4 km (2 mi) high, cut through ancient craters as they wind for hundreds of kilometers across the surface. It has been suggested that they were created by crustal shortening caused by the slow cooling and contraction of Mercury's iron core early in its history—a process comparable to the creation of wrinkled skin as an apple dries and shrinks.

Magnetic Mercury

If a pressure suit could be designed to prevent an astronaut from being fried or frozen alive, humans would find walking on Mercury much less arduous than on their home planet. With surface gravity little more than one-third that on Earth (though slightly higher than on Mars), a visitor weighing 64 kg (140 lb) would register a modest 24.5 kg (54 lb) on Mercury.

The reason that Mercury pulls above its size is that its density is extremely high, second only to the Earth in our planetary system. (Indeed, if the two planets were miraculously made the same size, Mercury would become the densest object in the Solar System.)

Before Mariner 10 arrived in 1974, the reason for this was not clear. Did Mercury have a massive iron core capped by a relatively thin silicate mantle and crust, or was it a simple mixture of these materials throughout, as in the case of some meteorites?

Thanks to the exploits of Mariner 10, we now know that, to all intents and purposes, Mercury may be regarded as an iron planet—a world whose massive iron core fills three-quarters of the entire body and is equal in size to Earth's Moon.

One explanation for this remarkable situation is that Mercury formed in a part of the solar nebula that was rich in heavy elements, such as iron. The planet may have melted and differentiated very quickly, allowing the metallic elements to sink to the center. The outcome was a dense core topped by a thin mantle and silicate crust. The core

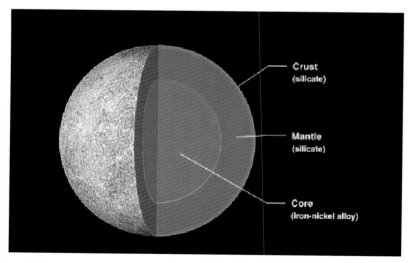

The Iron Planet—Mercury's metallic core has a diameter about three-quarters that of the entire planet. This accounts for its surprisingly high density, 5.44 grams per cubic centimeter, which is second only to the much larger Earth. The reason for this remarkably large core remains uncertain. However, it may help to explain the surprising strength of Mercury's magnetic field. (LPI)

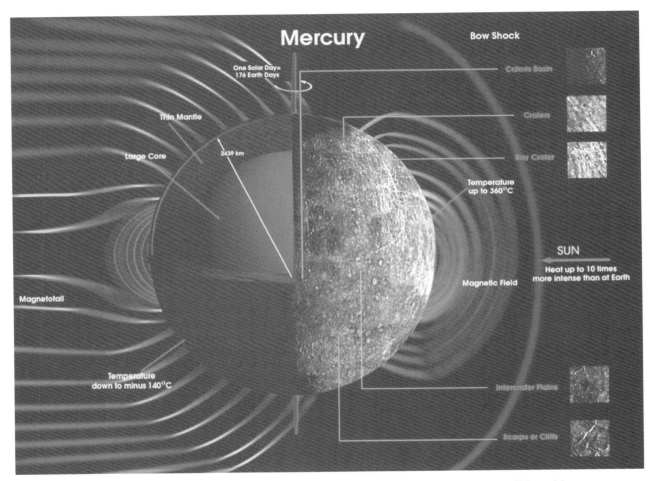

Magnetic Field—Despite its slow rotation, Mercury has a magnetic field that is stronger than those of Venus, Mars, or the Moon. Its strength at the equator is about 1% of Earth's magnetic field. Although the density of the solar wind is 10 times higher at Mercury than at Earth, the modest magnetic field is usually able to deflect most of the particles and create a teardrop-shaped magnetosphere which is 5% the size of the magnetic bubble surrounding the Earth. (ESA)

accounted for three-quarters of the planet's mass and occupied most of its interior. The surface then cooled and solidified, creating the smooth intercrater plains observed by Mariner.

Computer simulations show that a high-speed impact between the youthful Mercury and a large planetesimal may also have been a contributory factor. In a scenario reminiscent of the birth of Earth's Moon, the planet's silicate-rich outer layers may have been stripped away, leaving the iron-dominated remnant we see today.

Certainly, over the first few hundred million years of its existence, the newborn planet was subjected to a period of intense bombardment by numerous bodies up to asteroid size. These excavated thousands of craters, culminating in the Caloris Basin about 3.8 billion years ago. The origin of these cosmic bombs is uncertain—they may have been debris along the planet's orbit which was finally swept up by the planet's gravity, rather than planetesimals and comets originating in the asteroid belt or beyond.

There followed a period of dramatic volcanic eruptions as vast lava flows flooded the plains, infiltrating some craters and filling the Caloris Basin itself. Then, as Mercury cooled and contracted, the surface wrinkled and fractured, creating elongated cliffs. In the last 3 billion years little has changed apart from the occasional meteorite or comet impact that has sprayed bright rays around a handful of youthful craters.

Despite its iron heart, scientists were staggered when Mariner 10 data revealed that Mercury has a sizeable magnetic field. Magnetic fields are usually thought to be generated by a dynamo inside a rapidly rotating world. In the case of the Earth, currents circulating in the molten, iron-rich outer core turn the planet into a powerful magnet. Mercury, on the other hand, rotates very slowly and, since it is small and would normally cool quite quickly, it was presumed that its core had solidified long ago. The presence of a magnetic field suggests that this may not be the case.

One possibility is that at least part of the core is still liquid, perhaps because of some impurity such as sulfur that reduces its melting point. Alternatively, Mariner 10 may have detected a fossil field, one that was impregnated into

the planet during its formation and still lingers today. Until further observations by future spacecraft, the reason for Mercury's magnetic field will remain a mystery.

Whatever the truth, Mariner showed that the field appears to be an almost exact, miniature replica of the Earth's, although there are no radiation belts of trapped particles like those surrounding our world. It is aligned with Mercury's rotational axis, so that the magnetic poles correspond closely with the geographical poles.

Despite a field strength only one-hundredth that of the Earth, the shield can ward off the particles in the solar wind when the Sun is relatively inactive. It also diverts this stream of solar electrons and protons around the planet, creating a magnetic tail. Only when the Sun's outflow is operating at full blast do the particles directly strike Mercury's surface. Particles ejected into space by this peppering may then become trapped by the magnetosphere.

A World Without Atmosphere

Astronomers never expected tiny Mercury to have retained an appreciable atmosphere, so it was no surprise when measurements made by Mariner 10 in 1974 confirmed that, to all intents and purposes, Mercury is an airless world. Indeed, its gaseous blanket was shown to be even more flimsy than ground-based studies had suggested.

Minute quantities of hydrogen, helium, argon, and neon were detected, giving a surface atmospheric pressure, if that is the right word, of 0.000000000001 millibar, compared to just over 1,000 millibars for the Earth. This would easily qualify as a vacuum in most laboratories.

Then, in 1985, Andrew Potter and Thomas Morgan of NASA unexpectedly discovered the metallic elements sodium and potassium—not the sort of thing one expects to find in a planet's atmosphere.

Scientists speculated that they could come from meteoroids that strike Mercury, vaporize, and eject sodium and potassium into the atmosphere, both from themselves and from the surface. The solar wind raining down on the

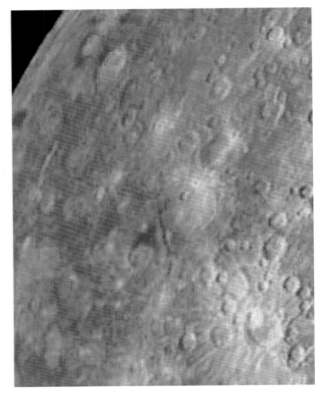

Surface Minerals—This false-color image, centered at 12°S 31°E, was created from Mariner 10 data. The small, bright Kuiper crater (lower right) shows fresh material excavated from rock that may have an unusual composition. The relatively dark and blue unit (top) is consistent with enhanced titanium content. The bright red areas nearby may represent primitive crustal material, while the dark brown plains (lower left) are probably covered with lava flows. (Mark Robinson, Northwestern University/NASA–JPL)

The second of these explanations is now favored. Ground-based observations show that minute traces of atomic oxygen, sodium, and potassium are mixed in with the atomic hydrogen and helium of the solar wind. It seems that these particles are soaked up by the dusty surface and slowly released to resupply continuously the tenuous atmosphere. The sodium and potassium are probably sprayed off the surface by the impact of high-energy solar wind particles and micrometeorites. Some of the hydrogen and oxygen may also come from ices that arrive in comets or meteorites.

Despite the dramatic successes of Mariner 10, many mysteries remain. In particular, the spacecraft was not

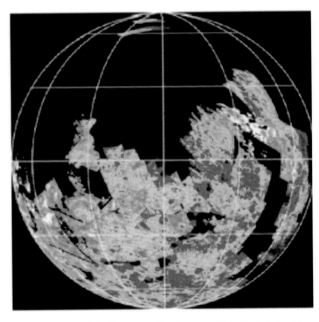

Elevation Map of the Southern Hemisphere—This digital elevation map, which covers about 25% of Mercury's surface, has been created by matching points on stereo images taken by Mariner 10. Elevations range from blue (lowest) through green, yellow, orange, red, and white (highest). (T. Cook & T. Watters, Smithsonian National Air and Space Museum)

surface could also release sodium and potassium into the atmosphere.

Similarly, helium, a light gas one would not expect to find on Mercury, could have been produced by radioactive decay of uranium and thorium in the crust, or from the charged particles arriving in the solar wind.

really equipped to determine the composition of Mercury's surface. Nevertheless, more than 20 years after the spacecraft's data were consigned to the archives, Mark Robinson (US Geological Survey/Northwestern University) and Paul Lucey (University of Hawaii) decided to undertake a new analysis of the unique images.

Using advanced computer and image-processing technology, combined with a better understanding of how light reflects off planetary surfaces, the pair have produced false-color images of Mercury that reveal significant differences in the materials that make up its barren terrain.

Although the computer-enhanced views do not allow a detailed analysis of the minerals that make up Mercury, they do reveal a complex surface of probable lava flows, volcanic ash and impact debris. Images also support suggestions that Mercury's crust (and underlying mantle?) contain much less iron oxide than the other rocky planets, Venus, Earth, and Mars, although the reason for this remains a mystery.

The pictures from Mariner 10 have also proved extremely useful in providing the first relief map of the little planet. By matching points on 1,709 pairs of stereo images obtained during the three flybys, Tony Cook and Tom Watters (Smithsonian Air and Space Museum) collaborated with Robinson to cobble together a mosaic that covers 25% of the Mercurian surface—mostly south of the equator. The map clearly shows the western highlands, revealed for the first time as Mariner 10 pulled away from the planet on March 29, 1974, and lower plains to the east of the Caloris Basin.

Meanwhile, the pre-Space Age names have been abandoned, and hundreds of surface features have been renamed by the International Astronomical Union. Almost 300 impact craters have been named, many of them in honor of renowned writers, musical composers, or artists (e.g., Stravinsky and Van Eyck). Valleys are named after famous ground-based radio astronomy observatories (e.g., Arecibo and Goldstone), while scarps recall famous ships from the golden age of exploration (e.g., Discovery and Santa Maria). The names given to Mercury by different cultures and languages are used for the planet's extensive plains (e.g., Tir and Budh).

The fixed reference point for the planet's zero longitude—Mercury's equivalent of the Greenwich Meridian on Earth—is drawn 20 degrees east of the small crater Hun Kal (which means "20" in the extinct language of the ancient Mayan civilization).

Revealing Radar

In recent years, tantalizing images of the landscape hidden from Mariner 10's cameras have been returned by modern, ground-based optical and radio telescopes. As far as we can tell, there is no reason to expect Mercury's unseen hemisphere to be very different from the one that was first revealed three decades ago.

The first fuzzy radar views of Mariner's dark side suggested that at least one large volcano was waiting to be unveiled, but subsequent observations showed the tell-tale signs of an imposing impact crater surrounded by

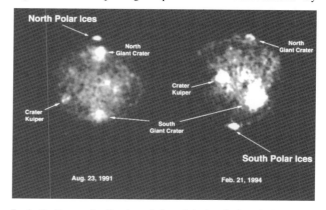

Ice in the Furnace?—Our knowledge of the hemisphere hidden from Mariner 10's cameras has been enhanced by bouncing radar signals off the planet's surface. The first radar images revealed bright features that might be giant craters. More detailed views obtained with the Arecibo radio telescope confirmed that these are fairly fresh impact craters. Radar-bright features at both poles may be deposits of water ice that have survived inside permanently shaded craters. (B.J. Butler, M.A. Slade and D.O. Muhleman, JPL/Caltech)

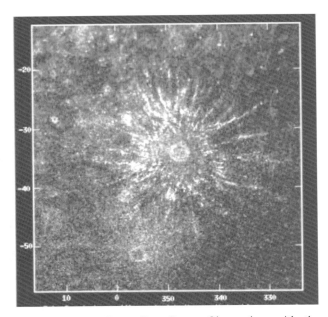

Radar Reveals a Large Ray Crater—Observations with the upgraded Arecibo radio telescope have shown that this feature, once thought to be a giant volcano, is actually an impact crater surrounded by a prominent ray system and a blanket of ejecta. (John Harmon, National Astronomy & Ionosphere Center, Arecibo Observatory)

prominent rays and blankets of ejected material. High-resolution radar observations of other bright reflective regions in the unphotographed hemisphere of Mercury indicate that they, too, are young craters surrounded by bright rays of ejected material, similar to Tycho on the Moon.

Radar has also enabled Duane Muhleman (JPL/Caltech) and colleagues to discover strange deposits close to the planet's poles. Bright radar reflections from circular areas near the north pole were first detected in 1991, using the 70-m (230-ft) Goldstone radio telescope in California in conjunction with the Very Large Array in New Mexico. Sharper images of both polar regions have since been obtained with the world's largest single radio telescope, the 305-m (1,000-ft) dish at Arecibo in Puerto Rico.

Unfortunately, Mariner 10 did not obtain any high-resolution images of the north pole, but there is no doubt that the radar-bright features at the south pole coincide with deep impact craters. The nature of the reflected signals favors the astounding conclusion that water ice survives in permanently shaded craters on the furnace-like surface of Mercury. Since the material is less reflective than pure ice, it seems likely that it is covered by a thin layer of dust or "soil."

Some calculations indicate that the deposits could be between 2 and 20 m (7 and 67 ft) thick—"true ice-rinks in Hell" as one scientist described them. (Similar suggestions have been made in recent years for the Moon, but Mercury may offer a much more extensive reservoir of ice, perhaps as much as several billion tonnes, or 1,000 times the amount on Earth's neighbor.)

If this is true, where did the water come from? One suggestion is impacts from comets, micrometeorites, and water-rich asteroids. Some water vapor derived from a suicidal comet crashing into Mercury could condense on the planet's cold night side and settle, more or less permanently, in deep, shaded craters, where the temperature never rises above −163°C.

Less likely, perhaps, is the possibility that water vapor outgassing from the planet's interior is frozen out at the poles. Whatever the true origin, water ice placed in such a deep freeze environment could remain stable over billions of years.

However, not everyone believes that the reflections represent ices, since similar reflections come from craters too small and too far from the pole to be cold enough for ice to exist. An alternative theory is that they are areas rich in sulfur, which is also efficient at reflecting radar waves.

One origin for the sulfur could be delivery by meteorite impacts over many aeons. Alternatively, it could have been deposited by early volcanic activity and then outgassed as a result of subsequent impact melting. The sulfur would linger in the tenuous atmosphere for some time, then be redeposited in the colder, shadowy, craters.

MESSENGER

Ironically, the remarkable success of Mariner 10 resulted in Mercury being largely ignored for the next three decades. No one seemed too interested in returning to a near-clone of the Moon when other, more exotic, objects beckoned. Proposals to return to the innermost planet were few and far between, and subsequently overlooked by space agencies with more spectacular and more popular projects to promote.

Only in the first years of this century were the pleas of planetary scientists eventually heard, to the extent that two ambitious Mercury orbiter missions are now underway on either side of the Atlantic.

MESSENGER—NASA's MESSENGER spacecraft is currently *en route* to Mercury. It will become only the second spacecraft to explore the planet from close range. One gravity assist from Earth, followed by two at Venus, will adjust its trajectory before going on to make three flybys of the innermost planet. It will finally enter orbit around Mercury in March 2011—almost seven years after launch. (JHU/APL)

The first to go was NASA's MErcury Surface, Space ENvironment, GEochemistry, and Ranging (MESSENGER) spacecraft, which received the go-ahead in March 2002 as part of NASA's lower-cost Discovery program. Delayed by concerns about cost overruns and shortage of time for testing and assembly, the mission eventually lifted off atop a Delta 2 rocket on August 3, 2004.

The 1,100-kg (1.2-ton) spacecraft carries seven scientific instruments, including cameras that will map the entire surface, a laser altimeter to gather topographic information, a magnetometer, and several spectrometers.

As with its venerable predecessor, Mariner 10, the key to MESSENGER's success will be the ability to match its speed with fast-moving Mercury. In order to slash the amount of fuel required to reach the planet, MESSENGER will take advantage of three gravity assists—one from Earth and two from Venus. Unfortunately, this extends the duration of its voyage across the inner Solar System to about 3½ years, compared with less than five months for Mariner 10.

During its roundabout journey, it will return to Earth's vicinity one year after launch, then skate past Venus in October 2006 and June 2007. The first Mercury flyby will take place in January 2008, when it will sweep to within 200 km (124 mi) of the planet's surface. Further close encounters in October 2008 and September 2009 will further slow the spacecraft in preparation for orbit insertion.

During these passes it will map most of the planet, filling in the blanks from Mariner 10's reconnaissance, in addition to studying the composition of the surface, the atmosphere, and the magnetosphere. These data will also be critical in planning the orbital phase of the mission. Only on the fourth approach in March 2011 will MESSENGER make history by firing its thrusters for 15 minutes before becoming the first spacecraft to orbit the planet.

The spacecraft will loop around the planet in a highly elliptical trajectory, which varies between 200 and 15,193 km (124 and 9,420 mi) above the surface. The plane of the orbit will be inclined 80 degrees to Mercury's equator, and the low point in the orbit will be reached at latitude 60°N.

Mercury's South Pole—One of the areas that will be examined by MESSENGER is Mercury's south pole. In this Mariner 10 view, the south pole is located inside Chao Meng-Fu, the 180-km (110-mi) wide crater on the planet's limb (lower center). Although the crater's floor is in shadow, sunlight is reflected from its far rim. Also visible is a double-ringed basin just above and to the right of the south pole, and a bright ray system of material ejected from a 50-km (30-mi) crater at upper right. The picture was taken from a distance of 85,800 km (53,200 mi). The horizontal line near the top was caused by a drop out in TV data. (NASA–JPL)

Its near-polar path will enable the spacecraft to map the entire surface for the first time as Mercury spins beneath it. The low-altitude orbit over the northern hemisphere will enable MESSENGER to conduct a detailed investigation of Mercury's geology and the composition of the giant Caloris impact basin.

In order to survive where solar radiation is up to 11 times more intense than on Earth, the spacecraft will carry a sunshield made of heat-resistant Nextel fabric—the same ceramic material that protects parts of the Space Shuttle—so that its instruments will be able to operate at room temperature. Exposure to reflected heat from Mercury will also be limited since the orbiter will pass only briefly over the planet's surface "hot spots."

MESSENGER's orbital reconnaissance will continue for at least 12 months (equivalent to two Mercurian days, from sunrise to sunrise, or four Mercurian years). The first "day" will mainly be focused on obtaining global maps and a broad planetary overview from the different instruments, while the second will be devoted to more specific, targeted, science investigations.

In particular, MESSENGER will try to answer six of the key outstanding questions about the nature of the little planet closest to the Sun:

- Why is Mercury so dense?
- What is the geological history of Mercury's surface?
- What is the structure of Mercury's core?
- What is the nature of Mercury's magnetic field?
- Do ices exist in craters at Mercury's poles?
- Where does the tenuous atmosphere come from and what is its composition?

"Exploring the many mysteries of Mercury will help us to understand all of the terrestrial planets, including Earth," said Sean C. Solomon, MESSENGER's principal scientific investigator. "We've had many exciting missions to Mars and Venus that yielded new theories about the processes that shaped the inner planets, and for 25 years now Mercury has clearly stood out as a place where major questions remain to be answered. Mercury is that last piece of the puzzle."

BepiColombo

MESSENGER will not offer the only opportunity to complete the detailed exploration of the Sun's closest planetary companion. Since 1994, the BepiColombo mission to Mercury has been a Cornerstone of the European Space Agency's long-term science program. In 2003, the Japanese Space Activities Commission approved that country's involvement in the pioneering mission. Unfortunately, a few months later, ESA's plans to deploy a small lander onto the planet's scorching surface were canceled due to budget constraints.

BepiColombo is intended to pursue in even greater depth the investigations undertaken by Mariner 10 and MESSENGER. It is named in honor of the Italian mathematician and designer who first explained Mercury's three to

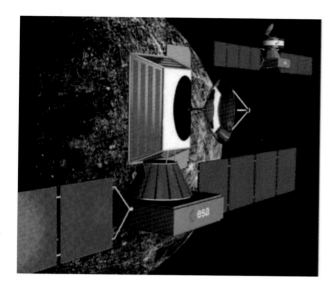

The BepiColombo mission—BepiColombo will be a dual spacecraft mission to explore Mercury. Scheduled for launch in 2012, it will follow a roundabout route lasting about four years. ESA is building the Mercury Planetary Orbiter (MPO), while the Japanese will contribute the Mercury Magnetospheric Orbiter (MMO). The MPO will study the planet's surface and internal composition, and the MMO will simultaneously study Mercury's magnetosphere. The mission is named in honor of Italian mathematician, Giuseppe Colombo. (ESA)

Giuseppe (Bepi) Colombo—Professor Giuseppe "Bepi" Colombo (1920–1984), a brilliant mathematician and engineer, was the first to explain why Mercury rotates three times during every two orbits of the Sun. In 1970, he realized that the period of Mariner 10's orbit, after its first flyby of Mercury, would be very close to twice the rotational period of the planet. He went on to explain that a second encounter with Mercury could be achieved with very little expenditure of fuel. Colombo also invented the concept of tethered satellites. (ESA)

two spin–orbit coupling and who later suggested the multiple flyby mission scenario for Mariner 10. The current baseline envisages the launch of both orbiters by a single Soyuz–Fregat rocket in 2013. Placed into an elongated orbit that allows them to be thrown toward the Sun by the Moon's gravity, the craft will follow the usual billiard ball trajectory involving gravity assists from Venus and Mercury.

Using ion thrusters powered by large solar arrays, BepiColombo will return to Earth for a gravitational kick, followed by encounters with Venus and Mercury, before it arrives at its destination after a trek lasting around four years. Once the solar electric propulsion system is jettisoned, the spacecraft will rely on more traditional chemical motors to brake into a 400 × 11,800 km (250 × 7,375 mi) orbit around Mercury.

The Japanese-built Mercury Magnetospheric Orbiter (MMO) will be separated at this stage and begin independent operation. From its highly elliptical polar path, the MMO will be ideally placed to study Mercury's magnetic field and investigate the interaction between the solar wind and the iron planet.

Meanwhile, the mother craft's chemical propulsion stage will fire for a second time, lowering the orbit before separation of the Mercury Planetary Orbiter (MPO). Swooping over the surface at an altitude of between 400 and 1,500 km (250 and 940 mi), the European-built contribution will be able to map the entire planet in detail as it rotates beneath the spacecraft once every 59 days.

The instruments on the MPO will consist of two cameras for high-resolution imaging and a suite of spectrometers for observing neutrons and radiation emitted by the planet's surface at infrared, ultraviolet, X-ray, and gamma-ray wavelengths. The spectra recorded will reveal the minerals and elements in the rocks, search for water ice inside polar craters, and confirm the constituents of Mercury's thin atmosphere. The MPO will also carry a laser altimeter to determine the highs and lows of the planet's rugged terrain, and two radio science experiments that should clarify Mercury's distribution of mass and internal composition.

Observations from both orbiters will continue for at least one Earth year. By the time BepiColombo's mission of exploration comes to an end, the aura of mystery surrounding Mercury will have all but evaporated. For the first time, scientists should have a much better understanding of why Mercury is mostly made of iron, why it is the only inner planet besides Earth with a global magnetic field, and whether it has preserved layers of ice in its polar craters.

However, even with our current limited knowledge, one thing seems certain. With such a hostile environment, Mercury is likely to remain for ever an uninhabited outpost as it speedily patrols the inner reaches of the Sun's domain.

Transit of Venus—This image of Venus on the eastern limb of the Sun was taken by NASA's TRACE spacecraft during the transit of June 8, 2004. Unlike airless Mercury, Venus is outlined by a faint, glowing ring during transits. Telescopic observations of such events provided early evidence that Venus has a thick atmosphere. Transits of Venus are quite rare because of the different inclinations of the planetary orbits. Venus will next be seen to cross the face of the Sun on June 5–6, 2012. (NASA)

3 VENUS: THE RUNAWAY GREENHOUSE

The Dazzling Beacon

Since time immemorial, humans have stared in wonder at Venus, shining like a brilliant beacon in the star-studded night sky. In pre-telescopic days, observers could not fail to be aware of the planet's frequent apparitions since, with the exceptions of the Sun and Moon, Venus appears as the brightest object in the heavens. Sixteen times brighter than the brightest star, the planet is a dazzling sight, so bright that, if one knows where to look, it can be seen in broad daylight with the naked eye. In favorable circumstances, the shimmering planet can even cast a noticeable shadow at night.

Not surprisingly, many ancient civilizations from Mexico to Babylonia regarded Venus as one of the most important objects in the night sky. The name that we use today comes from the Roman name for Aphrodite, the Greek goddess of beauty, love, laughter, and marriage.

From its apparitions within a few hours of sunrise or sunset, early observers soon established that there was a close connection between the planet and the Sun. However, they were initially confused by the appearance of an "evening star" that became ever brighter to the east of the Sun before disappearing in its glare, and a "morning star" which seemed to behave in reverse fashion to the west of the Sun. This led the Greeks to believe that they were twin planets, Hesperus and Phosphorus, until, in the sixth century BC, Pythagoras eventually realized that they were one and the same.

However, not until Copernicus produced his Sun-centered scheme for the Solar System in the fifteenth

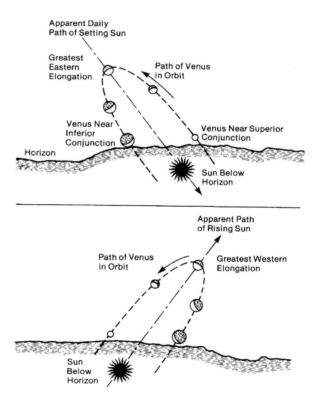

The Morning and Evening Star—As the result of its orbit around the Sun, Venus (like Mercury) appears as either as an evening (top) or morning "star" (bottom). This initially caused confusion among ancient sky observers, who thought they were two different planets. At its greatest elongation, Venus may rise or set up to five hours before or after the Sun. (NASA)

century did it become possible to explain the planet's bizarre behavior. As Venus moves toward us from the far side of the Sun (superior conjunction) it lies to the east of the Sun and appears as an "evening star." After it overtakes Earth on the inside orbital track at inferior conjunction, Venus moves to the west of the Sun and changes to a "morning star."

The theory seemed fine, but if the Copernican revolution was to overthrow the status quo, some hard evidence would be required. The man who provided the incontestable proof was the Italian Galileo Galilei, the most famous scientist of his day. Using his primitive "optick tube," Galileo became the first human to observe the phases of Venus and the changes in its apparent size as it approached and then receded from Earth.

As an "inferior" planet orbiting the Sun at an average distance of only 108 million km (61 million mi), Venus resembles Mercury in a telescope or binoculars by displaying phases, though these are much easier to observe since it is further from the Sun in the sky.* This means that the planet's appearance changes from a small full phase around superior conjunction on the far side of the Sun, to a large, but thin, crescent when nearing inferior conjunction. Venus appears at its brightest, reaching magnitude –4.3, when seen as a broad crescent near its greatest elongation of 48 degrees from the Sun in the sky. The planet then rises up to five hours before the Sun or sets up to five hours after it.

When at its nearest to Earth, Venus is invisible, since we can only see the dark side of the planet. However, on rare occasions, when the orbits are aligned correctly, Venus can be seen as a black "bullet hole" crossing the face of the Sun. Since they were first observed with a telescope in 1639, transits of Venus have been taking place in pairs at intervals of 8, 121.5, 8, and 105.5 years.

Those of 1761 and 1769 inspired expeditions around the globe in a (largely successful) endeavor to obtain an accurate measurement of the size of the Solar System.

The 1761 event also led to another dramatic revelation, when Russian astronomer Mikhail Lomonosov noted the appearance of a fuzzy halo around the planet. Lomonosov had discovered "an atmosphere equal to, if not greater than, that which envelops our earthly sphere."

The first transit of Venus to take place in the age of mass media occurred on June 8, 2004, when millions of people around the world were able to see it on TV and on the internet. After another transit in June 2012, the next Venusian Sun-crossing event will not take place until December 2117.

Transits of Venus are such novel events because the planet follows an orbit that is tilted more than 3 degrees to the Earth's orbital plane—the ecliptic. Since this inclination is steeper than those of all the other planets apart from Mercury and Pluto (which some deny to be a true planet), Venus usually passes above or below the Sun as seen from Earth.

The orbit is also notable for being closer to a circle than any other planetary paths. Since there is little variation in its distance from the Sun, its orbital velocity is also fairly constant. Traveling at an average speed of 35 km/s (22 mi/s), it completes one circuit of the Sun every 225 Earth days and laps the Earth every 584 days—an event known as inferior conjunction.

For some civilizations (e.g., the Maya of Central America), Venus was the object of greatest astronomical significance, more important than even the Sun. The brilliance of Venus immediately after inferior conjunction was so significant for these people that it was used as the basis for a calendar. It is also worth noting that, since five intervals of 584 days can be divided almost exactly by eight years of 365/366 days, apparitions of Venus in the sky repeat every eight years.

Removing the Veils

Generally speaking, the closer a planet is to us the more detail we can see and the better we can understand it. Based on these criteria, Venus should be an open book, since the gap between the two planets shrinks to 40 million km

---

* The Venus phases are also easier to observe because the planet is larger than Mercury and comes much closer to Earth.

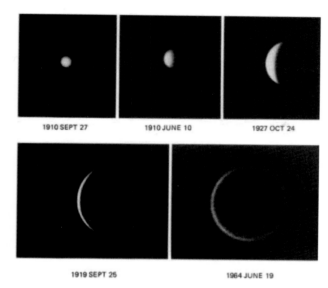

1910 SEPT 27   1910 JUNE 10   1927 OCT 24

1919 SEPT 25   1964 JUNE 19

The Phases of Venus—Venus appears fully illuminated, but very small, when it is at "superior conjunction" on the far side of the Sun from Earth. As Venus moves closer, it seems to grow in size and moves through its phases. By the time it approaches inferior conjunction, when the crescent is narrowest and Venus is nearest the Earth, it has grown in apparent size about seven times. (Lowell Observatory)

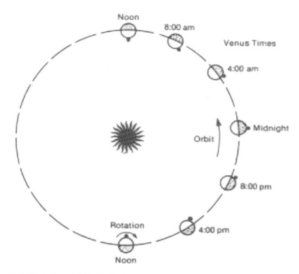

Axial Rotation: 243.1 Earth days
Orbital period: 225 Earth days (= Venus year)
Noon to Noon: 116.8 Earth days (= Venus day)
Venus year: 1.93 Venus days

Retrograde Rotation—Venus spins in a retrograde (east to west) direction, completing one full turn every 243.1 days. One orbit of the Sun takes 225 days. As a result, a day on Venus—from one sunrise to the next—lasts for 116.8 Earth days. (NASA)

(25 million mi) at inferior conjunction, when Venus is at its closest to Earth. Yet, until the last few decades, our celestial neighbor was reduced to a subject of speculation, the world most likely to attract lovers of mystery stories.

The one incontrovertible fact was that Venus is very similar in size to our Earth, so it was generally assumed that the two planets were twins. However, in the absence of a moon to tie down its mass, estimates of its bulk density ranged from 4.86 to 5.21 times that of water (compared with 5.52 for the Earth). Even such fundamentals as the length of the day and the nature of the surface remained unknown until the mid-twentieth century.

"No other object has caused more controversy and produced more varied testimony in the determination of its rotation period," wrote Edward Emerson Barnard of

Yerkes Observatory, in 1897. "This rotation controversy has raged for upwards of two centuries."

Most observers suggested rotations of about 23 hours, though Schiaparelli assumed that the planet completed one turn during each circuit of the Sun: in other words, the day and the year were the same—225 Earth days. American astronomer William Pickering added to the confusion by suggesting a spin period of 68 hours, with a highly inclined axis of rotation.

The problem arose because of the absence of any permanent, or even semi-permanent, detail on the visible surface of Venus. Instead of the solid surface displayed by the Moon and Mars, telescopes merely showed a bland blanket of yellowish clouds, occasionally etched with vague, shadowy markings. It is these clouds—which reflect 76% of incoming sunlight—together with the large size and relative proximity of Venus that explain the brilliance of the planet to the naked eye.

One technique that could assist was a measurement of the way in which the planet's rotation altered the wavelength of light reflected from Venus. Just as the pitch of a police car's siren is altered when a vehicle approaches and moves away, so this planetary Doppler effect enables astronomers to calculate a planet's spin. To everyone's consternation, observations made in 1956 by Robert Richardson (Mt Wilson and Palomar Observatories) suggested a very slow retrograde rotation. If true, this meant that Venus must be spinning from east to west—the opposite direction to the Earth and most other planets.

Further Doppler studies in the early 1960s by Bernard Guinot of Haute-Provence Observatory resulted in mixed messages. Although the data supported the "backward" rotation, the speed seemed to vary at different places on the planet's disk. The inconsistencies could only be explained by assuming that Venus was spinning remarkably quickly—once every four days!

Seeking a way around the apparent impasse, astronomers began studying the planet's cloud patterns. Although Venus revealed little in visible light, it was well known that fuzzy markings could be seen at ultraviolet wavelengths. The work of Guinot was confirmed by ultraviolet images obtained by French astronomer C. Boyer. Exposures taken only a few hours apart indicated that the planet's upper cloud levels were, indeed, racing from east to west in the remarkably short span of four days. But what about the hidden surface?

The solution finally came in 1961 with the advent of radar, a powerful new ground-based tool which enabled astronomers to pierce the cloud blanket and probe the planet's solid surface. By measuring the Doppler shift of the radio waves reflected from Venus, the secret of its spin could finally be revealed. In contrast to the superhurricanes raging above, the solid planet rotates very slowly, completing one retrograde revolution in 243 Earth days. This means that Venus has by far the longest rotation period of any planet in our Solar System.

Clearly something strange happened early in Venus's history. Current theory holds that Venus initially rotated in the same direction as most other planets. Then it simply flipped its axis 180 degrees. In other words, it still spins in the same direction it always has, just upside down, so that looking at it from Earth makes the spin seem backward. Scientists have argued that the Sun's gravitational pull on the planet's very dense atmosphere could have caused strong atmospheric tides. Such tides, combined with friction between Venus's mantle and core, could have caused the planet to flip.

Recently, French scientists have suggested that the rotation slowed to a standstill and then reversed direction. Taking into account the tidal effects of the Sun and other planets, the team concluded that Venus's axis could have shifted to a variety of positions throughout the planet's evolution, until it eventually tipped all the way over.

An interesting point here is that there is a definite relationship between the orbits of Venus and Earth and the rotation of Venus. At every inferior conjunction, when Venus lies directly between the Earth and the Sun, the planet presents the same side to us on Earth. (This, of course, is an obstacle to those who would like to characterize its entire surface with the aid of ground-based radar.) Some scientists have suggested that this implies some kind of tidal interaction between the two planets, though how these two widely separated worlds could influence each other in this way remains obscure.

Water World or Arid Desert?

The presence of a sizeable atmosphere on Venus has been known for centuries. The most obvious evidence was the presence of an all-embracing cloud deck that obscured any surface features. Early telescopic enthusiasts were also able to observe the fuzzy outer edge of the black planetary disk when Venus was near the limb of the Sun during one of its

Mariner 2—Mariner 2 was the first spacecraft to successfully complete a flyby of another planet. Based on the Ranger craft which were designed to crash into the Moon, the 204-kg (449-lb) spacecraft carried seven scientific instruments, but no cameras. The payload was designed to answer questions about the atmosphere, clouds, and temperature of Venus, and to explore the space environment. (NASA–JPL)

rare transits. They also noted an extension of the horns or cusps of the crescent Venus, a phenomenon ascribed to the scattering of light by the clouds.

But what about the temperature, composition, and depth of this atmosphere? Clearly, since Venus is only about three-quarters of Earth's distance from the Sun, it should be noticeably warmer than our planet. Exactly how warm remained a mystery until the mid-twentieth century, leaving room for considerable speculation about the nature of its surface and atmosphere.

Some scientists believed that Venus was an arid desert. Others, eagerly supported by science fiction writers, preferred to think of Earth's planetary neighbor as a "Carboniferous world," covered in tropical swamps like those tramped by the forerunners of the dinosaurs, more

than 200 million years ago. In 1918, the Swedish Nobel Prize winner, Svante Arrhenius, wrote, "The humidity is probably about six times the average of that on the Earth. We must conclude that everything on Venus is dripping wet. The vegetative processes are greatly accelerated by the high temperature; therefore, the lifetime of organisms is probably short."

The key to the puzzle was to discover the composition of the atmosphere. The first major advance came in 1932, when the spectral lines of carbon dioxide gas were recognized in the light reflected from the Venusian atmosphere. It was soon clear that most of the atmosphere is composed of this one gas—the figure accepted today is 97%, compared with 0.03% for Earth's atmosphere.

Carbon dioxide is notorious as a greenhouse gas that traps heat released by planetary surfaces, so, not surprisingly, the scientific argument swung in favor of those who believed that Venus is one vast, arid wasteland. This vision of an inhospitable furnace was supported by the first temperature measurements of regions below the clouds in the mid-1950s. Newly developed instruments that could detect microwaves emitted by the surface gave a figure of 325°C—comparable to conditions inside a red-hot domestic oven.

Not everyone was convinced, and the possibility of finding a more hospitable home for Venusian life seemed to improve in 1959 when two Americans, Commander Ross and C.B. Moore, detected water vapor during spectroscopic observations from a high-altitude balloon. The following year, theoretical studies by B. Warner in London indicated that free oxygen gas might exist there too.

Today these results are known to be highly inaccurate— the Venusian atmosphere contains only 0.1–0.4% water vapor, and a totally insignificant 60 parts per million of free oxygen—but at the time they persuaded some scientists to drastically revise their vision of the planet's surface conditions.

Instead of a barren, wind-swept desert that put the Sahara to shame, Venus was miraculously transformed into a water world, covered by a vast ocean that might contain primitive forms of life. The optimists in the scientific

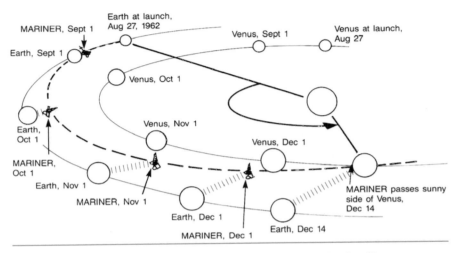

The Journey of Mariner 2—The best opportunities to launch a spacecraft from Earth to Venus occur once every 19 months. At such times, there is a window lasting a month or so when the energy required to send the spacecraft to the planet is at a minimum. Mariner 2 was launched by an Atlas–Agena rocket on August 27, 1962. After a journey lasting 3½ months, it flew past Venus at a distance of 34,833 km (21,645 mi). (NASA–JPL)

community argued that the clouds could be made of water ice crystals, which cannot be detected by a spectroscope. The merest detection of water vapor could then favor the interpretation that enormous quantities of supercooled water or ice shrouded the planet.

This Shangri-La world soon evaporated as Venus became a principal target for both the American and Soviet space programs. In 1962, NASA's Mariner 2—the first successful interplanetary spacecraft—flew past Venus at a distance of 34,833 km (21,645 mi). Although it carried no cameras, Mariner's microwave and infrared radiometers were sufficient to consign the water world to the dustbin of discarded theories. The spacecraft data indicated a surface temperature of at least 425°C, with little difference between the day and night hemispheres.

At last, the desert versus ocean controversy seemed to be settled. Then, five years later, confusion reigned once more when the Soviet Venera 4 atmospheric probe recorded a surface temperature of 265°C, in contrast to the searing

527°C measured during the flyby of Mariner 5. While the Soviets claimed that surface air pressure reached 22 Earth atmospheres, their US rivals (correctly) insisted that the figure was almost four times higher. Only after detailed analysis was it possible to reconcile the conflicting sets of data by recognizing that the Soviet craft had collapsed under the extreme heat and pressure at an altitude of 27 km (17 mi)—long before its remains reached the scorching surface.

It was now clear that Venus is the hottest of all the planets, surpassing even fiery Mercury. Today, the surface temperature is accepted as being between 450 and 475°C, dependent on altitude—well above the melting point of lead. Night gives no respite, since the dense atmosphere carries the unbearable heat of the day side right around the planet. Under such conditions any ocean would evaporate like a drop of water in a firestorm and no life as we know it could survive.

This dramatic difference between Earth and Venus can be explained if the two planets formed with similar compositions, but later evolved in different ways. The presence of liquid water is the key. Most carbon dioxide on Earth has been absorbed by the oceans and is now tied up in carbonate rocks. On Venus, gas injected by volcanoes remains free, creating a runaway greenhouse.

A Hellish World

At first glance, one could imagine that Venus and Earth are twins. Both were born in the same region of the inner Solar System, their diameters, masses, and densities are almost identical and both have thick, cloudy, atmospheres. On the

other hand, Venus has no natural satellites, it spins much more slowly than Earth in the opposite direction, and has a much denser atmosphere. What about conditions beneath the cloud blanket?

During the 1970s, as further American and Soviet spacecraft were despatched to probe the planet's secrets, the Venusian environment was found to be even more incredibly hostile than previously imagined.

Many of the early breakthroughs in our knowledge came from NASA's Mariner 10, which passed by the cloud-shrouded world in February 1974, *en route* to Mercury (see Chapter 2). Although the early pictures taken in visible light resembled views of a fog bank on Earth, images obtained at

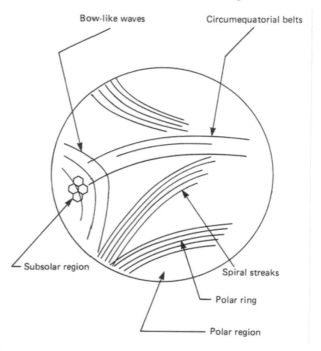

Venus from Mariner 10—This mosaic (left) was assembled from images taken in ultraviolet light. It was taken from 760,000 km (450,000 mi) by Mariner 10's TV cameras. Note the prominent swirl of cloud at lower right, above the south pole. Also apparent are C- and Y-shaped cloud features, and convection cells in the region where the Sun is directly overhead. The substance that causes these dark markings remains unknown. (NASA–JPL)

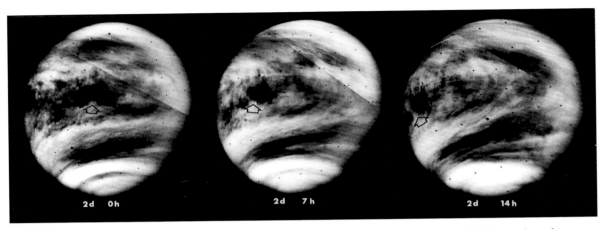

Shifting Patterns—This series of photomosaics was taken at seven-hour intervals on February 7, 1974, two days after Mariner 10 flew past the planet. They show the rapid rotation of light and dark markings at the top of Venus's cloud deck, as seen through ultraviolet filters. The bright feature marked with an arrow is about 1,000 km (620 mi) across. (NASA–JPL)

ultraviolet wavelengths revealed clearly defined cloud features that were only hinted at in ground-based views.

By studying pictures taken seven hours apart, it was possible to confirm that the upper clouds were speeding around Venus at 360 km/h (220 mph), circling the entire planet in about four days. This retrograde motion from east to west—the same direction as the planet's spin—was 60 times faster than the rotation of Venus. The only comparable winds on Earth are the jet streams, but they are limited to narrow zones, whereas the Venusian winds cover the entire planet. The source of energy for these high-velocity winds had to be the absorption of solar energy by the clouds.

Mariner's TV camera also spotted a mottling of cellular structures in the region of maximum heating, indicating that large convectional clouds were thrusting upwards from the main cloud deck on columns of rising air.

However, the most noticeable feature was a Y-shaped, symmetrical pattern centered on the equator. Broad bands of cloud diverged from the subsolar point, where the Sun was overhead, toward thick "hoods" over the poles. This suggested that, as on Earth, the upper air in equatorial regions spreads toward the poles, although the overall circulation on Venus is much simpler because of the small variation in temperature, the absence of oceans, and the slow planetary rotation. The overall motion may be compared to water swirling around a drain hole.

By studying changes in Mariner's radio signals as they passed through the atmosphere, scientists were able to detect at least two separate zones of cloud between 35 and 60 km (22 and 37 mi) above the surface. Above the main cloud deck were layers of tenuous haze.

Until the early 1970s, the composition of the clouds remained the subject of intense speculation. One of the favorite explanations was dust, stirred up by global, hurricane-force winds. Other, more exotic, suggestions included liquid mercury, ammonium nitrate, carbon suboxide, formaldehyde, nitrogen dioxide, hydrochloric acid, and even plastics!

Mariner's measurements, together with infrared spectroscopic studies made in 1973 from a Lear jet high in Earth's atmosphere, indicated that the yellowish clouds resemble the output from a hazardous chemical plant on Earth.

Together with traces of hydrochloric acid and hydrofluoric acid, large-scale reactions involving hydrogen, atomic oxygen, and sulfur were combining to form sulfuric acid. The tiny droplets (about 1 micron across) were in a concentrated solution of about 75% acid–water.

This discovery explained the extreme dryness of Earth's neighbor, since sulfuric acid is very effective in removing large amounts of water from the atmosphere. Such corrosive clouds can also exist over a much wider range of temperatures than the water clouds of Earth. Any sulfuric acid rain falling beneath the main cloud deck soon evaporates in the high temperatures, creating acid vapors and water vapor.

With the broad brush view of Venus now completed, it was clear that the beautiful planet that illuminates the twilight skies could no longer be regarded as Earth's twin. Instead of basking in a tropical paradise, astronauts foolish enough to visit Venus would rapidly expire from multiple assaults on their well-being from the totally hostile environment. Not only would they be crushed by the dense atmosphere, but roasted, dissolved and suffocated. No place yet discovered resembles more closely the popular vision of Hell.

Veneras Reveal Volcanic Venus

Soviet ambitions to explore the planets—and outdo their American rivals—began when the Space Race was still in its infancy. Even before Yuri Gagarin stunned the world by circling the globe in April 1961, two Soviet Venus probes had been launched from the secret cosmodrome at Tyuratum in Kazakhstan (officially known as Baikonur). The first—known as Sputnik 7—suffered a failure of the rocket's upper stage and failed to leave parking orbit around the Earth. Then Venera 1 headed into interplanetary space but communication was lost after only two weeks.

Six more missions launched in 1962 and 1964 met with similar fates—in contrast with the dramatic success of

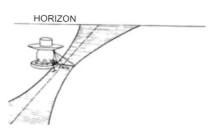

Venera 10 Views the Surface—Venera 10 was the second Soviet spacecraft to send back a "fish-eye" panoramic photo from the surface of Venus. It landed on October 25, 1975, near a volcanic region known as Beta Regio. Scientists suggested that the flat "plates" of bedrock could be eroded lava flows. The rake-like device (right of center) measured the density and composition of the surface. The distance to the horizon is probably about 100 m (330 ft). (NASA–Ames/USSR Academy of Sciences)

NASA's Mariner 2. However, with typical Soviet determination, the design bureaus led by the anonymous Chief Designer, Sergei Korolev, kept on trying. Their plan was to pierce the all-pervading clouds and parachute to the inhospitable surface.

Step by step, the goal edged ever nearer. In 1966, Venera 3 impacted with the planet, but contact was lost shortly before arrival. Finally, in October 1967—at the twelfth attempt—Venera 4 succeeded in transmitting data during its descent through the dense atmosphere. It recorded a temperature of 265°C and a pressure 22 times denser than at sea level on Earth, but Soviet claims that it had reached the surface had to be withdrawn after subsequent analysis.

The 1969 window saw Veneras 5 and 6 head toward Earth's nearest planetary neighbor, but although the probes were able to send back data for more than 50 minutes, they, too, were unable to withstand the alien conditions long enough to reach the surface.

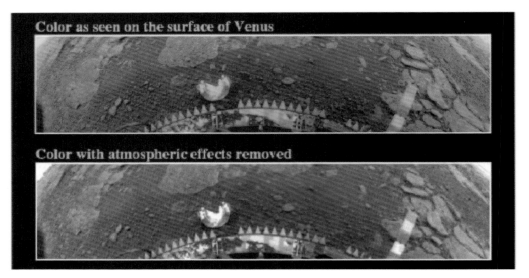

Venera 13 View of Venus—This panoramic image was returned by Venera 13, which landed at 7.5°S, 303.5°E on March 1, 1982. Sunlight filtering through the dense atmosphere gave the surface an orange tint. The lower image shows the true color—based on the color bar (right). The saw-toothed object at the bottom of each image is the lander's foot and the circular object nearby is part of the protective covering for the camera system. Chemical analyses indicate that the rocks are volcanic, and therefore black or gray. Pebble-sized pieces can be seen, but little soil. (James Head, Brown University/USSR Academy of Sciences)

Then, on December 15, 1970, came the red letter day. Equipped with a smaller parachute, Venera 7 made a more rapid descent on the planet's night side, eventually touching down at 5°S, 351°E. Scientists listening on Earth reported a barely discernible signal from the surface before it toppled over and contact was lost. Venera 7 had survived its hazardous voyage.

During the rest of the decade, the Soviets were able to build on this success. In 1972, Venera 8 landed on an upland plain not far from its predecessor. Fifty minutes of data from the surface confirmed the dense carbon dioxide atmosphere and blistering temperature. Measurements of natural gamma radiation indicated a surface made of granite-like rock. A light meter indicated that lighting was equivalent to pre-dawn illumination on Earth.

Three years passed before the next generation of Soviet landers was ready to breach the Venusian defences. Veneras 9 and 10 were equipped with a battery of instruments that could study every aspect of the planet, from its upper atmosphere and cloud decks all the way down to the rocky crust.

Both spacecraft touched down in a cloud of dust on the slopes of a radar-bright feature known as Beta Regio. In contrast to the hurricanes raging high above, the spacecraft recorded wind speeds of only 1.4–2.5 km/h (1–1.5 mph)—though in the thick carbon dioxide atmosphere, even these "breezes" would pack a powerful punch. Suggestions that Beta was a giant volcanic complex were supported by gamma-ray spectrometer readings, indicating a basalt lava composition. However, the most notable breakthrough came with the return of the first images from the surface.

The 180-degree black and white panoramic view from

Venera 9 showed a possible rock slide, with 30–40-cm boulders as far as the horizon, about 100 meters away. Venera 10 revealed an older, flat terrain covered with numerous rocky slabs, possibly evidence of eroded lava flows. Between the blocks were areas of weathered material, with pebbles, sand, and dust.

To the surprise of Soviet scientists, the powerful floodlights carried by each craft were not needed. It was subsequently realized that Venera 8 had landed near the day–night terminator, where the Sun was only 5 degrees above the horizon. For Veneras 9 and 10, which had landed in broad daylight, illumination was no worse than on a cloudy day on Earth.

To the dismay of geologists, the next two spacecraft in the series did not return any images of their landing sites, and few details of the surface conditions were returned. The most intriguing observations involved low-frequency radio bursts that were associated with extensive thunderstorm activity. (Intriguingly, Venera 9 had registered a brief glow in the night sky, suggestive of lightning.)

The best was yet to come. In March 1982, Veneras 13 and 14 landed about 1,000 km (625 mi) apart, to the east of an upland area known as Phoebe. Venera 13 set down on an ancient surface of rolling plains with a temperature of 457°C and a pressure of 89 atmospheres, while Venera 14 landed in a low-lying basin at 465°C and 94 atmospheres. They sent back the first (and so far, only) color pictures from the Venusian surface, revealing alien, rust-colored landscapes draped beneath a lurid orange sky. But there were subtle differences between the two sites.

Venera 13 landed with a tilt of about 8 degrees. Its 360-degree panorama showed a flat area with scattered angular rocks and pebbles of all sizes, while chemical analysis indicated the presence of a unique type of basalt that is usually associated with continental volcanism on Earth. Its companion came down on a tabular terrain marked by continuous layers of bedrock divided into slabs like a natural jigsaw puzzle. X-ray fluorescence analysis of a small sample drilled from a depth of 3 cm (1 in) indicated a composition similar to the fluid basalt lavas found on Earth's ocean floors.

Pioneer Venus

While the Soviets persevered with their pursuit of the perfect landing, NASA pressed ahead with its own ambitious program, involving a Venus orbiter and five atmospheric probes.

Unlike most American planetary missions, Pioneer Venus was managed by Ames Research Center in California, having inherited the program from Goddard Spaceflight Center. The drum-shaped orbiter, sometimes known as Pioneer 12, carried a dozen instruments that would explore every aspect of the planet, particularly its interaction with the solar wind, the composition of the mysterious clouds and the upper atmosphere, the shape of its gravitational field, and the topography of its hidden surface. Its most innovative experiment was a radar mapper that would reveal the true nature of the Venusian surface after centuries of speculation.

Pioneer Venus—Pioneer Venus comprised two separate missions. The Multiprobe mission (left) included the main Bus, which was not designed to survive atmospheric entry, and four atmospheric probes, each protected by a heat shield and pressurized capsule. The large Sounder Probe carried seven scientific instruments, compared with three on the North, Day, and Night Probes. The Pioneer Venus Orbiter (right) observed Venus for 14 years before it was incinerated during atmospheric entry. From a near-polar, elliptical orbit, it observed the atmosphere and mapped most of the planet's surface with radar. (NASA–Ames)

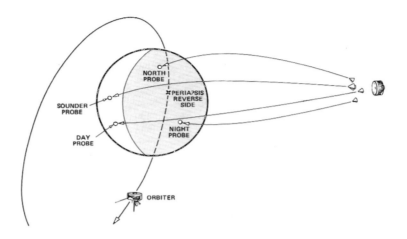

Pioneer Probes—The Multiprobe mission began on November 15, 1978, with the release of the Sounder Probe. Its three small companions followed four days later, when the Bus was still 9.3 million km (5.8 million mi) from Venus. They were targeted to investigate different regions of the planet's Earth-facing hemisphere—two on the night side and two on the day side. First to arrive, on December 9, was the Sounder Probe. All four probes survived, but only the Day Probe sent back any meaningful data after impact. (NASA–Ames)

The orbiter was expected to function for at least 8 months from a near-polar path. In contrast, the Pioneer 13 Multiprobe spacecraft were designed for a one-off kamikaze mission which just might culminate with a short-lived data return from the searingly hot surface.

The Pioneer probes could hardly have been more different from the large 660-kg (1,452-lb) Venera landers. Attached to the top of a 2.5-m (8.25-ft) diameter mother craft, known as the "bus," were four mushroom-shaped structures—a large probe weighing in at 316 kg (695 lb) and three small probes of 93 kg (205 lb) each. Once within striking distance of Venus, the probes would separate from the bus to complete their own individual voyages of exploration. Although the unprotected bus would soon disintegrate as it plunged into the upper atmosphere, scientists hoped that the heat shields on its former passengers would enable them to survive the shocking experience.

The orbiter set off from Cape Canaveral on May 20, 1978, following a long, looping trajectory that was designed to minimize the amount of onboard propellant required. Pioneer 13, on the other hand, took a faster route, so although the probes left Earth 2½ months after the orbiter, they slammed into the Venusian atmosphere only five days after Pioneer 12 arrived on December 4.

In a remarkable feat of endurance, the orbiter survived for no less than 14 years before it ran out of fuel and was condemned to a fiery death in the dense atmosphere. During its prolonged life, it was able to confirm that Venus has no magnetic field worth mentioning. As a result, Venus acts very much like a comet in its interaction with the supersonic particles of the solar wind. Like a boulder in a stream, the planet slows the solar wind and divides the flow, while the interplanetary magnetic field is draped around Venus to form a long magnetotail.

The absence of a magnetic shield means that the solar wind blasts into the upper atmosphere, splitting neutral atoms to produce a region of positively charged particles known as the ionosphere. The height of the ionosphere's upper boundary, the ionopause, varies with the strength of the solar wind, reaching as low as 250 km (155 mi) and as high as 1,500 km (930 mi). During the continuous battle with the solar wind, ions and electrons flow from the day side to create a night-side ionosphere as well as a long magnetic tail of ionized gas.

Other instruments on the orbiter detected radio emissions supporting Soviet suggestions that lightning is common on Venus, but the lack of a magnetic field meant that only feeble auroras were observed on the planet's night side.

One of the most intriguing results was the marked drop in sulfur dioxide levels during the mission, causing scientists to speculate that the orbiter's arrival had coincided with a massive volcanic eruption that had fed huge amounts of gas into the atmosphere.

The success of the orbiter was matched by the Multiprobe mission. The five-pronged assault began on November 15 with the release of the large Sounder Probe. Its three small companions followed four days later, when the bus was still 9.3 million km (5.8 million mi) from Venus.

The moment of truth came on December 9, when the Sounder Probe slammed into the atmosphere at 42,000 km/h (26,250 mph). Once its main parachute opened, the probe continued a more leisurely descent for 16 minutes before the chute was cut loose. The final free fall lasted 39 minutes, marked by a loss of contact on impact.

Traveling along diverging paths, one of the small probes plunged into the atmosphere on the planet's day side, while the other two sliced through the dark clouds of the Venusian night. Despite the lack of parachutes, their heat shields were able to protect them from the shock of entry, and by the time they struck solid ground the thick air had slowed them to a mere 35 km/h (22 mph) To everyone's surprise, the so-called Night Probe continued to transmit for two seconds after arrival, but the prize for endurance went to the Day Probe, which continued operating for 67 minutes.

Atmosphere

By combining measurements from the Pioneer Venus orbiter, the Multiprobe mission, and the Soviet Veneras, scientists were able to piece together a detailed profile of the planet's atmosphere.

Analysis of temperature data showed that the atmosphere of Venus is hot at the bottom and cold at the top, whereas the opposite is true for planet Earth. The soaring temperature of the lower Venusian atmosphere—the troposphere—was attributed to the massive carbon dioxide content, which resulted in a runaway greenhouse effect.

Since there is little free oxygen on Venus, there is no ozone layer comparable to that on Earth that is able to absorb solar radiation and heat the middle atmosphere. Instead, above 110 km (68 mi), the troposphere transitions directly into a region called the thermosphere. There, the relatively thin air is heated directly by absorption of solar ultraviolet radiation and the temperature increases slightly with altitude. A similar process occurs on Earth, although our planet's thermosphere is about 800°C hotter than that of Venus at an altitude of 300 km (186 mi).

Curiously—unlike Earth—there is no thermosphere above the unlit hemisphere of Venus. Despite the fast-moving winds that would be expected to transport heat from the sunny side, the upper atmosphere on the night side is the coldest place on Venus.

We are accustomed to high temperatures near the equator and freezing conditions in polar regions, but these variations do not exist on Venus. The most notable differences were found at the upper cloud deck, about 60 km (37 mi) above the surface. The North Probe, which plunged toward Venus at 60°N (equivalent to Anchorage in Alaska) measured temperatures 10–20°C colder than its counterparts, which entered nearer the equator. Concentrated solar heating at low latitudes accounts for the differences.

The Day and Night Probes also demonstrated that there would be little point in installing a thermometer in a

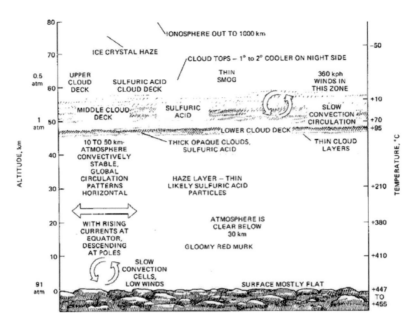

Cross-Section Through Venus's Atmosphere—The Venera and Pioneer Venus missions provided information on the horizontal structure of the planet's atmosphere. At high level are three layers of sulfuric acid cloud, which are swept by superhurricane-force winds. In the gloom below, the air becomes increasingly dense and hot, reaching a temperature of about 460°C. On the sweltering surface, air pressure is equivalent to that found almost 1 km (0.6 mi) down in Earth's oceans. Winds are sluggish but possess considerable momentum. (NASA)

Venusian weather station. The lower atmosphere is so massive that it retains almost all of its heat, even during the planet's prolonged night. As a result, the temperatures at any given altitude never differ by more than 5°C anywhere on the planet.

Spacecraft observations confirmed the presence of a sulfuric acid haze, which extends from about 90–70 km (56–43 mi) above the three main cloud layers. The upper cloud deck begins at 70 km and consists of tiny sulfuric acid droplets. Further layers of increasingly large acid droplets and particles of solid and liquid sulfur occur between 56 and 47 km (35 and 30 mi). At the level of the lower clouds, the temperature has already reached 95°C and atmospheric pressure is comparable to sea level on Earth.

Beneath the clouds the atmosphere is a little hazy, but below about 32 km (20 mi) visibility is good, except for the bottom 10 km (6 mi), which becomes increasingly murky due to scattering of light by the dense air, bathing the landscape with a reddish hue. An observer on the surface would not be able to see the Sun through the dreary blanket overhead.

As on Earth, convectional motion has been observed near the top of the Venusian clouds as huge bubbles of hot air thrust upward. Closer to the surface, where a barometer would register atmospheric pressure 92 times higher than on Earth (equal to the pressure an ocean diver would experience at a depth of more than 900 m or 2,970 ft), the fairly sluggish air currents generate only light breezes, in

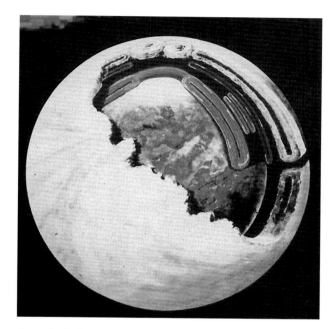

Atmospheric Circulation—This model of atmospheric circulation is based on observations by Pioneer Venus. The most significant movement is from east to west, causing the cloud tops to sweep around the planet in only four days. A much slower, north–south circulation—only a few meters per second—also occurs within the cloud layers. The atmosphere gradually spirals toward the poles, where it sinks and returns toward the equator. Several additional circulation cells ("Hadley cells") may also occur at lower levels. (NASA–Ames)

rushing downward into a giant hole in the cloud and returning at a lower level toward the equator. Several of these "Hadley cells" may exist at different levels. (Similar cells exist on Earth, but they only extend to mid-latitudes before being disrupted by the Earth's rapid rotation.)

The composition of the atmosphere was confirmed as 96.5% carbon dioxide, with 3.5% nitrogen and less than 0.1% water. Other minor constituents included sulfur dioxide, carbon monoxide, hydrogen chloride, and hydrogen fluoride. Since many of the gases are efficient absorbers of infrared (heat) radiation, the troposphere has been transformed into a sweltering greenhouse.

And yet it seems that Venus may not always have been this way. Measurements of the amounts of deuterium and hydrogen suggest that water was once much more plentiful. If the young Sun was 30% weaker than at present, liquid water could have survived for millions of years, possibly forming a global ocean. Only later, as the Sun brightened, did the oceans evaporate, creating a steam bath of a world where the temperature began to rise inexorably. Water vapor was carried to high levels where it combined with sulfur compounds or was split into hydrogen and oxygen atoms by ultraviolet radiation from the Sun. The light hydrogen gas then leaked into space while the oxygen combined with sulfur dioxide or oxidized surface rocks—a process similar to rusting on Earth.

Revealing Radar

Until the late 1970s, the study of the Venusian surface necessarily took second place to the study of its atmosphere, although pioneering work with ground-based radar provided some tantalizing results. These observations took advantage of the fact that radio waves can pass straight through the dense clouds and bounce back from the solid surface. The echoes received on Earth give important information about the roughness and composition of the terrain—the more the radar is scattered, the rougher the surface.

contrast to the 360-km/h (220-mph) hurricanes raging at the cloud tops.

Although these modest surface winds should have enough momentum to move dust and fine sand, the thick air will certainly inhibit saltation—the bouncing motion of sand grains across the surface—and the erosion of rocks by abrasion or sandblasting.

The most vigorous atmospheric circulation takes place at the level of the cloud deck. Resembling two giant conveyor belts, currents of warm air spread out from the equator toward the poles where they swirl around before

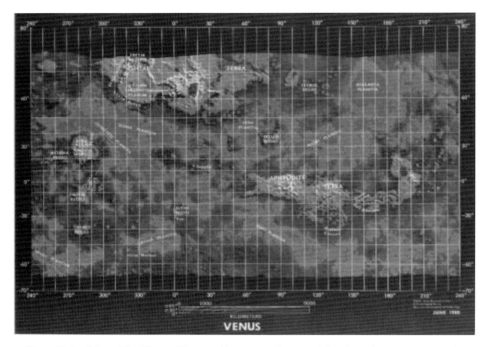

Pioneer Venus Radar Map—The Pioneer Venus orbiter mapped most of the planet between 73°N and 63°S, with a spatial resolution of 75 km (47 mi). Highest land is red and yellow, lowlands are blue. Apart from the uplands of Ishtar (top) and Aphrodite, Venus is remarkably flat and low-lying. 60% of the surface observed by the orbiter was within 500 m (1,650 ft) of the planet's mean radius. Zero degrees longitude is measured from a small feature known as Eve Corona, southwest of Alpha Regio. No data were obtained from polar regions. (NASA–Ames)

An early success for this radar technique was the discovery in 1962 that Venus rotates very slowly, so that one complete revolution takes 243 Earth days, in a retrograde (east to west) direction.

Unfortunately, this relationship means that Venus always shows the same side toward us at closest approach, so restricting the area that can be studied from Earth by radar observations.

Despite this handicap, it was possible to build up a picture of one hemisphere from the radar echoes picked up by the Arecibo radio telescope in Puerto Rico. By gradually refining the process over 20 years, it became possible to detect features only 10–20 km (6–12 mi) across on the

hemisphere centered on 320°E. Particularly notable were the isolated, radar-bright regions named Alpha, Beta, and Maxwell (in honor of British physicist James Maxwell). However, most of the surface seemed fairly smooth and featureless.

One limitation was the inability to accurately measure altitude. The breakthrough came through the radar altimeter on board the Pioneer Venus orbiter, which was capable of height measurements accurate to within 0.2 km (0.1 mi). Over a period of two years, the instrument was able to survey most of the planet between 73°N and 63°S, with a spatial resolution of 75 km (47 mi).

Analysis of these new data made it clear that Venus is a

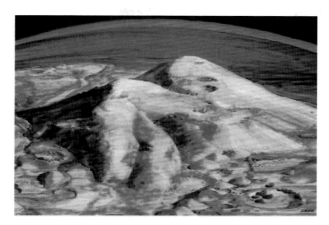

Artist's Impression of Beta Regio—A 1980 artist's impression of Beta Regio, based on Pioneer Venus radar data. One of the first features noted in ground-based radar images, it was originally thought to be a shield volcano with a central caldera. Pioneer Venus showed that it comprises two main volcanic constructs, Theia Mons (foreground) and Rhea Mons, split by a great rift valley. Beta Regio is located at 30°N 280°E. (NASA-Ames)

very smooth sphere, lacking extensive mountains and low-lying basins compared to Earth or Mars; 80% of the surface mapped by Pioneer Venus lay within 2 km (1 mi) of the planet's equivalent to sea level, indicative of widespread rolling plains. Although there are several "continents" which rise 2 km (1.2 mi) or more above the general level, these only cover 5% of the total area observed.

Nevertheless, there are extremes of relief, notably Maxwell Montes, whose peaks rise 11 km (6 mi) above the average height, dwarfing Mt Everest on Earth. On the other hand, the lowest point measured, the rift valley known as Diana Chasma, is only 2.9 km (1.8 mi) below the average level, which is only about one-quarter the depth of Earth's great ocean trenches.

The radar maps were dominated by two large upland regions. Ishtar Terra in the north is the size of Australia. At its core is the Lakshmi plateau, which is about the same height as the Tibetan plateau on Earth but twice the area. This continent is flanked to the north and east by mountains, the most spectacular of which are the Maxwell Montes.

Toward the east of the towering mountains is a 100-km (60-mi) circular feature that was thought to be a shield volcano like those found in Hawaii. The southern edge of the plateau appeared to drop very steeply, suggesting possible faulting and tectonic activity on a large scale. The rough surface seemed to be covered in loose debris.

The other "continent" was named Aphrodite and, though lower, it is twice the area of Ishtar and equal in size to Africa. (Apart from some features originally recognized in Earth-based observations, all features on Venus have been given female names.) Lying near the planet's equator, it, too, has mountains to the east and west, separated by a lower area.

Pioneer also found a huge rift valley system to the east of Aphrodite, 280 km (175 mi) wide and 2,250 km (1,400 mi) long. This was flanked by lines of hills, and contained the lowest point on Venus, while nearby was a chaotic region of parallel valleys and ridges. Scientists consider that these features are the best evidence found so far for crustal movements similar to those on Earth, although the surface of Venus seems to be one continuous slab, in contrast to the many mobile plates of Earth.

Whether or not there are active volcanoes on Venus remained a mystery, though the radar images seemed to indicate a number of giant volcanoes in addition to Maxwell. The most notable of these were two probable shield volcanoes, Theia Mons and Rhea Mons, which are 5 km (3 mi) high and located in the small upland region labeled Beta. Likely lava flows spread to east and west, the basaltic nature of which was previously recorded by two Soviet spacecraft which soft-landed nearby in 1975. Between the peaks was a large trough that might also be a rift valley. The radar-bright region called Alpha displayed another region of parallel fault structures, similar to the basin and range landscape of Nevada.

Large impact craters, so common on most of the other solid bodies in the Solar System, were largely absent, although the radar images did reveal a number of ring-shaped structures, some with central peaks, scattered across the plains. One such feature, located to the southwest of Alpha and about 200 km (125 mi) across, was named Eve. Such craters were absent on the continental areas, suggesting that they are much younger in origin than the plains. There are even larger circular basins such as those to the east of Ishtar and to the south of Aphrodite. However, one would expect most meteorites to burn up in the atmosphere, while any ancient craters should be affected and masked by billions of years of weathering and erosion.

Magellan

Even before the Pioneer Venus orbiter concluded its remarkable 14-year survey of Venus by plunging into the planet's atmosphere, both NASA and the Soviet Union had launched several spacecraft to map the surface in even greater detail.

First to arrive were the Soviet Veneras 15 and 16, each equipped with a side-looking radar that could reveal structures less than 2 km (1.2 mi) across. Unfortunately, their elliptical orbits meant that only the northern hemisphere above 30°N could be mapped.

The images clearly showed linear ridges and folds skirting the edge of the Lakshmi Planum plateau, with the Maxwell Montes dominating the eastern side of Ishtar. Most controversy focused on the various circular features. Were they craters of volcanic or impact origin? Some, dubbed "arachnoids" because of their spider-like shape, were a total mystery.

Clearly, images with higher spatial resolution were required to clear up the mysteries. The scientists' prayers were answered by NASA's Magellan spacecraft, which arrived in orbit around Venus in August 1990. Turning its powerful synthetic aperture radar toward the planet throughout three Venus rotations (roughly two Earth years), Magellan was able to map 99% of the surface with a resolution of between 120 and 250 m (400 and 825 ft). During repeat passes, it could also obtain stereo images taken from different viewing angles and search for changes on the surface. In September 1992, the spacecraft began a fourth observation cycle to enable scientists to map the planet's gravity and internal structure.

Magellan View of Venus Topography—This Magellan mosaic, centered at 180°E, was assembled from radar images taken during 1990–1994. Magellan imaged 99% of Venus's surface at a resolution of about 100 m (330 ft). Gaps in coverage were filled with images from the Earth-based Arecibo radar. Brown areas are high, generally rough terrain; the dark blue areas are low, typically smooth surfaces. To the west (left) is the "continent" of Aphrodite, with huge shield volcanoes such as Maat Mons in the "scorpion's tail" just right of center. (NASA–JPL)

By the end of its mission in October 1994, Magellan had returned more data than all previous US planetary missions and transformed our view of Earth's cloud-shrouded "twin." Perhaps the most notable discovery was the importance of volcanic activity on Venus. Wherever Magellan looked, there were volcanoes of all shapes and sizes. At least 100,000 small shield volcanoes, each less than 15 km (9 mi) across, were detected. Small pits at their summits marked the exit points for lava that had built up the volcanoes layer by layer. Many of the shield structures

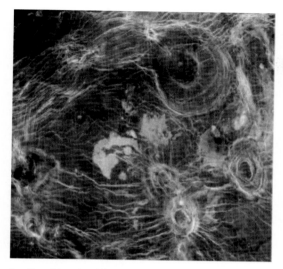

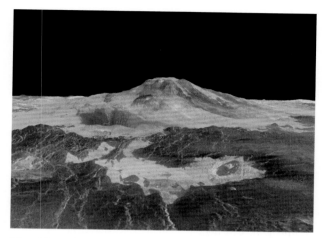

Maat Mons—Maat Mons is the highest and largest shield volcano on Venus, rising 8 km (5 mi) above the surrounding area. It may still be active. Lava flows spread out from the volcano for hundreds of kilometres. (NASA–JPL)

Arachnoids—"Arachnoids" are one of the more remarkable features found on Venus. They are seen on radar-dark plains in this Magellan mosaic of the Fortuna region. As the nickname suggests, they are circular or oval features with concentric rings and a complex network of fractures extending outward. They range in size from approximately 50 to 230 km (31 to 144 mi). They seem to be linked with an upwelling of magma that pushed up the surface to form cracks. (NASA–JPL)

gathered in swarms, possibly indicating a common origin from local hot spots beneath the crust.

Hundreds of larger volcanoes were also revealed, including some that resembled tick insects or pancakes, where thick, sticky lava had cooled close to the vent. In places, huge shield volcanoes dominate the landscape, with sunken calderas up to 100 km (60 mi) in diameter and lava flows hundreds of kilometers in length protruding like gnarled fingers across the surrounding terrain. Largest of all is Maat Mons, located in the "scorpion's tail" to the east of Aphrodite. Towering 8 km (5 mi) above the nearby plains, it rivals Mauna Loa, the largest volcano on Earth.

Although some pancake volcanoes are formed by sticky lava that solidifies before it can travel too far—despite the 460°C surface temperature—other landforms suggest the presence of lavas much more fluid than any on Earth. Long, sinuous channels snaking across the plains and sometimes ending in branching deltas, look suspiciously like water-eroded features. One of these, called the Hildr Channel, meanders for almost 7,000 km (4,375 mi), making it the longest feature of its kind in the Solar System. Magellan images show that each end is buried beneath younger deposits, so it was probably much longer in its youth. In the Venusian furnace, water erosion is hardly likely, so scientists suspect that the channels were carved by rivers of molten rock—possibly rare forms of lava, rich in sulfur or carbon, that can remain liquid at "room temperature" on Venus.

Other evidence of crustal motion and subsurface igneous activity is provided by unique features known as "coronae" and "arachnoids." Radar images show about 360 coronae—raised, circular or oval landforms that measure hundreds or thousands of kilometers across and are encircled by concentric rings of ridges and fractures. Scientists believe they formed as a result of liquid magma rising from below and deforming the surface.

Many of them are associated with major rift zones, such as the 1,000-km (620-mi) long Diana Chasma, and surface volcanism, such as the giant Artemis Corona to the south of Aphrodite. Arachnoids may be the predecessors of coronae. They exhibit circular fractures surrounded by numerous "legs" that seem to be created by dykes, where molten rock has seeped into cracks in the surface.

Another surprise was the relative absence of impact craters. About 1,000 of these were found, scattered randomly over the surface. The smallest of these are 3 km (2 mi) across, while at the top end of the scale, Mead Crater is 275 km (170 mi) in diameter. Erosion on Venus is very slow, so why are there so few craters on display? The most plausible explanation seems to be that they have been blanketed by volcanic material. But has this happened gradually or was the entire planet resurfaced by lava flows about 500 million years ago? The arguments still rage.

The overall picture that emerges is of a planet that has a very young surface, geologically speaking. This youth is related to processes taking place deep inside the planet, rather than wind or water erosion or plate tectonics. Whereas Earth's mountains and geological structures are not always directly related to convective currents rising and sinking in the semi-molten mantle, surface features on Venus seem to be much more closely linked to internal processes.

From Vega to Venus Express

Although the radar mapping of Venus took precedence during the 1980s and early 1990s, the Soviets (in collaboration with the French) were also prepared to have one last attempt at understanding the planet's atmosphere and landing spacecraft on its hostile surface.

The opportunity for the Vega missions came with the return of Halley's comet to the inner Solar System in 1986. (The name Vega comes from the first two letters in the Russian alphabet for Venus and Halley—there is no "H" in Russian.) En route to the icy intruder, the identical Vegas

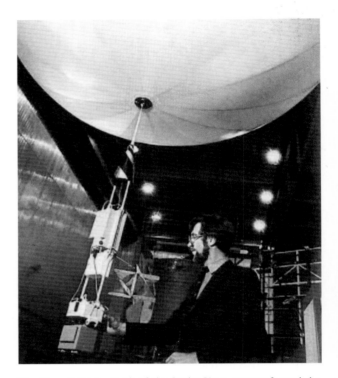

The Vega Balloon—Both of the Soviet Vega spacecraft carried a small helium-filled balloon, provided by the French space agency. Once the balloons were released from the Venus descent modules on June 11 and 15, 1985, they were inflated at an altitude of 54 km (34 mi). Swept toward the west at high speed, they traveled one-third of the way around the planet in just 46½ hours. The faint signals, sometimes as feeble as 2 watts, were picked up by 20 antennas scattered around the Earth. (CNES)

would each split into two parts: a Venus probe and a spacecraft that would examine the comet.

Each Venus entry module comprised a descent section similar to the previous Venera landers and a canister that contained a balloon. Once separated, the balloons were inflated with helium to a diameter of 3.5 m (12 ft). Carrying a small instrument package, they began to float in a region of convective clouds at an initial altitude of 53.6 km (33 mi),

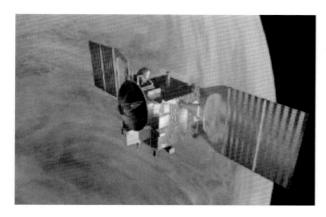

Venus Express—The European Space Agency's first mission to Venus was launched on a Russian Soyuz rocket in November 2005. Derived from the successful Mars Express mission, Venus Express reached the planet in just 153 days, then gradually maneuvered into its operational polar orbit. Over two Venusian rotations (486 Earth days), it will study many aspects of the atmosphere and hidden surface. Note the stubby solar panels, evidence of the plentiful solar radiation at Venus's distance from the Sun. (ESA)

where the air pressure was 535 millibars (just over half the sea-level pressure on Earth) and the temperature was 32°C. One balloon entered the atmosphere at 7°N, while the other was released at a comparable southerly latitude.

Beginning their flights near local midnight, they began to move rapidly toward the west, sending back intermittent signals at a mere 2 to 4 watts. By the time their batteries expired 46½ hours later, the first balloons to explore another world had traveled more than 11,000 km (6,875 mi) at an average speed of 240 km/h (150 mph) and penetrated far into the day hemisphere.

Scientists were fascinated by the ups and downs of their unprecedented voyages, especially the discovery of numerous downdraughts 10 times stronger than expected. When it approached Aphrodite, one of the highest regions on Venus, the Vega 2 balloon was buffeted by atmospheric waves that reached all the way to an altitude of 55 km (34 mi). As it dropped some 2.5 km (1.5 mi) the instruments recorded a pressure of 0.9 Earth atmosphere. Balloon 2

recorded an air temperature that was consistently 6.5°C cooler than for balloon 1. No lightning was recorded and balloon 2 recorded no breaks in the cloud. (The instrument failed on balloon 1.)

By the time contact was lost with each balloon, their respective Vega landers had long since succumbed to the hostile environment. Vega 1 landed on the Mermaid Plain north of Aphrodite, while its counterpart touched down 1,500 km (950 mi) to the southeast. Unfortunately, neither craft carried cameras, and the X-ray soil analysis instrument failed on Vega 1. However, Vega 2 found a volcanic rock known as anorthosite, which is rare on Earth but common in the highlands of the Moon. This sample appeared to be the oldest yet tested on Venus, and its high sulfur content suggested a chemical interaction with the atmosphere.

Intriguingly, the data sent back during the Vegas descent indicated the existence of two main cloud decks, rather than the three found by Pioneer Venus. One possibility was that atmospheric conditions had changed in the seven-year interval between the missions. The clouds were made of tiny droplets and said to resemble a thin fog. Among the constituents were sulfur, chlorine, and possibly phosphorus.

Almost five years later, the Galileo spacecraft passed Venus en route to Jupiter, giving scientists a brief opportunity to glimpse the planet from a distance of 16,100 km (10,000 mi), The solid-state imaging system obtained more than 80 near-infrared, ultraviolet, and visible images of the sulfuric acid clouds, showing less dense regions where heat is rising from below. The spacecraft also examined the interaction between Venus and the solar wind.

Similar opportunities arose in April 1998 and June 1999, when the Cassini spacecraft used the gravity of Venus to gain access to the outer Solar System. The most controversial result to come from these encounters was the failure to detect any evidence of lightning on Venus. Unable to explain the dramatic contradiction between results from Cassini, Venera, and Pioneer Venus, scientists could only assume that lightning on Venus is fairly weak and more difficult to detect than on Earth. "Because clouds over Venus are at very high altitudes of 40 km (25 mi) or more, it is likely that lightning at Venus, if it exists, is primarily cloud-to-cloud," commented Don Gurnett, leader of the Cassini Radio and Plasma Wave Science instrument team.

Clearly, despite decades of study, many questions remain about the dense, corrosive atmosphere that shrouds Earth's sister world. Scientists are hoping that new insights will be provided by the European Space Agency's Venus Express, which was launched in November 2005, and a Japanese orbiter that should fly later in the decade.

Derived from the successful Mars Express mission, Venus Express reached the planet in just 153 days, then maneuvered gradually into its operational orbit. The operational phase of the mission will last for two Venusian rotations (486 Earth days), during which the spacecraft will study many aspects of the atmosphere and hidden surface. Among the instruments is the Venus Monitoring Camera, which operates at visible, infrared and ultraviolet wavelengths to observe the clouds and search for volcanic activity below. However, its main task is to study the circulation and composition of the deep, dense atmosphere, and its interaction with the solar wind.

The Man in the Moon—As is the case with many planetary satellites, tidal effects have slowed the Moon's rotation so that it now spins exactly once during each orbit. As a result, we always see the same hemisphere. This near-side mosaic, taken by the Galileo spacecraft on December 7, 1992, shows the familiar dark "seas," now known to be lava-filled impact basins. Also visible is Tycho, a bright ray crater (bottom). The dark areas include: Oceanus Procellarum/Ocean of Storms (left), Mare Imbrium/Sea of Rains (left center), Mare Serenitatis/Sea of Serenity, and Mare Tranquilitatis/Sea of Tranquility (center), Mare Fecunditatis/Sea of Fertility (lower right) and Mare Crisium/Sea of Crises (right center). (USGS/NASA–JPL)

4 THE MOON: QUEEN OF THE NIGHT

The Lesser Light

The Moon has always inspired folklore and legends in many different cultures. For many centuries it was regarded as second only in importance to the Sun.

The reasons are readily apparent. Although it was obviously less luminous than the Sun, the Moon could still be seen in daylight hours and clearly dominated the night sky. Furthermore, apart from the Sun, it was the only body to reveal a visible disk. No one looking up at the star-studded celestial sphere could fail to be impressed by the yellowish, blotchy object that appeared to change shape, waxing and waning over a period of 29.5 days.

Not only would the Moon go through a complete sequence of phases, from invisibility at new Moon to a glorious, illuminated disk at full Moon, but it would sometimes pass in front of the Sun, a star or a planet, demonstrating that the "lesser light" was the closest object to Earth.

When it was no longer visible in the sky (the phase known as the new Moon), total or partial solar eclipses could occasionally be observed. At such times the black Moon passed in front of the brilliant Sun—demonstrating that the two objects have the same apparent size.

As we now know, this was the result of a unique coincidence. The Moon's average distance from Earth is 384,397 km (239,000 mi). Allowing for annual variations in the Sun–Earth distance, this means it is 380–400 times closer than the Sun. However, with a diameter of 3,476 km (2,160 mi), the Moon's diameter is 400 times smaller than that of the Sun—hence its surprising ability to obscure our much larger neighboring star.

Such a regular, predictable celestial cycle inevitably inspired ancient astrologers and timekeepers, with the result that astronomical calendars were based on the rotation of Earth (the day), the revolution of the Earth around the Sun (the year), and the revolution of the Moon around the Earth (the "moonth" or month). Even today, the Islamic calendar, for example, follows a purely lunar cycle. Each month begins with the first crescent-shaped sliver of the waxing Moon.*

Astrologers and scientists had no doubt about the links between the phases, the daily ocean tides and regular changes affecting the human body, including menstruation and mental state—hence the medical term "lunacy" (derived from the Latin "luna" or Moon). Extra staff were called into the asylums on the occasion of a full Moon.

As one might expect, the classical Greek observers managed to sort out much of what was myth and what was fact. The idea that the Moon physically changed shape from phase to phase was soon recognized as false, since the dark part of the lunar disk could often be seen faintly illuminated by earthshine—sunlight reflected from the Earth. Similarly, it was recognized very early on that the Moon was bright because it reflected light from the Sun. (In fact, its dark rocks do not make a very good mirror. Only 7% of incoming sunlight is reflected from the lunar surface—less than all the planets except Mercury.)

* Unfortunately, this lunar cycle does not exactly synchronize with the annual movements of the Sun in the sky, so the months slowly regress through the seasons over a period of about 33 years.

Galileo's Moon—These were the first published sketches showing the Moon at different phases, as seen through a telescope. They appeared in the *Siderius Nuncius* ("Starry Messenger"), written by Galileo Galilei (1564–1642). Clearly visible are large craters highlighted by shadows near the day–night terminator. Galileo correctly recognized that the indentations in the terminator indicated a mountainous surface. (Rice University)

In the third century BC, Aristarchus took the next logical step of explaining that the phases resulted from our changing view of the sunlit surface as the Moon orbited the Earth. He was also the first person to make a reasonable assessment of the Moon's size and distance by comparing the diameter of the full Moon with the size of the Earth's shadow during a lunar eclipse. His calculations suggested that the Sun was 20 times further than the Moon and seven times bigger than the Earth.

As time went by, these primitive estimates were revised. The first major breakthrough came about 225 BC, when Eratosthenes calculated the Earth's size with considerable accuracy. Then, around 150 BC, Hipparchus determined

that the Moon's distance was 30 times Earth's diameter (about 400,000 km or 250,000 mi). It was then a fairly simple calculation to work out the true size of Earth's satellite. (The modern figure is 3,476 km or 2,160 mi, so the Moon would comfortably fit in the Atlantic Ocean between Ireland and North America.) This makes the Moon one of the largest satellites in the Solar System in relation to its planet—leading to the common assertion that the Earth–Moon combination is actually a double planet system.

However, the nature of the Moon's surface remained uncertain, since the only features visible with the naked eye were the large, dark patches that make up "The Man in the Moon." One of the closest guesses was made by Democritus, as long ago as 400 BC, who suggested that it possessed mountains and valleys.

The invention of the telescope in the early seventeenth century delivered the *coup de grâce* to attempts to explain away the Moon's spots and to the religious ideal of the perfection of the heavens in general. The first person to seriously observe the Moon through the newfangled "optic stick" seems to have been an Englishman, Thomas Harriott. His drawings, however, remained unpublished.

It was the Italian, Galileo Galilei, who made the first reasonably accurate sketches of the pockmarked lunarscape when he first turned his improved spyglass toward the Moon in 1609. He was astounded at the rugged day–night terminator, marked by huge round craters and dark shadows cast by what appeared to be towering mountains. "It is like the face of the Earth itself which is marked here and there with chains of mountains and depths of valleys," he wrote.

Mapping Selene

Distracted by other scientific mysteries and problems, Galileo largely turned his attention away from the Moon, although in the 1630s he did observe lunar librations, a slight rocking in its motion that enabled parts of the far side to be seen. However, as time went by and instruments

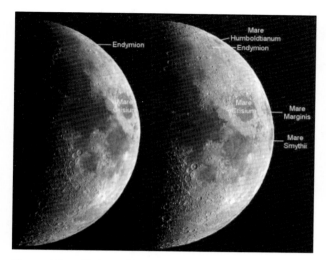

Near and Far—Various effects, generally known as libration, allow 59% of the Moon's surface to be seen from Earth at different times. One major factor is the elliptical orbit of the Moon. At the same time, the lunar equator is slightly tilted (1.5 degrees) from its orbital plane, while its orbit is inclined by 5 degrees relative to the Earth's orbit. (© David Haworth)

The Moon Eclipsed—This photo-montage shows the lunar eclipse of October 28, 2004, when the Moon passed through Earth's shadow. The images were taken over a period of 3½ hours by ESA's SMART-1 spacecraft, when it was about 290,000 km (180,200 mi) from Earth and 660,000 km (410,000 mi) from the Moon. The images of the Moon are shown in sequence, from left to right. The period of "totality," when it was completely inside the Earth's shadow, lasted about one hour. The Earth images were taken just before and after the eclipse. (ESA)

slowly improved, representations of the lunar surface gradually became more detailed and realistic.

By the mid-seventeenth century, observers such as Hevelius, a city councillor from Danzig, were giving geographical names to the most prominent features. For example, the crater now known as Plato was labeled "the Greater Black Lake." His system of nomenclature was widely used in Protestant countries until the eighteenth century, and Hevelius is recognized as the founder of the science of selenography (after Selene, the Greek goddess of the Moon).

However, it was the more elegant system chosen by Jesuit astronomer Giovanni Riccioli that has stood the test of time. On his map of 1651, based on the observations of his pupil Grimaldi, the prominent dark areas with relatively few large craters were given the Latin names for sea (mare), ocean (oceanus) and bay (sinus). Each was given a romantic description, such as Mare Tranquilitatis (Sea of Tranquility) or Mare Nubium (Sea of Clouds).

The lunar mountains were named after ranges on Earth (e.g., the Apennines and Alps), while the larger craters were given the names of famous people—usually astronomers. Since Riccioli was not a supporter of Galileo's view of the Universe, the great man was relegated to a small, insignificant crater in the Ocean of Storms, whereas Riccioli and Grimaldi were remembered by large walled basins. This system has since been extended to cover minor features discovered by modern instruments and spacecraft and the formerly unseen landmarks on the far side.

Despite the gradual improvement in instruments, many early observers believed that the Moon possessed air and water, and might support intelligent life. Even Sir William Herschel, the great eighteenth-century astronomer who discovered Uranus, was a believer in lunar inhabitants. This is surprising because there was no real evidence to support the idea. For example, when the Moon occults or passes in front of a star it disappears almost immediately, whereas if there was an atmosphere the star should flicker and fade before disappearing. Neither are there any signs of ice caps, clouds or dust storms, wind erosion or deposition.

During the nineteenth century, the first really reliable maps were produced, and the truth concerning the Moon's arid, airless environment was gradually accepted. However, this did not prevent science fiction writers from using their imaginations. For example, in his 1901 book *The First Men In The Moon*, H.G. Wells described a Moon inhabited by grotesque, insect-like creatures. Even in modern times there was considered to be enough of a chance for some primitive form of lunar life to cause the returning Apollo astronauts to be quarantined for 21 days.

Exactly how thin the lunar atmosphere was remained in dispute for many years. For example, W.H. Pickering, working in the late nineteenth century, obtained an estimate of 0.0005 times Earth's air pressure at sea level. Others saw this as too high, since the Moon's gravity is only one-sixth that of Earth and the escape velocity for any gases is only 2.4 km/s (5,400 mph).

Eventually it was concluded—correctly—that the Moon must have lost any atmosphere it had a long time ago, so that today there is an almost total vacuum. Instruments set up by the Apollo astronauts detected minute amounts of hydrogen, helium, argon, and neon, most of which are probably derived from the solar wind that impacts the surface.

There is a tremendous temperature variation on the Moon, due to the absence of an atmosphere or clouds to absorb and reflect solar radiation and to spread heat to the dark night side. Accurate figures for day and night temperatures were already available before spacecraft ever reached the Moon, one method being to measure the heat reflected to Earth. It turned out that temperatures at the lunar equator reach 120°C when the Sun is directly overhead. In contrast, a thermometer would fall to –170°C during the two-week-long night. Small wonder that the Apollo astronauts (none of whom had to endure the lunar night) were provided with a 17-layer thermal insulation suit and a water-cooled undergarment.

Despite its relative proximity, the Moon's peculiar synchronous rotation imposed severe limitations on the measurements that could be made. This resulted in considerable speculation about the nature and origin of the Moon's surface.

The accuracy of photographic atlases was limited by blurring caused by Earth's atmosphere, while the true shape and nature of features near the limb (visible edge) of the lunar sphere was difficult to discover due to foreshortening.

One particular area of dispute was the nature of the flat maria. A theory put forward by Thomas Gold in 1955 suggested that they were filled with soft dust many kilometers deep that would swallow any spacecraft attempting to land.

Another area of disagreement was the nature of the innumerable craters—some up to 250 km (156 mi) in diameter—that cover much of the surface. Some authorities considered that they were largely volcanic in origin, whereas others argued that huge meteorite impacts were to blame. The only way forward was to send robotic spacecraft and astronauts to survey the surface and bring back samples.

First to the Moon

If humanity was ever going to explore space, it would be necessary to develop powerful chemical rockets that could boost a spacecraft to the escape velocity of 11.2 km/s (25,050 mph). At that speed, the craft would be traveling so fast that it would break free of Earth's gravitational pull.

After the Earth-orbiting successes of the first Sputniks and Explorers, the next obvious challenge for Soviet and American rocket power was to reach out to our neighboring world, the Moon. The first faltering steps were marked by a series of ignominious failures, but by 1959 the communist superpower had moved into the lead.

A small breakthrough came on January 3, when Luna 1 swept past the Moon at a distance of 6,400 km (4,000 mi). Eight months later, the targeting had improved sufficiently to enable Luna 2 to smash into the Sea of Serenity, carrying metal spheres marked with the hammer and sickle. Its

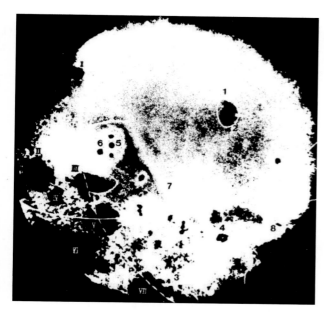

The First View of the Lunar Far Side—Luna 3 returned the first images of the Moon's far side on October 7, 1959. A total of 29 photographs were taken, covering 70% of the far side. This wide-angle view shows most of the far side (right of the dashed line), taken at a distance of about 65,000 km (40,600 mi). The dark patch at upper right (1) is Mare Moscoviense. The dark areas bordering the far side are Mare Marginus (III) and Mare Smythii (V). The small dark feature with a bright central peak is the crater Tsiolkovski (4). Feature 7 was named the Soviet mountains, but these were later found not to exist. The solid line is the lunar equator. (Novosti)

primitive instruments detected no sign of a lunar magnetic field or radiation belts.

The way was clear for a more advanced, camera-carrying craft to attempt the first photographic reconnaissance of the lunar far side. On October 6, 1959, after a two-day trip from Earth, the Soviet Luna 3 spacecraft passed over the Moon's southern hemisphere, then swung northward over the far side. At a distance of 65,200 km (40,500 mi) it began a 40-minute photographic session, taking 29 wide- and narrow-angle pictures using radiation-hardened 35-mm photographic film.*

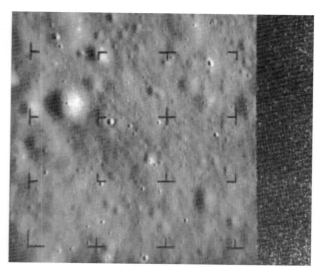

Ready for Impact!—This is the last image taken by Ranger 8 from a distance of 4.2 km (2.6 mi), two seconds before impact on February 20, 1965. The area shown is at 2.7°N, 24.55°E, about 60 km (37 mi) from the Apollo 11 landing site in the Sea of Tranquility. The area imaged is about 1.4 km (0.8 mi) across. The right side is missing because Ranger 8 crashed before completing transmission. Note the abundance of small impact craters. (NASA/NSSDC)

The spacecraft then resumed spinning while its onboard photo lab processed the film. As it headed back toward Earth, the pictures were scanned by a special light beam and relayed to Earth. By modern standards, they were noisy and lacking in contrast, but humanity's first fuzzy view of the lunar far side made headlines around the world.

The most obvious discovery was that the hidden side is largely devoid of the dark, lava-flooded maria ("seas") that make up the familiar face of the Man in the Moon. Two of the small exceptions to this rule were named the Sea of Moscow and Tsiolkovski (after the pioneering Soviet space theoretician). Another dark area, named the Sea of

---

* This was apparently "borrowed" from a captured US Genetrix spy balloon!

Ingenuity, was later found to be part of the much larger South Pole–Aitken Basin. One bright streak that was thought to be a mountain range was later found to be a long, linear ray from a crater named Giordano Bruno.

After the success of Luna 3, the next few years were full of failures and disappointment. However, after President Kennedy's May 1961 announcement that the United States was committed to sending humans to the Moon by the end of the decade, the Space Race with the Soviets increased in intensity.

In order to prepare the way for the manned Apollo program, NASA introduced a series of robotic probes that would enable engineers to gain experience of sending craft to the Earth's satellite and back, improve knowledge of the lunar environment, and provide the necessary information to select suitable landing sites.

First in line were the Ranger spacecraft, which, like Lunas 1 and 2, were designed for a simple suicide mission in which they would continue to send back images until they crashed into the Moon's surface. After a frustrating series of malfunctions and failures, the breakthrough came on July 31, 1964, when Ranger 7 smashed into the Moon on the northern rim of the Sea of Clouds, a region crossed by rays of material ejected from a large, fresh crater known as Copernicus.

The spacecraft's six TV cameras started snapping just 15 minutes before it smashed into the lunar surface, eventually transmitting 4,316 images that showed the barren surface in increasing detail. The last image, taken only 2.3 seconds before the spacecraft was destroyed, had a resolution of 0.5 m (1.6 ft) per pixel—1,000 times sharper than any images obtained from Earth.

The jinx had been broken. Six and a half months later, Ranger 8 returned 7,137 pictures over the last 23 minutes of its kamikaze plunge into the Sea of Tranquility (later to be the location of humanity's first footsteps in the lunar dust). This time, the Sun was lower in the sky, providing better contrast and information about the height and steepness of the mountains and craters. The final image taken before impact had a resolution of 1.5 m (5 ft) per pixel.

Ranger 9 starred in its own TV spectacular, after the sequence of spectacular approach images was converted for live broadcast. Altogether, it sent back 5,814 pictures over a period of 19 minutes, eventually rushing to its destruction inside the 112-km (70-mi) diameter Alphonsus crater. The final image, taken a quarter of a second before impact from an altitude of about 600 m (0.4 mi), had a resolution of 0.3 m (1 ft) per pixel. Navigation was also improving, enabling the spacecraft to plough into the crater floor only 6.5 km (4 mi) from its intended target.

Once again, the pictures demonstrated the pockmarked nature of our neighbor, with craters visible at all scales down to the limit of resolution. Some craters with dark haloes appeared to be volcanic in origin, although it seemed likely that cosmic impacts had excavated most of the larger basins. The surface generally seemed to be blanketed with fine dust or sand, but the Ranger images could not tell scientists whether it would behave like quicksand and swallow a spacecraft. That would have to await the next generation of soft-landing robotic craft.

One intriguing discovery made by analyzing Ranger tracking data was that the Moon's gravity was uneven, causing it to pull the spacecraft off course.

Robotic Surveyors

It was public knowledge that NASA's next step after the Ranger missions would be the delivery of soft-landers to the lunar surface. Not only was this maneuver vital for achieving a manned landing later in the decade, but it also promised to provide first-hand information about the nature of the alien terrain.

As was so often the case in those early years of the Space Race, the Soviets also had similar ideas, and they were determined to beat their capitalist rivals to the punch. After many failed attempts, they finally succeeded in placing Luna 9 in orbit around the Moon and delivering it to the Ocean of Storms on February 3, 1966. Five meters (16 ft) from the surface, a small, round capsule was ejected from the nose of the mother spacecraft. Protected from the jolt of

Surveying the Surface—This Surveyor 7 mosaic shows the lunar highlands north of a large, relatively young crater called Tycho. Surveyor 7 landed on January 10, 1969 at 40.88°S, 11.45°W and sent back about 21,000 pictures over a period of four weeks. The block in the foreground is about 0.5 m (1.6 ft) across and the small crater nearby is about 1.5 m (5 ft) in diameter. The hills on the horizon are about 13 km (8 mi) away. (NASA/NSSDC)

impact by shock absorbers, the metallic ball opened its four petals and activated its radio transmitter.

During seven radio sessions, which lasted a total of 8 hours 5 minutes, the historic first pictures from the lunar surface were relayed back to Earth. The Soviets were not anticipating any eavesdroppers, but Britain's Jodrell Bank radio observatory was listening in. Using a simple newspaper facsimile machine, the observatory was able to convert the signals into images. To the annoyance of the Soviets, the pictures were passed on to the press, providing the British media with an unexpected scoop.

Luna 9 remained in contact for the next three days, sending back views of small rocks nearby and a black horizon 1.4 km (1 mi) away. The first views showed that the capsule had landed at an angle of 16.5 degrees, apparently on the side of a small crater. In a subsequent panorama the capsule's tilt had increased to 22.5 degrees, suggesting that it had come to rest on a small rock or slightly unstable slope.

Four months later, it was NASA's turn to deliver a spacecraft to the lunar surface. Much more sophisticated than Luna 9, the Surveyors were designed to pave the way for the Apollo missions by demonstrating soft-landing capabilities on Earth's neighbor. A secondary objective was to confirm that the dusty plains were safe for a human landing.

After Surveyor 1 successfully touched down on the Ocean of Storms on June 2, 1966, it became the second spacecraft to send back images from the Moon's surface. Its TV camera scanned the local area, frame by frame, providing high-resolution views of a fine-grained material and occasional rocks more than 1 meter (3 ft) across. Panoramic mosaics and subsequent orbital surveys showed that it had arrived in a "ghost" crater that had been flooded with lava long ago.

Surveyor 1 sent back 10,732 images before the two-week-long lunar night began on June 14. It survived temperatures of –160°C before returning to life on July 6. Another 618 pictures were returned before the end of operations.

The Ocean of Storms was also the scene of the second Soviet landing. Once again, Luna 13 was ejected just before impact, arriving with a bump. Its images revealed a flat, fairly featureless landscape, relieved only by the eerie black shadow cast by the booms and antennas of the alien craft.

After the failure of Surveyor 2, the US program got back on track on April 20, 1967, when Surveyor 3 landed about 370 km (230 mi) south of Copernicus in the southeast section of the Ocean of Storms. The spacecraft bounced several times before coming to rest on a 14-degree slope near the rim of broad, subdued crater.* Dust kicked up by the thrusters caused some degrading of the images, and appeared to reduce mobility of the mirror used for imaging.

This time, in addition to the TV camera, the lander carried a sampling arm that was used to dig four trenches and test the properties of the "soil." Results indicated that it resembled a "fine-grained terrestrial soil," though the trenches also revealed a brittle crust up to 5 cm (2 in) deep.

* The crew of Apollo 12 visited Surveyor 3 in November 1969 and brought back parts of the spacecraft to Earth.

Lunar Landing Sites—This map shows where spacecraft have successfully soft-landed on the Moon. The eight Soviet Luna craft (red) explored many sites in the northern hemisphere, and sent back three small soil samples. Four of NASA's five Surveyor spacecraft (yellow) touched down close to the equator. The exception was Surveyor 7, which explored the vicinity of the fresh ray crater Tycho at 40°S. The first three Apollo landings (green) took place near the equator, but the other three were sent to more rugged highland areas. (NASA/NSSDC)

Attempts to crush a small white rock in the jaws of the scoop ended in failure.

After the loss of Surveyor 4, the fifth Surveyor made a hard landing about 25 km (15 mi) northwest of Tranquility Base, the site that would be made famous by the Apollo 11 crew's historic exploits less than two years later. TV images showed skid marks where the craft slid down a 20-degree slope, clear evidence that careful selection of landing sites would be imperative for the human explorers.

The most exciting aspect of the mission was the use of an alpha-scattering spectrometer to measure the elemental composition of the surface. Data showed it to be rich in oxygen, silicon, and aluminum—probably similar to basaltic lava on Earth.

The final two Surveyors were also resounding successes. After a touchdown in the Central Bay (in the center of the lunar near side) on November 10, 1967, Surveyor 6 became the first spacecraft to lift off from the Moon. A 2½-second burn of its thrusters enabled it to hop a couple of meters sideways, thereby providing an opportunity for stereo imaging.

Unlike its predecessors, the last of the series was despatched far to the south, close to the large, fresh, impact crater called Tycho. Although there were numerous small rocks, the area was surprisingly flat and smooth. The sampling arm dug seven trenches, turned over a rock and even saved the alpha-scattering experiment from failure by lowering it to the ground after it refused to drop on its own. With more than 87,000 pictures returned from five potential Apollo landing sites, the possibility of unpleasant surprises awaiting the first human visitors now seemed remote.

Orbiters

One of the key tasks undertaken during expeditions to unknown lands has always been the production of sketches and maps that will guide all those who follow in their footsteps. The Moon Race was no exception. Both the Soviet Union and NASA prepared the way for human

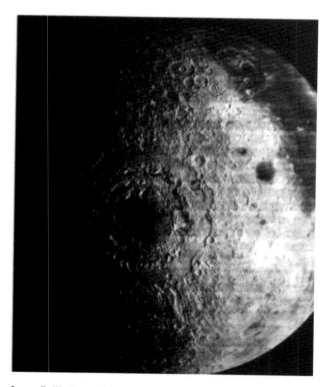

Lunar Bull's-Eye—This Lunar Orbiter 4 image shows the bull's-eye structure of the Mare Orientale (Eastern Sea) Basin. It lies on the western edge of the Moon's near side and is only visible at favourable times of libration. Three distinct rings can be seen. The outermost is the Cordillera Mountain scarp, almost 900 km (560 mi) in diameter. Orientale was formed by a giant impact early in the Moon's history, but, unlike most basins, it contains relatively little dark, basaltic lava. (NASA/NSSDC)

explorers by sending orbiters that would survey most of the lunar surface and seek out potential landing sites.

Building on the successes of the first Soviet flyby missions (Luna 3 and Zond 3) and the hard landings by NASA's Rangers, both superpowers prepared to conduct a detailed reconnaissance of the cratered world. Scientists eagerly awaited the results.

Today, it is accepted that almost all of the lunar craters and basins were excavated during a great meteorite and comet bombardment around 4 billion years ago. This ancient pattern has subsequently been overlaid by occasional impacts, most on a much smaller scale. However, in the 1960s there was considerable controversy over the nature of the innumerable circular depressions and mountains that produced the rugged lunar terrain.

Unfortunately, although the Earth must have endured a similar, violent past, there was little evidence of an era when huge incoming projectiles bored into the crust and excavated craters the size of European countries. On the other hand, scientists were familiar with the thousands of volcanic mountains and craters that spread molten lava and ash over the surrounding landscape. Which was the dominant process on the airless Moon?

Apart from occasional reports of "reddish glows" and "obscurations"—notably near the large crater named Alphonsus—there was no evidence that the Moon currently exhibited any volcanic activity. The most obvious volcanic features were small rounded hills with central pits. On the other hand, the craters were not randomly spread, as they should be if they were created by random cosmic impacts. The dark "seas" seemed to have been flooded by ancient lava flows that often covered or filled the oldest craters. Some craters appeared to line up in chains, and many of the larger basins were overlain or eaten into by smaller, younger craters.

Orbital spacecraft were the only way to answer these mysteries. As was so often the case in the early Space Age, the Soviets paved the way when Luna 10 braked into a 350 × 1,017 km (215 × 630 mi) orbit on April 3, 1966. Although it was not carrying a camera, the Moon's first artificial satellite was able to measure elevated levels of cosmic-rays, solar radiation, and micrometeorites. However, there were no radiation belts, no atmosphere, and no magnetic field worth mentioning. Orbital tracking confirmed the mysterious variations in the gravitational field hinted at by earlier missions.

Although the Soviets grabbed the headlines, it was the US Lunar Orbiters that revealed the true nature of the

Moon. Wide- and narrow-angle pictures taken by the Eastman–Kodak imaging system were exposed on 70-mm film, developed on board (rather like Polaroid instant cameras), then scanned by a high-intensity light beam and slowly transmitted back to Earth.

From an elliptical path that carried it to within 50 km (30 mi) of the surface, Lunar Orbiter 1 sent back 211 image pairs at medium resolution. They showed a relatively sparse population of craters on the Ocean of Storms, in contrast to the heavily pockmarked far side. However, it was the first ever picture of the crescent Earth above the lunar horizon that most captured the imagination.

Almost eight months passed before Luna 12 sent back the first Moon pictures from a Soviet orbiter. Unfortunately, they suffered in comparison with the US images, and only two were ever released to the media.

Lunar Orbiter 2 was devoted to scanning potential Apollo landing sites to the north of the equator, as well as large parts of the far side. However, the most memorable picture of the entire program was a sideways view of the

Copernicus—This oblique view of Copernicus, taken by Lunar Orbiter 2 on November 28, 1966, was described as "the picture of the century." Looking north across the 93-km (58-mi) diameter crater, it shows central peaks rising about 400 m (1,320 ft) above the crater floor, and a terraced rim. Copernicus is surrounded by bright rays that make it very prominent. One of the youngest large craters on the Moon, it probably formed less than 1 billion years ago. The most recent period of lunar history is known as the Copernican era. (USGS)

rugged terrain around the impressive "young" crater called Copernicus. Like its predecessor, the spacecraft was deliberately crashed onto the far side.

Mapping of the possible Apollo sites was completed by Lunar Orbiter 3, enabling mission planners to select eight prime landing areas. The impressive image resolution was confirmed when one picture actually showed Surveyor 1 on the surface.

The final two Lunar Orbiters became the first spacecraft to follow polar orbits around the Moon, which enabled them to carry out broad mapping and scientific surveys as the Moon rotated beneath them. During May 1967, Lunar Orbiter 4 photographed almost the entire near side and 75% of the far side. It was also able to send back the first reasonable images of the lunar south pole. Lunar Orbiter 5 completed the photography of the far side and collected medium- and high-resolution imagery of 36 preselected regions of scientific interest.

The two craft also provided valuable practice in preparation for the forthcoming manned lunar flights. Once again, careful tracking of their paths revealed significant gravitational anomalies that pulled the orbiters off course. It was determined that these anomalies were caused by mass concentrations (mascons)—subsurface piles of dense rock associated with the dark "seas."

Although NASA's robotic reconnaissance of the Moon was over, the Soviets continued to send occasional orbiters until 1974. In October 1971, Luna 19 carried a variety of environmental sensors and conducted panoramic mapping, while Luna 22 operated for three months, swooping to an altitude of 25 km (16 mi) in June 1974. Meanwhile, the US Explorer 35 carried out a low-profile mission to monitor the Moon's neighborhood from a 800 × 7,962 km (500 × 4,950 mi) orbit. Results made it clear that the absence of a magnetic field allowed the particles of the solar wind to impact the lunar surface. The modest little craft remained operational throughout the Apollo era.

Dress Rehearsals

Despite a handful of spectacular, but misleading, publicity extravaganzas by the Soviet Union, the American juggernaut inexorably overhauled its communist rival in the so-called Moon Race. Once rendezvous and docking were achieved by the two-man Gemini missions, all attention was diverted toward the Apollo program.

Despite a fire in January 1967 that killed the three occupants of the Apollo 1 capsule during a launch countdown test, NASA officials and contractors regrouped, eventually producing a much safer spacecraft. The show was back on the road on October 11, 1968 when Apollo 7 lifted off from Cape Kennedy, Florida, for the first manned flight of the Apollo Command and Service Module (CSM). Despite an irascible crew who were suffering from head colds and determined to vent their spleens on unfortunate ground controllers, the shakedown flight went according to plan.

Aware that the Soviets were in the running to launch a manned circumlunar flight before the end of the year, and plagued by delays in the development of the Lunar Module (the astronauts' landing vehicle), NASA bosses took the daring decision to send a relatively untried craft all the way to the Moon.

Millions gathered at the Cape to watch the launch of Apollo 8 on December 21, 1968. Three days later, everyone on Earth held their breath as the CSM occupied by Frank Borman, James Lovell, and William Anders disappeared behind the Moon. A failure of the main engine would strand them in lunar orbit, never to return.

Fortunately, all went according to plan, and the spacecraft entered an elliptical orbit that took them to within 111 km (69 mi) of the rugged surface. Two orbits later, a second engine burn circularized the orbit, providing a perfect vantage point to photograph and describe the craters, plains, and mountains that swept past the windows.

The most famous of their TV broadcasts came on Christmas Eve when the crew read from the book of

Earthrise—This famous view of the rising Earth greeted the Apollo 8 astronauts as they came from behind the Moon after the lunar orbit insertion burn on December 24, 1968. The craters and mountains in the foreground are near the eastern limb of the Moon as viewed from Earth. The lunar horizon is approximately 780 km (485 mi) away, and the width of the visible horizon is about 175 km (110 mi). On the Earth, 384,400 km (240,000 mi) away, the sunset terminator bisects Africa. (NASA)

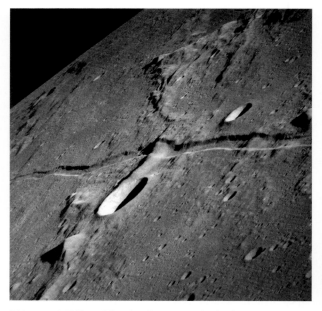

Ridges and Rilles—The Apollo crews obtained many stunning orbital views of the Moon, including some potential landing sites. This Apollo 10 image shows the 240-km (150-mi) long Rima Ariadaeus, a classic example of a straight rille caused by crust dropping between two faults. The rille runs between the flat plains of the Sea of Tranquility (off picture to the right) and the Sea of Vapors (top left). (NASA)

Genesis. "In the beginning, God created the heaven and the earth...." However, it was a picture of the small, half-illuminated Earth rising above the arid lunar horizon that particularly affected people back home and helped to generate a new awareness of our home planet's fragility.

After becoming the first humans ever to study Earth's neighbor in intimate detail, the crew of Apollo 8 headed for home on December 25, eventually splashing down in the Pacific Ocean a little over two days later.

By March 1969, the Lunar Module was ready for its first test flight around the Earth. Once again, the craft performed beautifully, despite its ungainly looking appearance. All that remained was to try out all of the Apollo hardware during a full dress rehearsal above the Moon.

Launched on May 18, 1969, Apollo 10 carried the second crew to travel to another world. After orbit insertion, Tom Stafford and Eugene Cernan powered up the Lunar Module "Snoopy" and separated from the CSM

"Charlie Brown." During the next hour, the men swept to within 14 km (9 mi) of the Sea of Tranquility, the primary site for the first landing. Numerous close-up photographs of the Moon's surface, in particular the planned Apollo landing sites, were taken. Once again, the geologists on Earth were having a field day.

Despite observing a plethora of boulders, enough "to fill up Galveston Bay," the Apollo 11 target area seemed "very smooth, wet clay, like a dry river bed in New Mexico or Arizona."

The jettison of the descent stage almost ended in disaster when the ascent stage went into a sudden spin, but Stafford managed to bring it under control, and the return trip to Charlie Brown was completed without further problems.

After 31 spectacular orbits of the Moon, the trusty main engine fired once again to blast them out of the Moon's gravitational grasp. Apollo 10 splashed down in the Pacific on May 26, after a flight of 8 days and 1.3 million km (830,000 mi). The "dry run" had been completed with no major hiccups. The world waited in anticipation for the historic "giant leap" that would see humans from planet Earth set foot for the first time on another world.

Tranquility Base

In 1961, President John F. Kennedy committed the United States to land a man on the Moon and return him safely to Earth by the end of the decade. Eight years later, that commitment was realized when the courageous crew of Apollo 11 completed humanity's first expedition to the surface of another world.

The adventure began on July 16, 1969, when the huge Saturn V rocket lifted off from Cape Kennedy in Florida, cheered on by more than one million people who crammed onto the beaches and roadsides, and by millions more watching TV around the world.

The three-day journey to the Moon in the cramped Command and Service Module Columbia was also flawless, and at 75 hours 41 minutes into the flight, radio contact was lost as the three-man crew of Neil Armstrong, Edwin "Buzz" Aldrin, and Michael Collins disappeared behind the Moon. When Apollo 11 reappeared, the spacecraft was following an elliptical path that took the men to within 113 km (70 mi) of the rugged surface. A few hours later, the orbit became almost circular, clearing the way for the historic excursion to the Sea of Tranquility.

On July 20, Armstrong and Aldrin floated into the Lunar Module, Eagle, then separated from Columbia. The first step was successfully negotiated when Eagle's descent engine fired to alter its flight path, bringing it to within 16 km (10 mi) of the surface. Then came the moment of truth

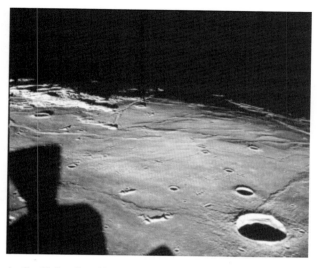

Apollo 11 Landing Site—This oblique photograph, looking west over the Sea of Tranquility, was taken from the Apollo 11 Lunar Module (LM) one orbit before descent, while Eagle was still docked to the Command Module. The landing site was to the right of center, near the day–night terminator. Although craters, ridges, grooves, and lava channels are visible, the region is relatively smooth. The crater Maskelyne in the right foreground is 23 km (14 mi) in diameter. The Apollo 11 landing site is 210 km (130 mi) from the center of Maskelyne. The black object (lower left) is a LM thruster in the camera's field of view. (NASA)

as the crew headed for the lunar plains, 102 hours 33 minutes into the mission.

For the first part of the descent, Eagle flew with its head down, but Armstrong eventually turned the craft around to enable the landing radar to lock on. On a number of occasions, flashing alarms on the primitive computer warned of an information overload, but the astronauts pressed on regardless. Aldrin's deadpan voice told the tale as he relayed the altitude and speed of forward motion.

Back on Earth, controllers in Houston were becoming concerned as the fuel gauge swung inexorably toward "empty." Time was running out. They received some assurance when Aldrin announced that he could see the faint shadow of one of Eagle's footpads through the

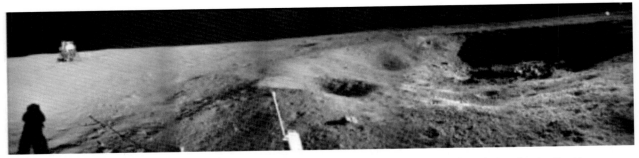

Tranquility Base—Neil Armstrong is almost completely absent from the pictures taken during the Apollo 11 moonwalk. However, the commander's long shadow is visible (left) in this panoramic view taken on the rim of East Crater, so named because it is 60 m (200 ft) east of the Lunar Module. Armstrong had to guide Eagle over this 30-m (100-ft) crater during the final approach. The images were taken near the end of humanity's first walk on the lunar surface. The white object in the foreground is a stereo camera. (NASA, Brian McInall)

window. Now only about 12 m (40 ft) above the ground, the LM was "picking up some dust." A few seconds later, the engine stopped. They had made it with 20 seconds of fuel to spare.

Eventually, a delighted Armstrong described their surroundings. Tranquility Base was "a relatively level plain cratered with a fairly large number of craters of the 5- to 50-foot variety and some ridges, small, 20 or 30 feet high ... and literally thousands of little, one- and two-foot high craters." The commander also noted the presence of some angular blocks perhaps 0.6 m (2 ft) across and a hill about 1 km (0.6 mi) away.

The crew were too excited to sleep, so humanity's first excursion onto the pristine lunar surface was brought forward a few hours. TV viewers around the globe were treated to the gray, ghost-like image of Armstrong awkwardly clambering down the LM ladder and dropping gingerly onto the dusty "sea."

"That's one small step for man, one giant leap for mankind," he declared. Armstrong was soon describing the fine gray material that adhered to his boots. Tread marks, a few millimeters deep, left clear evidence that humans had finally set foot on Earth's closest neighbor.

After about 20 minutes, the eager Aldrin joined him outside the safety of Eagle's cabin, and the pair set to work erecting the Stars and Stripes and collecting samples of local material. Hammering core tubes into the fine dust proved quite a challenge. Armstrong took charge of the photography, with the result that the first man on the Moon appears in only a handful of images taken during that momentous geological field trip.

They also deployed three science instruments: a 1.2-m (4-ft) long rectangle of aluminum foil that would collect gather from the solar wind collector during their sojourn on the surface; a retroreflector made of 100 prisms that would reflect laser beams and enable the accurate measurement of the Earth–Moon distance; and a solar-powered seismometer to measure moonquakes.

As the 150-minute mark approached, the men, now covered in gray grime, hustled to load their boxes of precious samples on board the Eagle. Then it was time to seal themselves in and repressurize the cabin. A few hours later, they opened the hatch once more to dump their rubbish, and after a final verbal grilling from ground control, they were able at last to grab some well-earned rest.

Twenty-one and a half hours after their dramatic arrival, the men prepared for another first—blast-off from the Moon. If the ascent engine failed, they would be stranded 384,000 km (239,000 mi) from home and condemned to a lonely death on an alien world. Fortunately, the engine ignited on time, sending a blast of exhaust gas that set the flag flapping violently. Then they were safely on their way to a rendezvous with Michael Collins in Columbia.

After a successful docking, the crew vacuumed themselves and the sample boxes before passing them through to Collins in the Command Module. Analysis back on Earth showed that the 21 kg (46 lb) of extraterrestrial material were igneous (volcanic) rocks that had formed 3–4 billion years ago, long before all but the oldest rocks on Earth.

The crew and some others who had come into contact with the unique samples were subjected to three weeks of quarantine, but all emerged unscathed. As expected, Earth's arid neighbor appeared to be most unsuitable for the evolution of even primitive bacteria.

Science on the Hop

After the headline-grabbing exploits of humanity's first landing on another world, Apollo 12 was something of an anticlimax. After a near-abort due to a lightning strike during launch on November 14, 1969, the spacecraft headed for the Oceanus Procellarum (Ocean of Storms). Although scientists unanimously rejected the selected site, mission planners ignored their advice in favor of a more media-friendly landing alongside an old friend, the unmanned Surveyor 3.

So it was that Pete Conrad and Alan Bean set the LM Intrepid down on the edge of Surveyor Crater. Conrad described humanity's second lunar landing site as "a place a lot dustier than Neil's ... like an undulating plain." When Bean joined him on the surface, he reported seeing "little shiny glass" in the rocks.

Before setting off on their first exploratory walk, the crew set up a sophisticated surface science package, powered by a small nuclear generator. This included a seismometer to measure moonquakes, a magnetometer and various instruments to measure the solar wind, and any traces of a lunar atmosphere. They then strolled over to

Surveyor Crater—Charles Conrad Jr, Apollo 12 commander, examines the unmanned Surveyor 3 spacecraft during the mission's second moonwalk. The Lunar Module "Intrepid" (top right) landed on the Moon's Ocean of Storms only 180 m (600 ft) from the dead spacecraft. The television camera and several other components were removed from the Surveyor and brought back to Earth for scientific analysis. Surveyor 3 soft-landed on the Moon on April 19, 1967. (NASA)

nearby Shelf Crater, observing bedrock at the bottom and boulders resting inside the rim. On the return leg, they described two unusual mounds.

"It looks like a small volcano," said Bean, "only it's just about 4 feet high and about 5 feet across the top. It slopes down to a base with a diameter of 15 to 20 feet."

During the second moonwalk, they followed a heart-shaped traverse which took them past various unremarkable craters where they took photos, collected rocks, and extracted core samples. At one point, Conrad kicked the dark surface, revealing a lighter, cement-colored layer. Near the end of their trek, they stopped to cut away pieces of the Surveyor—now burned brown by the Sun—for analysis back on Earth.

When Intrepid blasted off from the Moon after 31½ hours on the surface, it was carrying 34.3 kg (75 lb) of lunar rock and soil, which was subsequently found to include a

Shepard's Rickshaw—Apollo 14 Commander Alan Shepard assembling a double core tube as he stands by the Modular Equipment Transporter (MET). Nicknamed the "rickshaw," the MET was a cart for carrying tools, cameras, and sample cases over the lunar surface. The photo was taken about 170 m (560 ft) northeast of the Lunar Module during the second moonwalk on February 6, 1971. (NASA)

4.45-billion-year-old rock that was unusually rich in radioactive elements. Scientists were surprised when the Moon continued "ringing like a bell" for 51 minutes after the LM was deliberately sent to its destruction on the lunar surface.

To the delight of the scientific community, the third Apollo landing mission was targeted at a more interesting region of undulating hills, about 50 km (30 mi) north of the Fra Mauro crater. The highlands were believed to be made up of an ejecta blanket blasted out by the huge impact that created the Mare Imbrium (Sea of Rains).

Unfortunately, the number 13 proved to be unlucky for the three-man crew. On April 13, 1970, when the spacecraft was more than three-quarters of the way to the Moon, an explosion tore the heart out of the Service Module. The landing was immediately abandoned and the crew were forced to use the Lunar Module as a lifeboat. They could only stare through the windows of Aquarius and snap photos of the rugged world that they were destined never to set foot upon.

Fortunately, Apollo 13 was able to limp home, but there was obviously a serious safety issue to solve, and it was not until nine months later that another crew was sent beyond Earth orbit. In the meantime, the Soviets took advantage of their rival's discomfiture to carry out the first successful robotic sample return. Luna 16 brought back 101 g (3.5 oz) of soil from the Sea of Fertility.

Eventually, NASA decided that Apollo 14 would complete the previous aborted mission. So it was that Alan Shepard, Stuart Roosa, and Ed Mitchell—a remarkably inexperienced crew with only 15 minutes of space time between them—set off for the lunar highlands on January 31, 1971. Almost five days later, Shepard and Mitchell headed for Fra Mauro in the Lunar Module, Antares. Despite a number of technical problems, including a radar altimeter that was feeding back incorrect information, they eventually touched down on a hummocky plain about 1 km (0.6 mi) from Cone Crater, the main geological objective of the mission.

Although one foot of the LM was inside a small crater, causing it to tilt by 7 degrees, Antares was quite stable and the two moonwalks were able to go ahead as planned. During the first excursion, which lasted 4 hours 47 minutes, the men collected samples in the local area and deployed a surface science package that included a seismometer, a solar wind collector and a laser reflector.

Mitchell strung out a line of sensors that were to be used in an active seismic experiment. It was only a partial success, since the hand-held "thumper" device set off only 14 of 21 small explosive charges. There was also a small grenade launcher that would be fired after lift-off in order to calibrate the Apollo 12 and 14 seismometer network. (It was actually deployed too close to the other instruments and never used.)

The prime objective of the second excursion was to collect samples from the rim of Cone Crater, about 1.6 km (1 mi) away. However, the lack of visual clues, tiredness, and slow progress as they trudged up the slope with their hand-held rickshaw caused them to turn back—unaware that they were just a few meters short of the rim. Nevertheless, the booty from their 4 hour 41 minute walk included about 22 kg (48 lb) of rock and soil samples, as well as hundreds of photographs.

Unable to roll rocks down the slopes of Cone Crater, Shepard resorted to using a "genuine six-iron" on the handle of a contingency sample return container to send a golf ball "miles and miles" down the bumpy, barren lunar fairway. After 33½ hours on the lunar surface, the weary crew of Apollo 14 eventually headed for home with 42.8 kg (94 lb) of precious samples. Once again, the Lunar Module was sent plummeting into the surface in a final suicide mission.

Mountains of the Moon

The near-disaster of Apollo 13, coupled with the delay in returning to flight and tight budgetary constraints, led to NASA announcing the cancelation of Apollos 18 and 19 on September 2, 1970.* Time was running out if a reasonably broad scientific exploration of the lunar near side was to be undertaken before humanity returned to more parochial concerns in Earth orbit.

By the summer of 1971, three successful manned landings led to increasing confidence that the remaining three missions could venture to more rugged, remote regions away from the equator. In order to fulfill the potential of these expeditions, increased mobility was required, so a fold-up roving vehicle was carried on the Lunar Module. The spacecraft was also upgraded to survive on the surface for up to 72 hours, while improved spacesuits also allowed longer excursions by the astronauts.

The first mission to benefit from this was Apollo 15, which was targeted at a geologically interesting site nestled between the Apennine Mountain Front and a huge gash in the surface known as Hadley Rille. Within a fairly small area, the crew could explore a basaltic plain, a probable collapsed lava "tube" and mountains more than 3,000 m (10,000 ft) high.

The Lunar Module, Falcon, skimmed over the Apennine range and made a steep descent before a successful

Hadley Rille—Apollo 15 astronaut James Irwin working alongside the rover at the edge of Hadley Rille. The picture was taken looking northwest from the flank of St George Crater. The rille measured about 1 km (0.6 mi) across and 350 m (1,150 ft) deep. It has been widened and made shallower over time by movement of wall material downslope. Large blocks are visible on the walls and floor. (NASA)

touchdown on July 30, 1971. Within a couple of hours, the eager Commander, David Scott, opened the top hatch for a quick 360-degree scan of the spectacular site.

"All the features around here are smooth," he reported. "There are no sharp, jagged peaks or no large boulders apparent anywhere. To the east of (Mount) Hadley Delta ... there appear to be lineaments running, dipping, through to the northeast."

The first surface exploration began the next day, when Scott became the seventh man to walk on the Moon. After a struggle to deploy the rover, he and Jim Irwin set about testing their new vehicle. To their disappointment, the front wheel steering was not functioning; they would have to rely on rear steering only.

* Apollo 20 was canceled on January 4, 1970, as scientists were gathering in Houston to discuss the results from Apollo 11.

Roving near Mount Hadley—James Irwin, Apollo 15 Lunar Module pilot, working alongside the Lunar Roving Vehicle during the first moonwalk at the Hadley–Apennine landing site on July 30, 1971. The shadow of the Lunar Module "Falcon" is in the foreground. This view is looking northeast, with Mount Hadley in the background. Apollo 15 was the first mission to use such a roving vehicle. (NASA)

The drive over the undulating terrain proved to be a "rock 'n' roll ride," spewing out "rooster's tails" of dust as they reached a speed of around 8 km/h (5 mph). In a mere 25 minutes they reached Elbow Crater and nearby Hadley Rille, having driven a distance of 3 km (2 mi)—more ground than the Apollo 14 crew were able to cover in an entire moonwalk.

At each stop, ground controllers in Houston were able to command the rover's TV camera to pan across the scene, showing the two men at work, enthusiastically taking pictures, picking up rocks, and using a rake to sift the soil for small pebbles.

One kilometer (0.6 mi) wide and 350 m (1,150 ft) deep, Hadley Rille was particularly spectacular, sharply etched in bright sunlight and black shadow. Standing close to the steep edge of the sinuous valley, Irwin noted several large blocks that had apparently rolled downslope onto the flat floor.

The men were so absorbed with the first part of their planned route that mission control ordered them to omit stop three and return to the Falcon. On arrival, they unloaded the nuclear-powered Apollo Lunar Surface Experiment Package (ALSEP) and began setting up the experiments about 110 m (350 ft) from the Lunar Module. Included in the ALSEP were a seismometer, solar wind collector, magnetometer, laser reflector, dust detector, and ion detector.

There were also two sensors that would measure heat from the interior, but it proved to be extremely difficult to drill the deep holes for these probes in the compacted surface material. Extraction of a core from a 2.4-m (8-ft) deep hole also proved to be a major headache, but scientists back on Earth thought it was worth the effort when they found 58 distinct layers beneath a mixed "topsoil."

The second Moon drive headed toward a cluster of craters at the foot of Mount Hadley Delta. Discovered among the scattered boulder fields were two objects of interest. The first was a large rock coated with greenish volcanic glass that was found to be rich in magnesium. The once-molten glass had rapidly solidified after being sprayed from a lava fountain 3.5 billion years ago.

Then the men came across the "Genesis" rock, a small, white specimen sitting on top of a mound. Later analysis showed that this modest fragment was a long-sought remnant created some 4.5 billion years ago, when the Moon (and the Solar System) was in its infancy. About 400 million years later, this ancient piece of the early lunar crust had been subjected to a major impact shock during a period of massive bombardment by asteroids and comets.

After the seven-hour exploratory drive on August 1, the crew were restricted to a much shorter trip the following day. This time, they headed west, straight for Hadley Rille. Once there, the TV camera clearly showed layers of supposed lava flows topped by numerous boulders. Pieces of bedrock were hammered free for subsequent analysis in Houston.

By the time the Falcon lifted off from the surface, the

crew had broken all records. During 67 hours on the Moon, they had taken part in three excursions, driven 27.9 km (17.3 mi), collected almost 78 kg (170 lb) of rock and soil samples, and taken 1,152 photographs. Meanwhile, orbiting up above, Al Worden had launched a mini-satellite to explore particles and magnetic fields, and conducted an automated survey of the terrain over which his spacecraft passed.

Desolate Descartes

With only two missions remaining, NASA managers were keen to explore sites of exceptional scientific interest— which meant heading for the highlands. However, they were torn between a routine visit to a near-equatorial region, or a more dangerous journey to Tycho, a "fresh" ray crater far to the south. In the end, pragmatism won out: the site selected for the penultimate mission was a fairly flat plateau close to Descartes Crater and southwest of the Apollo 11 landing.

The chosen site had two basic types of terrain: the smooth or undulating Cayley Plains, and the Descartes formation—made of hilly, highland plateau material. It was hoped that this region, probably the result of ancient volcanism, would be far enough from the Imbrium basin to have been unaffected by material ejected during the huge impact that created it 3.8 billion years ago.

Apollo 16 lifted off from Cape Kennedy on April 16, 1972, carrying John Young, a veteran of the Apollo 10 landing rehearsal, Charlie Duke, and Thomas Mattingly. Four days later, despite a six-hour delay due to technical problems, Young and Duke safely set down the Lunar Module, Orion, close to the target zone. When the men took a look around, they sighed with relief with the realization that Orion's rear footpad was no more than 3 m (10 ft) from the rim of a 9-m (30-ft) deep crater they had overflown during the final seconds of the descent.

The pair were raring to go. Unfortunately, Young's first footsteps in the Descartes dust were not recorded for

Plum Crater—Apollo 16 Lunar Module pilot Charles Duke is photographed collecting lunar samples at Station No. 1, during the first extravehicular activity at the Descartes landing site on April 21, 1971. This picture, looking eastward, shows Duke standing at the rim of Plum Crater, which measured about 40 m (130 ft) wide and 10 m (33 ft) deep. The parked Lunar Roving Vehicle can be seen in the background. (NASA)

posterity due to a balky antenna on the lunar module. However, once he was joined by Duke, they unfolded the roving vehicle, installed its TV antenna, and set up the experiment package (which included an ultraviolet camera and a seismic experiment involving a grenade launcher that would be fired by remote control after the crew left for home).

All went according to plan, apart from an unfortunate accident when Young's boot caught in a power cable, pulling it away from the connector and cutting off the power supply to the heat flux probes that they had so diligently lowered into the surface. There was no hope of a repair.

Once the ALSEP was deployed, they were free to explore. Their first trip in the battery-powered rover took them about 1.5 km (1 mi) due west to the 300-m (1,000-ft)

Driving over Descartes—The distant Apollo 16 Lunar Module "Orion," as seen by Charles Duke from the approaching lunar rover. The picture was taken on April 23, 1971, as Duke and John Young were returning from their third lunar field trip. The color television camera mounted on the rover is in the foreground, with part of the umbrella-like, high-gain antenna at top left. (NASA)

The target for the second EVA was a cluster of five craters near Stone Mountain, a rounded hill about 4 km (2½ mi) to the south of the landing site. Once again, they had a great time driving the rover across the boulder fields and undulating, cratered terrain. From the slopes of Stone Mountain, Young and Duke could see into South Ray Crater. Although most of the debris around it was a brilliant white color, there were also occasional streaks of dark material.

On arrival, the men meticulously took magnetic readings, raked the soil, and filled their sample bags in full view of the geologists on Earth, but the predicted lava flows failed to materialize. Instead, the scene was dominated by unremarkable breccias. It was beginning to look as though the hills in the area were all composed of debris ejected during ancient impacts.

There were the inevitable glitches, including the breakdown of the rover's pitch indicator. At one point during the EVA, Young caught his hammer on the right-rear fender, tearing off part of the dust guard. From then on, the men were showered with fine dust that sprayed from the wheels.

The third and final EVA was shortened by a couple of hours in order to give them more time to prepare for lift-off. Their objective was the 1-km (0.6-mi) wide North Ray Crater, close to a sizeable hill dubbed Smoky Mountain. As they drove away from the South Ray ejecta blanket and the number of boulders dropped markedly, they made rapid progress.

On arrival, the men collected more samples and posed beside a huge, black boulder, labeled "House Rock," which had probably been blasted out of North Ray Crater during its impact formation. Once again, subsequent analysis showed that it was composed of the ubiquitous breccia, rather than something of more interest.

The return leg included a single stop at another large, black boulder which had a south-facing overhang and a deep hollow underneath. The men made great efforts to reach into the hollow with a shovel to retrieve samples of soil protected from the solar wind.

By the time Orion lifted off from Descartes on April 23, Young and Duke had collected 94 kg (207 lb) of samples

wide Flag Crater and its smaller neighbor, called Plum. With the Sun low in the sky behind them, long shadows were cast over the undulating landscape.

Young had to drive slowly in order to avoid the numerous rocks and small craters in the broad ejecta blanket that spread out from South Ray Crater, a huge hollow about 6 km (9 mi) to the south. In addition, the low Sun added to the basic problem of estimating sizes and distances.

To everyone's disappointment, the boulders around Flag and on the lower slopes of a nearby hill named Stone Mountain turned out to be breccias—rocks produced by the melting and mixing of the lunar surface materials after one or more large impacts. There was a distinct absence of volcanic rocks.

and driven 26.7 km (16.6 mi) during 20 hours and 12 minutes on the surface. Once again, their colleague, Thomas Mattingly, had made his own contributions to lunar science by releasing a small satellite and conducting a survey of the surface during his long, lonely spell in orbit.

The Grand Finale

After the disappointing geological results obtained by Apollo 16, scientists were desperate to take full advantage of the final manned landing. NASA fell in with their demands by reassigning geologist Harrison Schmitt to the crew of Apollo 17. The only professional scientist to set foot on the Moon, Schmitt was given the opportunity to explore at first hand a narrow valley set into the mountainous eastern rim of the Serenitatis Basin.

Bounded on three sides by the towering Taurus Mountains, the selected site promised a spectacular finale to the Apollo program. Although the valley floor was essentially flat, the adjacent highlands meant that the difference in height across the Taurus–Littrow valley measured about 2,750 m (9,000 ft), or roughly half as deep again as the Grand Canyon.

Not only was the valley coated with some of the darkest (and freshest?) material on the Moon, but it also appeared to contain a cluster of cinder cones. Geologists speculated that the dark layer might be volcanic ash spewed from these cones—maybe within the last half billion years.

The mountains themselves were thought to have been created by the enormous impact that created the Sea of Serenity almost 4 billion years ago. In contrast, the Taurus–Littrow valley was partially filled by basaltic lavas that flooded the area a few hundred million years later.

Watched by half a million spectators at the Cape and millions more on TV, Apollo 17 lifted off from Florida at 33 minutes past midnight, local time, on December 7, 1972. Fittingly illuminated by a trail of fire, the final human mission to the Moon headed out across the black Atlantic.

Three days later, the spacecraft braked into lunar orbit.

Immediately, Schmitt launched into an enthusiastic commentary on the rugged landscape that swept across their view, including a description of a "bright little flash" from a meteorite impact on the flank of the large crater, Grimaldi.

The next day, he joined commander Gene Cernan in the LM Challenger, and headed for the lunar surface. Swooping over the rounded peaks that flanked their landing zone, they eventually touched down less than 200 m (660 ft) from the target zone. Although Cernan had to guide Challenger over a boulder field on the final approach, there were generally few rocks visible among the "subtle hummocky-like craters" at the landing site.

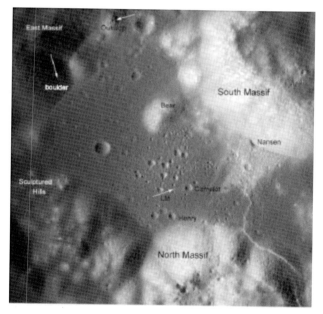

Taurus–Littrow Landing Site—The landing site of Apollo 17 is nestled near the center of the narrow Taurus–Littrow valley, among several 100-m (330-ft) craters. Located on the southeastern rim of the Sea of Serenity, it is bounded on three sides by high mountains. To the northeast are the dome-shaped Sculptured Hills. The second EVA included traverses along the southern rim of Camelot Crater (right of the landing site). North Massif and the Sculptured Hills were visited during the third moonwalk. (NASA, Ron Wells)

Schmitt at Split Rock—Apollo 17 scientist–astronaut Harrison Schmitt is standing at Station 6 during the third EVA at the Taurus–Littrow landing site on December 13, 1972. The huge, broken boulder was found to be a breccia, a rock composed of fragments of other rocks. Scoop marks on the side of the boulder show where Schmitt collected a sample. South Massif, 8 km (5 mi) away, forms the right half of the skyline; East Massif is visible to the left. (NASA–LPI)

As on previous missions, the first moonwalk began with deployment of the rover and the nuclear-powered experiment package. However, the final ALSEP was very different from its predecessors, with instruments to measure the composition of the sparse atmosphere, to study dust particles derived from impacts and to search for cosmic gravity waves. The package also included an advanced seismic experiment and the heat flow experiment that had been prematurely terminated on Apollo 16.

Once again, drilling holes in the fine, compacted soil for the heat flow sensors and core samples proved a major headache, causing Cernan's pulse and oxygen consumption to soar. As a consequence, their first field trip was restricted to a short drive to Steno Crater. Nevertheless, Cernan accidentally damaged the rear fender, resulting in the men being sprayed with clinging gray dust. The next day, the men made a makeshift repair, using Moon maps, tape, and lighting clips.

Their second field trip—a record-breaking 7 hour 37 minute drive to Nansen Crater at the foot of South Massif—was highlighted by what seemed to be a momentous discovery. After spending several hours happily collecting samples of dark and light surface material, the men were heading back when they made a brief stop at a modestly sized crater called Shorty. As Schmitt began to photograph the surface, he suddenly noticed "orange soil."

Schmitt enthusiastically set to work digging a trench, while speculating that the peculiar exposure might be related to a volcanic vent. Scientists back on Earth spoke about "one of the most important finds on the Moon" and began to speculate that subsurface water might also be involved.

Not until the samples were analysed back on Earth was it discovered that the soil was made of glass beads created by a "fire fountain" of lava about 3.5 billion years ago. The orange color was derived from high concentrations of titanium and iron.

The third excursion, which lasted almost as long, took the men to the foot of North Massif. There they explored the numerous boulders that had rolled or slid down the steep slopes thousands, perhaps millions, of years ago. Their most spectacular find was a huge split rock, now broken into five massive slabs. Although it was perched on a steep slope, the men were able to clamber around it and collect samples that showed it to be a complex breccia that had been modified by a series of large impacts that led to partial melting.

The final EVA concluded with a number of PR events and a brief, emotional speech by mission commander Cernan. During their three expeditions, the men had spent more than 22 hours on the lunar surface—longer than any other mission. During this time, they had driven 36 km (22 mi), traveling up to 7.4 km (4.5 mi) from Challenger, and collected 110 kg (242 lb) of rocks.

The human exploration of the Moon ended on December 14, 1972, with Challenger's dramatic departure in a cloud of dust. With its precious cargo unloaded, the LM was deliberately crashed into the South Massif. Further seismic data were provided by the detonation of explosive charges left on the surface. On the homeward leg, Ron Evans grabbed some of the limelight by stepping outside to collect film cartridges from the Service Module cameras. The brief, but glorious, Apollo era was over.

The New Moon

Although the $25-billion Apollo program had been launched largely for national prestige and as a demonstration of US technological superiority, there was, inevitably, a considerable scientific return. In less than a decade, the Moon had been scrutinized by numerous robotic craft, and 12 humans had walked or driven over six separate sites on its dust-laden surface. In addition to collecting some 379 kg (840 lb) of samples, the men conducted 41 different experiments on the surface and from orbit.

Over the last three decades dozens of research laboratories and hundreds of scientists around the world have studied the Apollo lunar samples. As a result of this intensive activity, the post-Apollo Moon is a very different world from the cold, dead satellite envisaged by scientists in the sixties.

On a global scale, the Moon was found to exhibit various unexpected anomalies. Analysis of spacecraft flight paths showed that they arrived a little later than anticipated. Studies showed that the lunar center of mass is slightly displaced toward the Earth, while the near side is about 3 km (2 mi) further from Earth than expected.

A logical explanation was that the crust on the far side is substantially thicker than the 56 km (35 mi) indicated by seismometer readings in the near side's Ocean of Storms. This theory was also supported by the almost complete absence of dark "seas" on the far side, suggesting that asteroid impacts had only occasionally penetrated its thicker crust.

Then there were the mass concentrations that caused lunar orbiters to temporarily increase speed as they pass overhead. These "mascons" were found to be associated with the largest circular basins that formed the familiar "seas" on the near side. Apparently the extra gravitational pull was caused by oceans of lava that had welled up from below, filling the huge impact basins.

Apollo seismic experiments showed that there are frequent moonquakes, but these are smaller and less common than their Earthly counterparts. The Moon was found to possess a thick crust (typically 60 km or 37.5 mi) on top of a 1,200-km (750-mi) deep mantle. The upper part of this mantle was rigid and rocky, while the lower zone was probably warm enough to be partly liquid.

Long-term laser ranging indicated that a small, liquid iron core lingered at the center, but the absence of a global magnetic field indicated that any fluid dynamo had shut down long ago. The only magnetic readings came from ancient rocks that still retained some remnant magnetism.

The Moon's mighty outer armour has effectively prevented any large-scale faulting or crustal movement over billions of years. The impenetrable crust also prevents any magma seeping to the surface and causing a volcanic eruption.* On the other hand, it is ideal for preserving the lunar impact record.

---

* There have been occasional reports of brief, sudden flashes or glows near large craters such as Alphonsus and Aristarchus. The later Apollo orbiters also detected emissions of radon gas.

Orange Soil—Some of the most famous lunar samples came from the "orange soil" found near Shorty Crater during the second Apollo 17 moonwalk. This picture shows the 89-cm (35-in) long trench dug in the orange outcrop to obtain samples. Analysis back on Earth showed that it was made of tiny beads of orange glass, formed in a volcanic "fire fountain" over 3.5 billion years ago. The gnomon (left center) was used to indicate surface slope, solar position, and color to aid scientists in studying lunar surface photographs. (NASA)

The Apollo Lunar Surface Experiments Package—This photograph, taken on April 21, 1972, shows Apollo 16 commander John Young standing in front of the Apollo Lunar Surface Experiments Package that was deployed during their first moonwalk. The lunar surface drill is to Young's right, with the drill's rack and bore stems to the left. Beyond the rack is the Surface Magnetometer. The dark object (right background) is the Radioisotope Thermoelectric Generator (RTG) nuclear power source. Between the RTG and the drill is the Heat Flow Experiment. A part of the Central Station is on the right edge of the picture. (NASA)

Crater counts, studies of overlapping craters, and dating of Apollo rock samples enabled scientists to piece together an outline lunar history. Born at about the same time as the Earth, about 4.6 billion years ago, the molten Moon quickly separated, with the denser material sinking and the lighter minerals rising to the surface.

Over a period of 650 million years, the young world suffered a continuous, cataclysmic bombardment by asteroids and comets. The result was a surface scarred by impact craters and basins of every size, and the widespread creation of breccias—composite rocks made from material crushed and melted by the collisions.

After the main basins had formed, there followed a period of tremendous volcanic activity. For some 700 million years, lava poured into the circular lowlands on the near side, often overflowing from one basin into a neighbor. This volcanic resurfacing was largely absent from the far side, where the ancient highlands prevailed.

The eruptions ceased as abruptly as they started, leaving a Moon marked by dark "seas." Over the last 3,200 million years, volcanic activity faded and disappeared, while impact activity dropped off dramatically. One of the last major collisions, which created the crater Copernicus, took place about 1 billion years ago.

Scientists were hoping that Apollo would solve the mystery of the Moon's origin, but the evidence remained ambiguous. Studies of lunar samples showed that our cosmic neighbor was born in the same region of space as the Earth. However, there were notable differences in composition. In particular, the Moon was highly depleted in iron and volatile elements that are needed to form atmospheric gases and water. The Moon appeared to be completely dry, arid, and lifeless.

Lunar rocks could generally be divided into three types. The dark, mare basalt that makes up the lunar "seas"; a highland rock known as KREEP—so called because of its unusually high content of potassium (K), rare-earth elements (REE), and phosphorus (P); and a common highland volcanic material called anorthosite. All of these were once molten and formed in the Moon's violent youth.

Broad similarities between the lunar composition and the material that makes up the Earth's crust and mantle suggested to many scientists that the Moon contains a lot of terrestrial material. It is widely accepted today that the Moon was created by a Mars-sized body that collided with the young Earth. As a result of the colossal impact, debris from both objects was blasted into orbit around the Earth, where it came together to form our planet's solitary satellite.

Soviet Robots

When NASA announced its human Apollo program, US and European scientists eagerly looked forward to receiving numerous samples of lunar "soil" and rock from different locations on the Moon's near side. After their return to Earth, it would be possible to conduct the first analysis of unique materials from another world using state-of-the-art instruments in terrestrial laboratories.

Although it was late in entering the Moon Race, the Soviet Union mounted a two-pronged challenge by simultaneously attempting to land a man on the Moon and using robotic craft to explore the surface. Unfortunately, the manned program hit innumerable technological obstacles, while the first sample return attempts failed to deliver, the most famous of these being Luna 15, which raced Apollo 11 to the Moon in July 1969 and then crashed ignominiously on the Sea of Crises.

Not until four astronauts had left their footprints in the Moon did Luna 16 make the breakthrough by touching down on the Sea of Fertility on September 17, 1970. After the sample arm extended over the surrounding "soil," the hollow drill was driven 35 cm (10 in) into the ground until it struck solid rock, and 101 g (3.5 oz) of material was then deposited in a return capsule, ready for take-off.

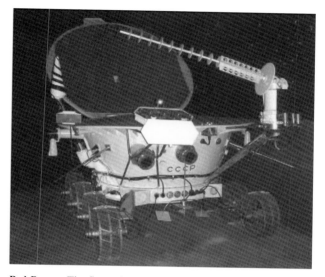

Red Rover—The first robotic rover to trundle over the Moon was the Soviet Lunokhod 1, which Luna 17 delivered to the Sea of Rains on November 17, 1970. After rolling down ramps onto the surface, the vehicle covered 10.5 km (6.5 mi) and sent back more than 20,000 TV images over a period of almost 11 months. Note the wire wheels, the forward-facing stereo cameras, and the solar cells on the underside of the lid. The hexagonal box holds a French laser reflector. (Peter Bond)

After four days on the surface, the ascent stage engine fired, sending the upper section of the spacecraft, complete with its precious cargo, onto an Earthbound trajectory. When it was 48,000 km (30,000 mi) from home, the spherical capsule was released for a ballistic entry into the atmosphere. Slowed by a parachute, it eventually landed on the steppes of Kazakhstan on September 24.

After its delivery to Moscow, the sample was unsealed in a sterile chamber filled with helium. Analysis showed that, as expected, it was composed of volcanic basalt similar to that brought back by Apollo 12. The core varied from fine material at the top to larger grains—possibly rock fragments—near the bottom. It also included a tiny bead of volcanic glass. Three grams were eventually swapped for similar amounts of Apollo 11 and 12 samples.

Luna 20 Return Capsule—The first successful robotic sample return was completed by Luna 16, which drilled into the Sea of Fertility on September 20, 1970. A sealed capsule with 101 g (3.5 oz) of "soil" was later delivered back to Earth. The picture above shows the sample return capsule from the second mission, Luna 20, which brought back a 55-g (2-oz) sample of ancient highland material. The third successful sample return (Luna 24 in 1976) was also the last Soviet/Russian Moon mission. (Novosti)

Bearing in mind the recent, near-fatal debacle with Apollo 13, the Soviets tried to emphasize the advantages in terms of human safety and cost in using robotic craft for lunar exploration. The official line fed to the western media was that they had never intended to send men to the Moon.

The next advance came on November 15, 1970, when Luna 17 delivered the first automated roving vehicle to the Sea of Rains. Steered by a team of controllers back on Earth, Lunokhod 1 resembled a round, metallic bath tub, fitted with a lid and eight independently powered wire wheels. On the underside of the lid were solar cells for charging the batteries, while a radioactive power sourced provided warmth during the long, cold night.

After trundling down the forward ramps onto the slopes of an ancient crater, the rover set to work. Over the next 11 lunar days it traveled 10,540 m (6.5 mi), sending back more than 20,000 TV pictures and over 200 TV panoramas. It also conducted more than 500 tests of lunar soil density and mechanical properties. Other information on rock compo-

sition came back from an X-ray spectrometer. Also on board were an X-ray telescope, cosmic-ray detectors, and a laser-reflecting device for measuring the Earth–Moon distance. Operations officially ceased on October 4, 1971, the anniversary of Sputnik 1, although it had probably shut down some time earlier.

After a failed sample return mission with Luna 18, the Soviets returned to winning ways with Luna 20, which conducted a repeat performance of the Luna 16 mission after a successful landing on February 18, 1972. This time, a camera captured pictures of the soil collection site, but the upper stage returned with a rather disappointing 55-g (2-oz) sample of ancient highland material from the rim of the Sea of Fertility. Analysis showed that the fine grains were rich in aluminum and calcium oxides, pointing to a volcanic rock called anorthosite.

The next Soviet invasion of the Moon began on January 12, 1973, when Luna 21 delivered Lunokhod 2 to the vicinity of the 55-km (35-mi) crater Lemonnier, on the eastern edge of the Sea of Serenity.* Although the upgraded rover was designed to travel farther and faster than its predecessor, Lunokhod 2 only survived for about 4 months in the harsh environment.

It eventually covered 37 km (23 mi), rolling over the foothills of the Taurus Mountains and along the rim of a small rille called Fossa Recta. At one point, the rover had to be reversed after it sank up to its axles in deep dust. Altogether, 86 panoramic images and more than 80,000 TV pictures were transmitted. The surface was found to be relatively poor in iron compared with the Lunokhod 1 site, and a weak magnetic field was detected. An unexpected bright glow above the horizon was attributed to fine dust suspended in the lunar vacuum. Unfortunately, after operations ended on June 4, the parked position of the rover meant that the laser reflector could no longer be used.

---

* Only a month earlier, the crew of Apollo 17, the final manned Moon mission, had lifted off from a site about 180 km (110 mi) south of Luna 21.

The two remaining Soviet sample return missions experienced mixed fortunes. Luna 23 crash-landed on the Sea of Crises in November 1974, but the hoped-for soil sample was subsequently retrieved by Luna 24—the last craft to soft-land on the Moon. This time, 170 g (6 oz) of soil was stored inside a flexible 2-m (6.6-ft) tube in order to avoid spillage. The layered mixture—presumably the result of successive deposits—was composed of fine powder mixed with centimeter-sized pebbles. Altogether, the three samples brought back by Soviet automated craft weighed about 320 g (11.3 oz)—extremely modest compared with the 384 kg (847 lb) of lunar material collected by the Apollo crews.

Lunar Prospecting

By the mid-1970s, many of the questions about our neighboring world seemed to have been answered, and ambitions to send human explorers far from home were replaced by more routine missions in low-Earth orbit.

For the next two decades, the Moon merely served as a useful target for calibration of instruments on passing spacecraft, or as a gravitational slingshot for ships heading for deep space. In 1973, for example, Mariner 10 obtained the first good views of the Moon's north polar region as it headed toward Mercury. Four years later, the first complete family portrait of the Earth and Moon was taken by Voyager 1 en route to Jupiter. In 1982, the ISEE-3 spacecraft was diverted and accelerated toward comet Halley during five lunar flybys.

Two Earth–Moon encounters were included in the tortuous trajectory that enabled NASA's Galileo spacecraft to reach Jupiter. On December 8, 1990, it sent back detailed images—including the first infrared map of the lunar far side—that provided information on the Moon's surface composition. The basaltic maria or "seas" were shown to be rich in iron, unlike the metal-poor highlands.

Similar studies were undertaken in December 1992, although this time the ultraviolet spectrometer was used in

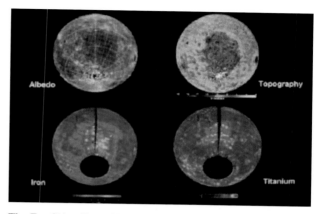

The Far Side—Four Clementine views of the lunar far side are dominated by the South Pole–Aitken Basin, the largest, deepest impact basin in the Solar System. The basin is about 2,500 km (1,560 mi) wide, and up to 13 km (8 mi) deep. The albedo map shows that the basin is markedly darker than the surrounding highlands. Its floor shows enhanced iron and titanium concentrations—a very different composition from the rest of the far side and the lunar highlands in general. (Lunar and Planetary Institute)

an unsuccessful search for water ice inside craters at the north pole.

Meanwhile, Japan launched two spacecraft toward the Moon in 1990, becoming only the third nation to do so. Since both were designed as technology demonstrators, science took a back seat. The larger Hiten mother ship, carrying only a dust detector, carried out a series of navigational maneuvers around the Earth and Moon, while the tiny Hagoromo satellite entered orbit around the Moon but failed to send back any data.

Further evidence of a renaissance in lunar studies came with a unique collaboration between the US Ballistic Missile Defense Organization and NASA. Originally intended as a military mission to test optical sensors beyond Earth orbit, the Deep Space Program Science Experiment was revamped to track and study the Moon, along with one or more near-Earth asteroids. Developed in less than two years at a cost of $75 million, it was regarded as the prototype of a new "faster, better, cheaper" approach to space exploration.

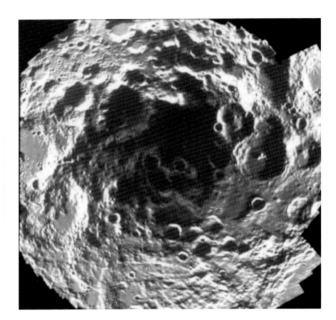
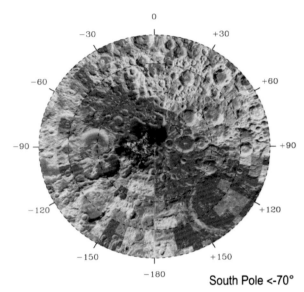

South Pole <-70°

Water Ice at the South Pole?—This mosaic of about 650 Clementine images (left) shows the Moon's south polar region, from 80°S to the dark pole in the center. The permanently shaded region near the pole marks a depression inside the much larger South Pole–Aitken Basin. Radar results from Clementine first indicated that this could contain deposits of water ice. The map (right), based on neutron spectrometer data from Lunar Prospector, confirms the presence of hydrogen (dark blue)—probably associated with surface water ice. (LPI/LANL)

Renamed Clementine,* the spacecraft was launched from Vandenberg Air Force Base on January 25, 1994. Almost four weeks passed before it finally braked into a polar orbit around the Moon. Over the next 10 weeks, it swept to within 430 km (265 mi) of the lunar surface, relaying between 4,000 and 6,000 images back to Earth each day.

Altogether, more than 30 gigabytes of science data were returned, including nearly two million images of the Moon, as well as gravity field, topographic, and other measurements. This flood of data enabled delighted scientists to produce the first global digital map of the Moon made in 11 different colors (including ultraviolet and infrared wavelengths), and the first global topographic map. This

revealed the full extent of the difference in relief between its two hemispheres, highlighting in particular the 2,500-km (1,560-mi) diameter South Pole–Aitken Basin—the largest and deepest impact basin in the Solar System—that dominates the far side of the Moon.

However, it was the apparent discovery of water ice in permanently shaded craters at the lunar south pole that grabbed the headlines. An *ad hoc* radar experiment that was dreamt up after Clementine reached the Moon was used to study the darkened craters. As the spacecraft passed

---

* Like the girl in the popular old song, it would seek out riches and then be "lost and gone for ever."

overhead, its radio signals were bounced off the ground so that the echoes could be studied by telescopes on Earth. Before the remarkably successful mapping mission was concluded on May 4, the experiment produced a reflection that suggested the presence of ice.

Not everyone was convinced that Clementine had detected an ice signature. For example, ground-based measurements made with the giant radio dish at Arecibo favored a rough, reflective terrain over an icy surface. The only way to resolve the argument was to send another spacecraft.

NASA's low-cost Lunar Prospector was launched on January 6, 1998, and arrived in lunar orbit five days later. Included in its payload was a neutron spectrometer that was designed to detect minute amounts of hydrogen, a possible tell-tale sign of surface water ice. The instrument did, in fact, pick up signs of significant hydrogen enrichment, particularly over the north pole. This led mission scientists to estimate that the lunar poles may contain up to 6 billion metric tons of water ice. The analysis indicated the presence of discrete, near-pure water ice deposits buried beneath a possible 40 cm (18 in) of dry regolith or "topsoil."

Lunar Prospector also provided the first global maps of the Moon's elemental composition and localized magnetic fields—including the two smallest known magnetospheres in the Solar System. Each about 160 km (100 mi) across, their magnetism was strong enough to prevent the solar wind from striking the surface. Other data led to the first precise gravity map of the entire lunar surface, and the discovery of seven previously unknown mass concentrations. Three were located on the Moon's near side and four on its far side. The mission ended on July 31, 1999, when the spacecraft was deliberately crashed into a crater where ice was thought to exist, but no signature of water vapor was detected. Lunar Prospector also made history by delivering the ashes of the distinguished planetary geologist Eugene Shoemaker to the surface of the world he had studied so avidly over many decades.

On the Road to Mars

The possibility that pockets of water ice exist on the supposedly arid Moon—even if they are at the rugged poles—renewed interest in lunar prospecting and led to increased speculation about future human bases. During the 1990s, several space agencies produced reports outlining long-term programs that would culminate in astronauts returning to the Moon—this time for good. However, these reports seemed destined to remain mere paper studies without any political commitment.

The situation suddenly changed on January 14, 2004, when US President George W. Bush announced his Vision for Space Exploration, in which he outlined a 30-year effort to send humans to the Moon and then on to Mars. In order to pave the way for this momentous movement beyond low-Earth orbit, NASA would be required to send a series of robotic missions that would identify lunar resources and test new exploration technologies.

Bush's announcement came at a time when NASA was floundering after the loss of the Shuttle Columbia and the grounding of the three remaining orbiters. Without the heavy-lift capability of the Shuttle, there was no way to complete construction of the International Space Station. Even more embarrassing was the reliance on the Russians for the transportation of crew and cargo to the orbiting outpost.

However, there were also external factors at play. When, in October 2003, China became the third nation to send one of its citizens into orbit, there was even the long-term possibility that the next people to walk on the Moon would hail from outside the United States. Other spacefaring countries had already made clear their intentions to explore the Moon with sophisticated robotic missions.

First to do so was the European Space Agency with its SMART-1 spacecraft, which launched during the night of September 27/28, 2003. Primarily designed as a technology demonstration mission, SMART-1 relied upon a revolutionary solar-electric propulsion system (often called an ion engine) to fire a high-speed stream of xenon gas into space.

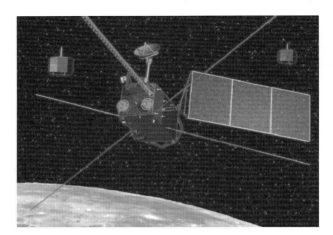

Moon Base 2020—NASA's Vision for Space Exploration foresees a series of robotic missions that will pave the way for humans to return to the Moon by 2015–2020. This artist's impression depicts a human-tended lunar base designed to test the technologies required for a long-term human occupation. Note the rocket in the background, the large rover, and the automated base station. (NASA)

Selene—Japan's Selene mission, scheduled for launch in 2007, will carry out the most ambitious lunar exploration since Apollo. The largest of the three orbital spacecraft will carry experiments to identify the composition of the lunar surface, map its topography, and study the magnetic field, particles, and radiation. Once in orbit, it will release a relay satellite that will act as communications link with the Earth. Another small satellite will take part in a radio experiment to precisely measure the orbit of the Moon. (JAXA)

Gently propelled by this almost continuous jet of particles, the spacecraft consumed just 60 kg (132 lb) of xenon fuel as it slowly spiralled around the Earth 331 times. After each loop, it moved a little closer to its target, until it was finally captured by lunar gravity on November 15, 2004. Over the next three months it edged closer to the surface, until it reached a polar orbit of 400 km × 4,000 km (250 × 2,500 mi).

SMART-1 carried six scientific experiments to image the surface, determine its composition and study the interaction with the solar wind. The most unusual experiment was an X-ray spectrometer designed to identify chemical elements on the surface by detecting the way they fluoresce or glow when struck by X-rays from the Sun.

Following its modest success with Hiten, Japan lined up two very different robotic projects. Lunar-A was to be the first mission to fire two penetrators into the lunar surface—one on the near side and one on the far side. Crammed

inside the dart-shaped probes would be two seismometers and sensors to measure heat flow from the Moon's interior. The data obtained by the instruments would be transmitted to Earth via the Lunar-A mother spacecraft, orbiting about 200 km (125 mi) overhead. Originally scheduled for launch in 1999, the mission ran into severe technical and budgetary problems, and it is possible that Lunar-A will never fly.

Prospects for the second mission, Selene, seem much brighter, with launch currently scheduled for summer 2007. One of the largest robotic missions ever sent to the Moon, Selene will comprise two 50-kg (110-lb) spacecraft piggybacking on the main orbiter. After a five-day trip, Selene will enter orbit around the Moon, and, as it spirals closer to the surface, the octagonal relay satellite and Very Long Baseline Interferometry Radio (VRAD) satellite will be released.

The Selene orbiter will produce global maps of elemental abundance, mineralogical composition, surface relief and magnetic field, and measure the charged particle environment. As its name suggests, the relay satellite will act as a communications link with the main craft. Careful tracking of the satellites will also enable scientists to learn more about gravity anomalies on the far side, while radio experiments involving the VRAD will refine our knowledge of the Moon's orbit and rotation.

The race to return to our neighboring world is soon to be joined by two new space powers. India's Chandrayaan-1, slated to be launched by 2008, will map the lunar terrain for minerals and conduct scientific experiments. The full scientific payload has yet to be announced, but it is expected to include a 20-kg (44-lb) "impactor" that will be fired into the surface.

China's lunar program is even more ambitious. The one-tonne Chang'e-1 spacecraft, expected to launch in 2007, will carry an altimeter, gamma and X-ray spectrometer, microwave radiometer, and space environment monitor system. If this is successful, the Chinese have announced their intention to follow up with a rover by 2012 and a robotic sample return mission by 2020.

Although political differences have minimized US cooperation with India and China in recent years, there seems little doubt that Europe and Japan will be keen to join NASA in returning to the Moon. However, before astronauts can once again leave their footprints in the lunar dust around 2015–2020, a new crew transportation system must be built and tested in Earth orbit.

Meanwhile, the US has expressed its intention to send a series of increasingly sophisticated robotic missions to characterize its surface (including a detailed search for water ice) and to test technologies that will enable humans to survive for long periods in its hostile environment, and subsequently on Mars. The first of these robotic precursors, called Lunar Reconnaissance Orbiter, is scheduled for launch in 2008.

Earth and Mars Compared—Mars is the outermost of the four terrestrial (rocky) planets. With a diameter of 6,794 km (4,333 mi)—little more than half that of Earth—Mars has a surface area comparable to all the land masses on our planet. Whereas Earth is largely covered by oceans, Mars is currently in an ice age. Note the patchy ice clouds over the Martian highlands. (NASA)

5 MARS: THE RED PLANET

The Mysterious Red Planet

Every 26 months or so, a reddish "star" appears in the night sky, moving fairly rapidly against the background constellations and increasing in brightness until it outshines everything apart from the Moon, Venus, and Jupiter.

Although it was not the brightest of the "planets" or wandering stars, the object's ruddy glow and prolonged stay among the lesser stellar lights gave it a special significance to the first astrologers and stargazers. Most ancient civilizations associated this distinctive interloper with fires, conflicts, plagues, and disasters. Not surprisingly, since it was generally associated with deities of death and misfortune, the planet eventually ended up being named Mars, after the Roman god of war.

By the fourth century BC the Chaldean astronomers of Babylon had calculated that Mars (or Nergal as they knew it) was at its brightest every 779.9 days. We now know that that is the time interval between oppositions, when Earth and Mars are at their closest.

Unfortunately predicting its position at other times was much less straightforward. One of the planet's most perplexing characteristics was that it performed a clear reversal of direction when at its brightest. (Jupiter and Saturn followed suit, but in much less spectacular fashion.) For many millennia, this retrograde motion defied all explanation. The most widely accepted solution was provided by Aristotle (384–322 BC), who proposed a system of small circular orbits (epicycles) as Mars and the other planets circled around the dominant Earth at the center of the planetary system.

Over the centuries, this arrangement became increasingly complicated, without satisfactorily accounting for the vagaries of planetary motion. The first breakthrough came in 1543, when Nicolaus Copernicus published a short pamphlet called *De Revolutionibus*, in which he moved the Sun to the center of the Solar System and placed the planets in circular orbits around it. Since Mars was the fourth planet from the Sun and Earth was the third, the apparent retrograde motion could easily be explained by our world catching up with and passing Mars, like a runner overtaking on the inside track. (Earth is closer to the Sun, so it moves more rapidly around its orbit than Mars.)

Predicting planetary positions still proved problematical, until Johannes Kepler realized that these wanderers follow elliptical, rather than circular, paths. Subsequent calculations showed that Mars has one of the most elongated orbits among the planetary family. This means that its distance (and brightness) vary considerably at each opposition. At its closest, Mars may approach to within 55 million km (about 34 million mi) of Earth. During the least favorable oppositions, it comes no closer than 100 million km (62.5 million mi).

Although, with the exception of Venus, Mars is Earth's nearest planetary neighbor, it proved to be a disappointment to early telescopic observers. To Galileo and his successors, Mars remained a featureless red blob which was clearly much smaller than Jupiter. Modern measurements show that its diameter is 6,794 km (4,333 mi)—little more than half that of Earth. Despite this modest size, the surface area of Mars is still comparable to all the continental land masses on our planet.

In order to learn more about the Red Planet, it was

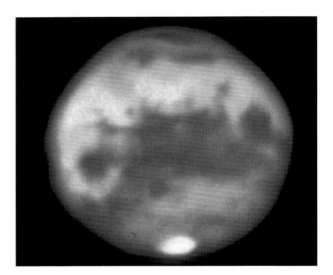

Ground-based View of Mars (0°N, 43°W)—This 1988 image was obtained at the start of the southern spring, when the southern polar cap was prominent. The southern highlands appear much darker than the northern lowlands. The Valles Marineris canyon system is the narrow feature to the left of the central dark region, known as Acidalia Planitia. Atmospheric turbulence means that Earth-based telescopes can usually resolve features no smaller than 300 km (185 mi) across, even when Mars is at its closest. (Leonard Martin/Lowell Observatory/LPI)

necessary to observe features in its atmosphere or on the surface. The first person to produce a reliable sketch of a Martian surface feature was a Dutchman, Christiaan Huygens (who was already famous for discovering Saturn's largest satellite, Titan, and the nature of Saturn's rings). The Africa-like feature drawn by Huygens in 1659 was Syrtis Major, the most prominent dark marking on the orange–red surface.

By 1672, Huygens was able to announce the presence of a white spot covering the planet's south pole. By carefully timing the reappearance in his eyepiece of particular markings, he was also able to report that Mars rotated once every 24 hours. This figure was later refined by Giovanni Domenico Cassini and others. Today, we know

that a Martian day lasts for 24 hours 37 minutes—a little longer than a day on Earth. The tilt of its rotation axis is very similar to that of the Earth, so Mars, too, has seasons.

One hundred years later, another renowned observer, William Herschel, used his superior, home-made telescopes to speculate that the white spots over each pole were ice caps—similar to those on Earth, but much thinner. Not only did the polar caps grow rapidly in the winter and shrink equally quickly each summer, but significant changes were also seen in some of the blotchy, dark regions. Occasionally, clouds and dust storms were visible—yet further evidence that Mars was a smaller version of our benign world. Little wonder that Herschel concluded, "Its inhabitants probably enjoy conditions analogous to ours in several respects."

Although there was considerable circumstantial evidence that Mars was the most Earth-like of all the planets, finding hard evidence was far from easy. Maps produced in 1830 by Berlin banker Wilhelm Beer, in collaboration with Johann Heinrich Mädler, showed a complex, mystifying pattern of dark and light patches. Uncertain of their true nature, the observers considered the reddish-ochre areas to be deserts separated by blue–green seas.

Although many of the dark features were given names representing seas or oceans, the first efforts to find water and oxygen proved inconclusive. Then, in 1860, Emmanuel Liais, a French astronomer living in Brazil, put forward the astounding suggestion that the shifting patches were actually vegetation fed by water from the melting polar caps. This theory rapidly gained support, and, until well into the twentieth century, many astronomers continued to believe that they were observing primitive plants that spread in the spring and died back in the autumn. But what about intelligent life?

The Canal Conundrum

A new era in Mars studies opened during the extremely favorable opposition of 1877, when the Red Planet was

Percival Lowell—Millionaire businessman and amateur astronomer Percival Lowell (1855–1916) built his own observatory at Flagstaff, Arizona, to observe the planets. He was particularly fascinated by Mars. His drawings of the Red Planet included intricate networks of "irrigation canals" that he believed were constructed by intelligent Martians. He drew similar linear markings on sketches of Mercury, Venus, and the moons of Jupiter. (Smithsonian Air & Space Museum)

almost at its closest to Earth. In America, Asaph Hall took advantage of the opportunity to find two small moons in orbit around Mars, but his discoveries were largely overshadowed by the furore that enveloped Giovanni Schiaparelli, director of the Brera Observatory in Milan.

With the aid of a recently installed 22-cm (8.6-in) refractor, Schiaparelli began to sketch the most detailed

map of Mars yet produced. The outcome was a beauty to behold, and, updated during future oppositions, it became a standard reference for all planetary observers.

Many of the names he gave persisted until the Space Age. A light spot in the southern hemisphere, for example, he called Nix Olympia, "the Snows of Olympus." (Now recognized as the largest volcano in the Solar System, it has been renamed Olympus Mons.) Other features were given picturesque labels such as Syrtis Major (the "Great Marsh"), Elysium, Cydonia, Tharsis, and Thyle.

However, the most striking characteristic of Schiaparelli's map was a network of linear markings that crisscrossed the "deserts," linking one dark area to another. Schiaparelli described this strange spider's web as "canali" (channels).* Convinced that they carried water from the poles to arid regions at lower latitudes, he named them after famous rivers, both fictional and real—Gehon, Hiddekel, and Phison from the rivers in the Garden of Eden; Lethes and Nepenthes from the mythological underground realm of Hades; and Ganges, Euphrates, and Nilus from planet Earth.

Unfortunately, no one else noted the presence of any Martian channels during the 1877 apparition. The controversy was reignited two years later, when Mars once again shone brightly in the night sky. On this occasion, Schiaparelli noted that some of the channels had doubled— a process he called "gemination." For a while, he remained a lone voice in the wilderness, but, in 1886 two French astronomers announced that they, too, had seen the elusive "canali."

As the controversy raged over the existence and nature of the linear markings, a wealthy American amateur astronomer named Percival Lowell was poised to make his own sensational contribution. The millionaire from Boston stepped into the gap by building his own, state-of-

* A few straight, channel-like features had previously been sketched by Beer and Mädler. Italian astronomer Pietro Secchi was the first to use the term *canali* to describe two fine lines he had seen on Mars in 1859.

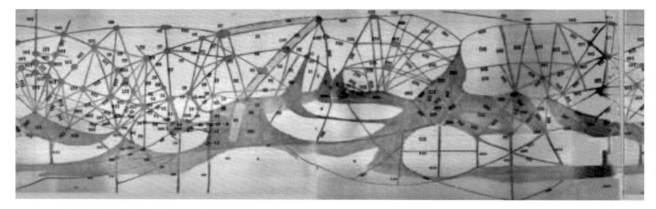

Lowell's Canals—This map of Mars was drawn by Percival Lowell and published in 1895. Apart from the famous dark patches, it is dominated by numerous straight, linear features, some running parallel to each other. Lowell described these as strips of irrigated land alongside artificial "canals" which had been constructed by an intelligent civilization to carry water from the polar ice caps across the planet's arid deserts. Note that some of the so-called canals cross broad, dark areas that were supposed to be regions of cultivation. (From *Mars*, by Percival Lowell, 1895)

the-art observatory beneath the clear skies of Flagstaff, Arizona.

Between 1894 and his death in 1916, Lowell studied Mars at every opportunity, producing thousands of drawings and three books. Not only did he claim to have seen Schiaparelli's original network, but he added many more linear markings—which he deliberately termed "canals" and "oases." By the time that *Mars as the Abode of Life* was published in 1906, he had mapped no fewer than 437 canals—51 of which were double—and nearly 200 oases. In his mind, there was no doubt that these were artificial features constructed by intelligent Martians.

According to Lowell, the technologically advanced aliens had built a huge irrigation scheme that carried water from the polar ice caps, in a desperate attempt to hold back the advancing deserts and save their civilization. Bearing in mind the limitations of ground-based telescopes, the linear markings had to be at least 48–64 km (30–40 mi) across. Aware that the existence of such enormous canals was exceedingly unlikely, Lowell considered that the features probably comprised a fairly narrow, water-filled channel flanked by cultivated land, like an artificial version of the Nile valley.

The debate over Lowell's observations and interpretation raged long after his death and inspired a whole genre of science fiction—notably H.G. Wells's *The War of the Worlds*, published in 1898. Some observers failed to see any canals at all. Others insisted that they were natural features or optical illusions. Perhaps studies of the Martian environment could provide an answer.

Measurements showed that, despite its greater distance from the Sun, Mars was far from a frozen world. Although night temperatures plummeted to around –70°C (and possibly –100°C at the poles), summer days in the tropics were calculated to reach a balmy 35°C.

If temperatures seemed reasonably favorable for liquid water, what about the thickness of the atmosphere? (Water evaporates at lower temperatures when air pressure decreases.) Until the early 1960s it was generally agreed that air pressure on Mars was between 80 and 100 millibars (about one-tenth of the sea-level pressure on Earth.) At such a modest pressure, water would boil at around 45°C. Any surface water would evaporate rapidly, reducing the likelihood of rain or snow. Most precipitation would probably take the form of hoar frost or dew.

Another question mark concerned the ice caps. In 1947, Gerard Kuiper detected a substantial amount of carbon dioxide in the atmosphere. The following year, Kuiper detected water ice in the polar caps, scotching suggestions that they might be composed of frozen carbon dioxide. However, there was obviously very little oxygen or water vapor in the atmosphere. Any clouds appeared as wispy, cirrus-like features—presumably made of water ice crystals.

Although minimal amounts of liquid were available from the air or the polar caps, the possibility still remained that vegetation could be fed by water that was stored in underground aquifers and seeped to the surface through springs. Ground-based studies could find no evidence for the tell-tale signature of plant chlorophyll, but this was not conclusive. Lichens on Earth managed to survive without chlorophyll. Perhaps these hardy specimens were the Martian life that everyone so eagerly sought.

The First Mariners to Mars

A number of factors combined to make Mars a prime target for early space missions. Not only was it one of Earth's closest neighbors, but it held the potential for astounding scientific discoveries—including the possibility of finding life beyond our planet.

Since Mars reached opposition approximately every 26 months, the launch windows for missions to the Red Planet came around at two-yearly intervals. However, since the orbit of Mars is quite elongated, the best launch opportunities occur about 16 years apart, so 1971, 1988, and 2003 were particularly favored.

In order to use the minimum energy and allow the heaviest possible payload, a spacecraft would typically follow a curved transfer orbit. Over a period of six or seven months, it would gradually catch up with the slower-moving Mars before flying past or braking into orbit around the planet. However, the first objective was to reach the escape velocity of 11.2 km/s (25,000 mph) and leave Earth behind—something the Soviets had failed to do on

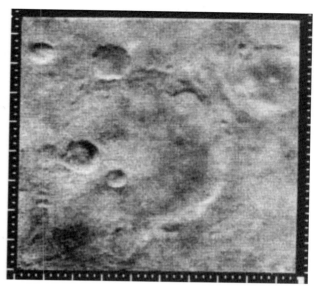

Mariner Crater—Mariner 4 was the first spacecraft to get a close look at Mars. From a distance of 9,846 km (6,118 mi), it revealed a heavily cratered surface that showed little evidence of liquid water and no sign of life. Dominating this image is a 151-km (94-mi) diameter impact feature at 35°S, 164°W, later named Mariner Crater. A linear trough, part of Sirenum Fossae, runs from lower left through the southern part of the crater. The image was taken from 12,600 km (7,875 mi) and covers an area of approximately 250 sq. km (155 sq. mi). (NASA–JPL)

five out of six launch attempts in 1960 and 1962. The sole exception was Mars 1, which lost contact at a distance of 106 million km (65.9 million mi).

NASA entered the competition to be the first to reach the Red Planet with a double header known as the Mariner–Mars 1964 project. The plan was to launch two identical spacecraft, each weighing 261 kg (575 lb), on trajectories that would enable them to fly over Mars in a diagonal path from the equator to the south pole.

A simple TV system would return 22 images that, it was hoped, would be comparable in detail with the best ground-based views of the Moon. Six more experiments would measure radiation, solar particles, magnetic fields, and

Mariners 6 and 7 Mosaic—Images taken by Mariners 6 and 7 are attached to a globe of Mars that shows the vague dark markings seen in ground-based telescopes. Mariners 6 and 7 were intended to study the equatorial regions and the southern hemisphere. Altogether they sent back 55 close-up pictures during flybys on July 31 and August 5, 1969. This was the first real attempt to begin detailed mapping of the Red Planet. (NASA–JPL)

micrometeorites in interplanetary space, while the characteristics of the thin atmosphere would be determined by studying radio signals from the Mariners as they passed behind the planet.

The program got off to a bad start on November 5, 1964, when Mariner 3's protective shroud failed to separate, condemning it to a quick, silent death. The mission was over almost as soon as it had begun. After hasty modifications, Mariner 4 followed on November 28, and this time all went according to plan.

After a trek lasting 7½ months, the pioneering craft made the first successful flyby of Mars on July 14, 1965. Forty minutes before closest approach, the TV system began its preprogrammed picture-taking sequence, and for the next 25 minutes it scanned a series of 21 overlapping scenes (plus 22 lines of a 22nd image) that stretched in a diagonal line from approximately 37°N to 55°S.

Mariner 4 then swept behind the right-hand limb of the planet, as seen from Earth, allowing scientists to monitor radio signals that probed the thin blanket of Martian air. The data indicated that surface pressure was a mere 5 millibars—200 times less dense than on Earth—while the daytime temperatures were estimated to hover around –100°C. There was no sign of a magnetic field or regions of trapped radiation like Earth's Van Allen belts.

The impression of a surprisingly non-Earthlike world was reinforced when playback of the images stored on the tape recorder began. Transmitted at a snail-like pace of 8⅓ bits per second, it took 8 hours 20 minutes to piece together a single image. Eight days later, scientists were able to analyze the first close-up views of Mars. To everyone's dismay, the fuzzy pictures were dominated by impact craters ranging from 5 to 120 km (3 to 75 mi) in diameter. With its thin atmosphere and ancient, pockmarked surface, Mars seemed a much closer relative to the Moon than to the Earth. Lowell's canals and vegetation were nowhere to be seen.

The one consolation was that Mariner 4 had imaged only 1% of the Martian terrain. Perhaps there were "oases" where conditions were more hospitable. NASA decided to send two larger, more sophisticated, spacecraft to find out. With a launch mass of 413 kg (909 lb), Mariners 6 and 7 carried narrow- and wide-angle TV cameras attached to a movable scan platform, and various instruments to study the atmosphere and surface conditions. The maximum rate of data return was 16.2 kbit/s (2,000 times faster than Mariner 4).

Once again, the Mariners were directed to survey swaths that stretched from the equator across the southern hemisphere. Mariner 6 was launched on February 24, 1969, followed by Mariner 7 on March 27. After a 100-million-km (62-million-mi) voyage, the twin craft swept past the planet on July 31 and August 4, 1969 respectively, at an altitude of about 3,550 km (2,200 mi).

Closest approach by Mariner 6 was just south of the equator. Once again, air pressure was very low (6.5

millibars), with carbon dioxide confirmed as the main constituent and an absence of nitrogen. However, the temperature at noon was a balmy 16°C, even if it did plummet to –73°C at night, with a low of –125°C over the south pole (low enough to support frozen carbon dioxide). A series of 26 near-encounter images included some intriguing views of a polar ice cap with a rugged border.

The advantage of a dual mission was demonstrated when Mariner 7 was reprogrammed to take more pictures of the southern ice cap, but 33 near-encounter images showed a lack of craters in the Hellas region, craters filled with frost, and some ice-free areas within the limits of the polar cap. Fine particles were clearly visible at 15–40 km (9–25 mi) above the horizon in some images.

A long-range view of the entire planet showed the circular Nix Olympica, but it was not possible to identify the strange marking as a giant shield volcano. So although Mariners 6 and 7 showed that Mars was, after all, a much more dynamic world than the Moon, it would take a little longer to uncover the planet's geological marvels and remarkable past.

Emerging from the Dust

Even before Mariners 6 and 7 got off the ground, a more ambitious follow-up program was well under way. This time, the pair of Mariners scheduled for launch during the favorable 1971 launch window were intended to provide the first orbital survey of the Red Planet.

Each was assigned a different objective. Mariner 8 would enter a near-polar orbit in order to perform a 90-day reconnaissance that covered a large portion of the planet's surface. Its companion was intended to take a closer look at "time-variable" features, such as clouds, haze layers, ice caps and dust. Mariner 9 was to follow an elliptical orbit, inclined at 50 degrees to the equator, that would bring it to within 885 km (550 mi) of the surface.

Meanwhile, the Soviets were preparing their own ambitious onslaught on Mars. The original flotilla was to comprise two orbiter–lander combinations. However, the temptation to cock a snook at their capitalist rivals caused them to introduce a simple orbiter that would arrive before Mariner 8, thus becoming the first spacecraft to enter orbit around Mars (or any other planet).

The campaign began badly for both protagonists. A malfunction in the guidance system of the Atlas–Centaur rocket sent Mariner 8 plummeting into the Atlantic Ocean on May 9, 1971. The following day, the hastily assembled Soviet orbiter—subsequently dubbed Cosmos 419—became stranded in Earth orbit after a programmer inserted the incorrect time for engine firing. Two down and three to go.

Soviet prospects seemed to improve when Mars 2 and 3 successfully lifted off from Baikonur on May 19 and 28.

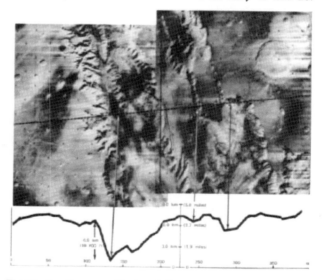

The Martian Canyons—Mariner 9 discovered an enormous canyon system that was later named the Valles Marineris (Mariner Valleys). This rift valley reached one-fifth of the way around Mars, dwarfing the Grand Canyon on Earth. It was found to be 4,000 km (2,500 mi) in length—enough to stretch across the Atlantic Ocean—with a maximum width of 600 km (375 mi) and a depth in places of 6 km (almost 4 mi). This profile was calculated from ultraviolet spectrometer readings along the dashed line. (NASA–JPL)

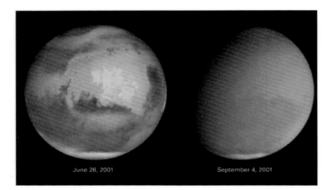

June 26, 2001 September 4, 2001

A Global Dust Storm—Major dust storms regularly arise on Mars during the southern summer. When Mariner 9 and a pair of Soviet spacecraft began to orbit the planet in November 1971, the surface was totally obscured by a global dust storm that had been raging since September. Fortunately, Mariner was able to image the surface after the storm began to abate in January 1972. However, landers deployed from Mars 2 and 3 crashed during the storm and images sent back from the orbiters were largely useless. These pictures from the Hubble Space Telescope show another global dust storm that arose in 2001. (STScI/NASA)

Back in Florida, engineers scrambled to modify Mariner 9 before the May 30 launch date. After the loss of its twin, the remaining spacecraft would have to map the planet from a compromise orbit inclined at 65 degrees to the equator.

Despite the shortage of time, the launch of Mariner 9 was successful. Following a faster route than its Soviet rivals, the 1-tonne spacecraft made history by braking into orbit around Mars on November 13. Mars 2 and 3 were close on its heels, arriving on November 27 and December 2. However, the planet was not prepared to give up its secrets easily.

A dust cloud spotted from the ground on September 22 began to take on a life of its own; spreading rapidly it combined with another storm and the entire planet was enveloped in an opaque blanket. All that could be distinguished were four dark spots in the Tharsis region and the south polar cap. The first, featureless pictures led one scientist to jokingly observe that they must have arrived at Venus by mistake. His comment was not well received.

Frustrated by the dust storm, scientists reprogrammed

Mariner to turn its TV cameras on the moons, Phobos and Deimos. Originally considered hardly worthy of study, the satellites were found to be irregular, cratered objects, possibly main belt asteroids that had been captured by Mars long ago.

While the Mariner team searched for objects to study, the Soviets did not have the luxury of waiting. Their spacecraft were preprogrammed to release the landers (each carrying a tiny mobile rover) into the atmosphere on arrival at Mars. Unfortunately, the angle of entry of the Mars 2 lander was too steep, preventing deployment of the parachutes. Humanity's first robotic ambassador to another planet disintegrated on impact.

Mars 3 survived its passage through the raging storm and began to transmit an image of the landing site. However, signals ceased after 20 seconds, and all that was received was a plain gray picture. Soviet engineers speculated that a discharge of static may have shorted all the craft's circuits.

The Soviet orbiters survived long enough to send back some images of the exposed Martian surface, along with information on surface temperature, air pressure, and the nature of surface rocks. It was determined that the temperature of the northern polar cap was below $-110°C$, and that the concentration of water vapor was 5,000 times less than on Earth.

However, it was Mariner 9 that grabbed the headlines with a series of unexpected discoveries that revolutionized theories about the planet's evolution. After the storm had subsided, the mapping program began on January 2, 1972. It was immediately clear that the dark spots poking through the dust blanket were actually four giant shield volcanoes. The largest of these (formerly known as Nix Olympica) was named Olympus Mons. An estimated 25 km (15 mi) high and 600 km (370 mi) across, it was the largest volcano ever seen, dwarfing anything on Earth. The other three towering mountains were aligned on top of the newly discovered Tharsis bulge.

On January 12, Mariner unveiled part of a mammoth canyon system located southeast of the volcanoes. Subsequent mapping showed that this huge gash in the landscape stretched for 4,800 km (3,000 mi)—one-fifth of the way around the planet. In places it was up to 600 km (375 mi) wide and 6 km (4 mi) deep. The rift valley complex was named Valles Marineris (Mariner Valleys).

Most intriguing of all were sinuous, often braided, channels that snaked across the surface and appeared to have been carved by running water. The largest of these dry "river beds" were 1,500 km (930 mi) long and 200 km (125 mi) wide. The implications were staggering. At some time in the past, the Martian atmosphere must have been much warmer and wetter, allowing huge quantities of liquid water to flow across the surface in the form of catastrophic flash floods. There was speculation that gulleys on the sides of impact craters had been produced by falling rain.

By the time the mission ended in late October, Mariner 9 had imaged the entire planet with a spatial resolution of 1–2 km (0.6–1.2 mi). Scientists were eagerly awaiting their opportunity to explore the surface of the "new Mars" and search for signs of primitive life.

Viking—The 1976 Viking mission to Mars involved two orbiters and two landers. The orbiter (bottom) had four solar panels, a main antenna and various instruments. Its initial job was to survey the planet for a suitable landing site. Later, it studied the planet's surface and atmosphere in unprecedented detail and acted as a relay station for lander data. The lander was stored inside the clam-shaped bioshield (top), which was jettisoned soon after launch. On the surface of Mars, the lander surveyed the soil, wind, and atmosphere, and conducted numerous experiments to search for evidence of past or present life. (NASA–JPL)

The Viking Invasion

Mariner 9 had revealed a dynamic planet that had undergone—and perhaps was still undergoing—dramatic changes during its 4.5-billion-year lifetime. Although there was no evidence of liquid water on the present surface, the pictures of meandering channels provided powerful circumstantial evidence that primitive life could have evolved in the distant past, when conditions were more favorable. A lack of impact craters on the large shield volcanoes indicated that lava may have erupted on the planet in the geologically recent past.

The next step in unraveling the Martian mysteries would be to map the planet at higher resolution from orbit while searching for ground truth with two advanced landers. The twin orbiter–lander missions, known as Viking, were approved in December 1968, with launches envisaged for 1973. Unfortunately, budget cutbacks soon obliged NASA to put back the mission by two years.

Although the delay increased the mission cost in the long term, it turned out to be blessing in disguise, since at least two of the technologically challenging lander experiments would probably not have been ready for the 1973 launch date.

The fact that each Viking comprised two large spacecraft also added to the complexity. Despite the reuse of Mariner 9 technology, it was still necessary to enlarge the orbiters to enable them to carry the sophisticated soft landers, a larger engine, and extra fuel. The Vikings were eventually found to be the most expensive NASA planetary mission ever undertaken, with an overall cost of roughly $1 billion at 1975 rates.

Each Viking spacecraft weighed around 3 tonnes at launch, of which more than half was fuel. The lander measured 2.1 m (7 ft) in height (including the main antenna) and 3 m (10 ft) wide. After descending under a parachute, the touchdown was to be controlled by three variable thrust engines. Three legs, arranged in a triangle, gave the lander a ground clearance of 22 cm (8.7 in).

Carl Sagan with the Viking Lander—Scientist, space popularizer, and award-winning author Carl Sagan (1934–1996) with a model of the Viking lander that was used during simulations of the Mars environment in Death Valley, California. A leading supporter of the theory that life was common throughout the Universe, Sagan was disappointed when the Vikings found no compelling evidence for the existence of Martian organisms. Born in Brooklyn, New York, the son of a Jewish garment worker, Sagan graduated in astronomy and astrophysics before going on to teach at Harvard. In 1968 he moved to Cornell, where he became director of the Laboratory for Planetary Studies. During the 1970s he worked on most of the US planetary missions, but his most remembered contributions are the messages for aliens placed on Pioneers 10/11 and the Voyager spacecraft. (NASA–JPL)

This extremely sophisticated craft was equipped with two computers that could store commands for 60 days of operation, as well as a recorder with 198 m (653 ft) of tape for storage of up to 40 Mbits of data. Images and other results from the surface could be directly transmitted to Earth for about two hours each day, or relayed via the orbiters. Seventy watts of electricity—less than the amount used by a normal domestic light bulb—was provided by two nuclear generators powered by heat from the radioactive decay of plutonium-238 oxide. Four rechargeable batteries were available as backup in peak periods.

One of the main concerns was to avoid contamination of the Martian surface by organisms transported from Earth. The landers were, therefore, sterilized before launch by heating them in an oven to 113°C for 40 hours. They were then sealed in a bioshield, which was only jettisoned after the spacecraft were safely on their way to Mars.

The first countdown got off to a bad start when problems with the Titan–Centaur launcher and a drained spacecraft battery forced engineers to swap the Vikings. Consequently, the second pair of Viking spacecraft were first to leave Earth on August 20, 1975, followed by their twins on September 9. An unusually long flight path took them more than half way around the Sun—a distance of about 815 million km (505 million mi).

Anticipation was high when Viking 1 finally fired its main engine for 38 minutes to slow into Mars orbit on June 19, 1976. Three days later, its orbital period was adjusted to enable the spacecraft to fly over the planned landing site in Chryse Planitia each day but unfortunately, the "golden plain" did not live up to its glittering description. The touchdown site, which had taken years to select, turned out to be on the floor of what looked like a deeply incised river bed. The plan to celebrate the US bicentennial with a landing on July 4 would have to be binned.

Over the next few days, scientists desperately tried to find a more suitable site before Viking 2 arrived to complicate matters even more. After waiting for new pictures and agonizing over contradictory radar measurements of surface roughness, they finally plumped for a new landing area on Chryse, about 575 km (365 mi) west of the original site. The prolonged discussions paid off when Lander 1 completed a successful descent on July 20. It was just 28 km (17 mi) off target.

The mission team were faced with a similar dilemma when Viking 2 arrived in orbit on August 7. Once again, the primary landing zone—this time further north on Cydonia—proved unsuitable, with polygonal cracks, small craters and plentiful rocks. As the backup site near the Alba Patera shield volcano was also far too rough, the scientists eventually decided to set their explorer down on Utopia Planitia, half a world away from its companion.

Lander 2 eventually set down on September 3, about 200 km (125 mi) from Mie Crater and 6,400 km (4,000 mi) from Lander 1.

With their primary responsibility safely discharged, the Viking orbiters were able to concentrate on mapping the surface and studying the atmosphere. Over the next few years they imaged the entire surface of Mars at a resolution of 150–300 m (500–1,000 ft), showing details as small as 8 m (26 ft) across in selected areas. They far exceeded their planned lifetimes. Viking Orbiter 2 was powered down on July 25, 1978, after 706 orbits, while Viking Orbiter 1 shut down on August 17, 1980, after 1,400 orbits.

The landers displayed even greater longevity. Viking Lander 2 ceased communications on April 11, 1980, while Lander 1 succumbed on November 13, 1982, after operating for more than three Martian years. Between them they transmitted over 1,400 images and performed 125 soil-sampling sequences at the two sites.

Mars from Orbit

If Mariner 9 had whetted scientists' appetite for more, the Viking orbiters certainly gave them more than they had dared hope. From their polar orbits, the spacecraft were able to map and study almost the entire surface, witnessing the dramatic seasonal changes that take place during a Martian year. A grand total of 52,000 images were returned before they were finally shut down.

The orbiters carried three experiments: a TV system, an atmospheric water detector, and an infrared thermal mapper. The science team was also able to learn about the atmosphere by monitoring changes in radio signals as the spacecraft passed behind the planet.

On arrival in Mars orbit, one of their first tasks was to send back detailed pictures of preselected landing sites to ensure that the piggybacking landers would be able to complete their surface mission successfully. Once the landers were released, the orbiters were able to concentrate on their planetary surveys.

On a global scale, it was clear that Mars is divided into two very different hemispheres: the fairly flat, low plains in the north, and rugged, cratered highlands in the south. The boundary between the regions was very steep and distinct, cutting across the equator and rounding the Tharsis bulge.

Much attention, of course, was given to the dramatic landforms discovered by Mariner 9. Hundreds of shield volcanoes of various sizes were revealed, in addition to many small, steep-sided cones that were apparently formed by more explosive eruptions. However, by far the most impressive edifices were the giants of Tharsis, the youngest volcanic region on Mars.

It was impossible to describe Olympus Mons without resorting to superlatives. The huge, gently sloping cone contained more lava than the entire Hawaiian mountain chain. Its central caldera was shown to comprise a series of separate craters formed by repeated collapses after underground magma drained away during eruptions. Thousands of individual lava flows could be seen running down its flanks. After pouring over the near-vertical cliff at its edge,

Clouds Around Olympus Mons—This painting is based on a mosaic of black-and-white Viking orbiter images of Olympus Mons, the largest volcano in the Solar System. It shows the 24-km (15-mi) high shield volcano protruding above thin clouds of water ice. The huge structure measures about 700 km (430 mi) in diameter and the complex caldera in the center is 80 km (50 mi) across. In places the steep flanks form a cliff 6 km (4 mi) high. (NASA–JPL/Gordon Legg)

they spread out for hundreds of kilometers over the surrounding plains.

About 1,600 km (1,000 mi) to the southeast was a row of three volcanoes—Arsia Mons, Pavonis Mons, and Ascraeus Mons—that crowned a huge crustal bulge more than 10 km (6 mi) high. To the north of Tharsis, beyond a chain of more modest volcanic cones, was the immense Alba Patera, an ancient volcano that measured roughly 1,600 km across, making it even wider than Olympus Mons. However, despite its size, Alba Patera was found to be very flat, reaching only about 3 km (2 mi) above its surrounding plains.

Associated with Alba Patera and many of the other giant volcanoes were numerous straight, linear faults that resembled dozens of railway tracks running parallel across the barren landscape. It seemed certain that tremendous stretching and cracking of the planet's crust had coincided with the formation of the Tharsis bulge.

The most spectacular creation of this crustal extension was the great gash known as Valles Marineris. This huge complex of equatorial canyons apparently grew as the crust began to pull apart, forming an enormous rift. Subsequent erosion—either by glaciers or by running water—and landslides then widened and shaped the various parallel chasms.

The Vikings also turned their attention to the mysterious polar caps. Winter temperatures at the south pole plummeted to around –140°C, cold enough for carbon dioxide gas to freeze out onto the surface. The north pole was only a little less welcoming. However, although a thermometer would still register a frigid –60°C in the northern summer, this was much too warm for carbon dioxide ice. The residual northern ice cap had to be made of the more familiar water ice. Curiously, this did not seem to apply to the southern cap, where temperatures remained marginally below the freezing point of carbon dioxide.

Although the build up of carbon dioxide ice caused both ice caps to grow dramatically during the winter, the southern one showed the most dramatic expansion, reaching all the way from the pole to 55°S. In contrast, their shrunken summer versions were disrupted by curved, ice-

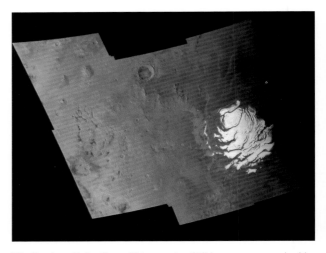

The Southern Polar Cap—This mosaic of 18 images was acquired by Viking Orbiter 2 on September 28, 1977. The south pole is located off the upper right edge of the ice cap. The picture was taken during the southern summer, when the polar cap was nearing its final stage of vaporization and retreat. The residual cap is approximately 400 km (250 mi) across. Although the exposed surface consists mainly of carbon dioxide frost, its permanent lower layers are water ice. Irregular pits with sharp-rimmed cliffs (left) appear "etched," presumably by wind. The large crater near the top (named "South") is about 100 km (62 mi) across. (USGS, NASA–JPL)

free valleys that cut into the permanent, residual cap. Viking close-ups revealed layers of ice and dust on the valley walls, material deposited during a repetitive seasonal cycle that stretched back for millennia.

Widespread sand dunes and frequent dust storms—including two that almost covered the entire planet—revealed Mars to be a desert world. Wind was obviously the main agent of erosion at the present time, but the Vikings also found numerous channels that appeared to have been carved by running water. Some of the networks were braided and cut through pre-existing craters. Occasionally, teardrop-shaped "islands" revealed the direction of flow.

The largest of these valleys, known as outflow channels, seemed to have been created by flash floods in which huge volumes of water poured from a subterranean source.

Others, discovered by Mariner 9, took the form of sinuous channels that were often fed by tributaries, but the Viking orbiters found even smaller valley networks that were presumably once fed by springs.

If the planet was once wet, where had the water gone? Some must have escaped into space, but where was the remainder? One clue came from impact craters that were surrounded by aprons of what looked like solidified mud flows. Heat released by the impacts had melted the deep-frozen permafrost and turned the surface into a slurry.

Viking Vistas

On July 20, 1976, the team in mission control waited with bated breath as they received the telemetry from Lander 1. Would their billion-dollar baby survive the traumatic landing sequence? They did not have long to wait. Immediately after touchdown, Camera 2 began a 5-minute scan of the ground on which the craft was resting. Strip by strip, the scene assembled on the screens in Pasadena, and, for the first time in history, human eyes could see what the surface of the Red Planet really looked like.

The black-and-white close-up image revealed numerous pebbles and small rocks resting on a fine-grained surface. In the right-hand corner, the surface under Footpad 2 seemed fairly solid, although some sediment had collected inside its concave bowl.

A second, panoramic picture elicited even more excitement from the waiting team at JPL. Over a period of half an hour, Camera 2 sent back a 300-degree view of an undulating landscape that bore a close resemblance to a rocky desert on Earth. Many of the most prominent boulders were given descriptive names by the imaging team—"Volkswagen," "House on the Hill," and "the Midas Muffler Rock." Subsequent images revealed that the largest rock in the vicinity, nicknamed "Big Joe," was about 3 m (10 ft) wide and 1 m (3 ft) high—large enough to have destroyed the lander if it had been unlucky enough to set down on top of it.

Chryse Planitia from Viking Lander 1—The Viking 1 landing site on Chryse Planitia displays numerous orange–red sand dunes and rocks, apparently overlying patches of exposed, darker bedrock. The 1-m (3-ft) high rock at the extreme left was nicknamed "Big Joe." Shallow trenches dug by the lander's sampler arm are visible in the foreground. The reddish color was thought to be caused by limonite (hydrated ferric oxide). Such weathering products form on Earth in the presence of water and an oxidizing atmosphere. The reddish sky color is caused by dust suspended in the lower atmosphere. (NASA–JPL)

The next landmark was the first color picture. On July 21, under pressure from the media, the imaging team rushed out a beautiful image of a pale blue sky above a brown surface scattered with grayish rocks. Five days later, after they had time to calibrate properly the images taken through blue, green, and red filters, a revised version was issued. This time, to many people's consternation, the sky had turned salmon pink, while the Chryse plain looked more like the orange–red planet of legend. Scientists blamed scattering of light by tiny particles of airborne dust for the alien hue. The red stain of the rocks was explained by the presence of rust (hydrated iron oxide).

However, the science team had more important things to worry about. The protective "cage" on the Lander 1 seismometer remained stubbornly in place, preventing the instrument from picking up any Marsquakes. Fortunately, its counterpart on the second lander worked perfectly, although it only detected two distant rumbles, each measuring about 3 on the Richter scale, during its lifetime.

Lander 2's mission almost came to an abrupt end when one footpad came to rest on a sizeable rock, causing it to tilt 8½ degrees to the west. As a result, the first panoramic views showed a sloping horizon, even though, in reality, the landscape was extremely flat. Apart from the plethora of large boulders and the absence of sand dunes, the scene at Utopia was not dissimilar to Chryse.

Over the coming weeks and months, the meteorology instruments sent back the first measurements of weather on another world. Summer conditions were quite monotonous and repetitive, with midday temperatures soaring to −14°C

Utopia Planitia from Viking Lander 2—Viking Lander 2 touched down about 6,400 km (4,000 mi) from its twin. Utopia Planitia was located further north than Chryse Planitia, but the sites looked broadly similar. The boulder-strewn field of rust-red rocks reaches to the horizon nearly 3 km (2 mi) from the lander and fine particles of red dust have settled on spacecraft surfaces. Color calibration charts for the cameras are visible at three locations on the spacecraft, to the right of the Stars and Stripes. The circular structure is the high-gain antenna (top), the main communication link with Earth and the orbiters. (NASA–JPL)

in Chryse, but plunging to −77°C just before dawn. Lander 2 experienced more extreme conditions, especially in winter, when the predawn temperature bottomed out at −120°C, about the freezing point of carbon dioxide. A thin layer of water frost coated the ground during each winter at Utopia. However, when occasional midwinter dust storms cast gloom over the landscape, diurnal temperatures remained almost steady.

Studies of atmospheric composition confirmed that 95% of the atmosphere is made of carbon dioxide. To everyone's surprise, nitrogen was recorded for the first time, at about 2.7%. The third most plentiful gas was argon.

Wind and pressure sensors detected passing fronts, cyclones, and anticyclones. Gentle breezes, generally blowing downhill from nearby uplands, were occasionally interrupted by gusts reaching 48–65 km/h (30–40 mph). The thin atmosphere reacted to seasonal changes in the sizes of the ice caps. When the southern cap was at its largest, air pressure dropped to around 7 millibars at both sites, rising to about 10 millibars as the carbon dioxide ice reverted to a gas.

The most exciting, and controversial, results came from three biology experiments designed to look for possible signs of life. Each was designed to search for a chemical reaction from primitive organisms in the Martian soil.

The first one looked for evidence of photosynthesis. Scientists hoped to see evidence that organisms were taking up carbon dioxide and carbon monoxide "labeled" with radioactive carbon-14. Unfortunately, after a weak positive response on the first attempt, the results could not be repeated.

A second experiment used carbon-14 as a tracer inside a sterile "organic soup." It was hoped that Martian organisms would breathe out radioactive gas as they ate the broth. After an initial surge in activity, the gas production process ground to a halt—behavior untypical of multiplying bacteria. However, strong heating seemed to destroy whatever was causing the unusual reaction.

In the third experiment, an extraordinary "outburst" of oxygen occurred when the soil was moistened, but very little happened when a nutrient broth was added. Since a similar release of oxygen occurred when the soil was sterilized, scientists concluded that a strong oxidizing agent in the soil, such as hydrogen peroxide, could be to blame.

Another key piece of evidence was provided by a fourth experiment which detected no organic (carbon-rich) material in the soil. Scientists reluctantly concluded that the Vikings had found no convincing evidence of life on Mars (although biology team member Gilbert Levin still continues to dispute this conclusion). The general consensus was that the Martian environment is self-sterilizing. Living organisms are prevented from existing on the surface of Mars today by the combined effects of strong ultraviolet radiation from the Sun, extreme aridity and an oxidizing soil.

Phobos and Deimos

Although Schiaparelli's "canali" grabbed most of the headlines during the close approach of Mars in 1877, another astronomer on the other side of the Atlantic was also determined to take advantage of the favorable opposition. Asaph Hall set himself the task of finding out whether the Red Planet had any companions.

Since the early years of telescopic studies, astronomers had speculated that Mars may have two small moons. In those days, when the symmetry and perfection of the heavens was generally accepted, it seemed appropriate to assume that if Venus had no satellites, Earth had one and Jupiter had four, then Mars should possess two.

In 1726 Jonathan Swift took this unfounded belief one step further by making a point of mentioning the mythical Martian moons in his novel, *Gulliver's Travels*. Swift's hero learned that the Laputans had discovered two satellites "which revolve around Mars; whereof the innermost is distant from the center of the primary planet exactly three of his diameters, and the outermost five; the former revolves in the space of ten hours, and the latter in twenty-one and an half." The predictions proved to be remarkably accurate.

Using the 0.66-m (26-in) refractor at the United States Naval Observatory—at the time the largest telescope of its kind in the world—Hall began a systematic, meticulous search of the skies around Mars. Starting far from the planet, he gradually moved inward. Weeks passed without success, but, reassured by his wife, the former Angelina Stickney, he pressed on.

Then, on the night of August 11, his dedication and perseverance paid off with the discovery of a small satellite. After a break forced by bad weather, Hall impatiently returned to his observations and promptly found a second satellite even closer to the planet. He named the inner moon Phobos (Fear) and its outer companion Deimos (Panic), after the companions of the Greek war god, Ares.

It did not take long to calculate their orbits. The average distance of Deimos from the center of the planet was found to be 23,460 km (14,580 mi)—only one-tenth the Earth–Moon distance. It followed a near-circular path, tilted at an angle of 1.8 degrees to the planet's equator. With an orbital period of 30 hours 21 minutes, Deimos took a little over one Martian day to complete a single circuit. Consequently, an observer on Mars would see it track slowly westward across the sky.

Phobos was even closer to Mars, just 9,380 km (5,830 mi) from the planet's center—so close that the satellite would be permanently below the horizon and out of sight for anyone at a latitude higher than 69°S. In order to avoid crashing into Mars, the moon swept around the planet once every 7 hours 39 minutes (about one-third of a Martian day). This meant that someone fairly close to the Martian equator would see Phobos rise above the western horizon roughly every 11 hours. It would cross the sky "backward" (from west to east) in just 4½ hours. During this brief appearance, it would go through more than half its cycle of illuminated phases.

Close studies of their orbits showed that Phobos is slowly accelerating and spiraling in toward Mars. It seems that, in about 100 million years from now, the little moon will end its existence in a cataclysmic explosion as it ploughs into Mars. (In 1959, Russian astronomer Iosif Shklovskii suggested that this could be explained if Phobos was hollow inside—possibly an alien space station!)

Images taken by various spacecraft showed that both

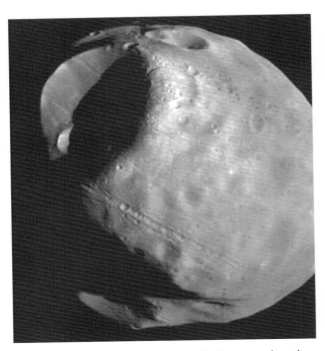

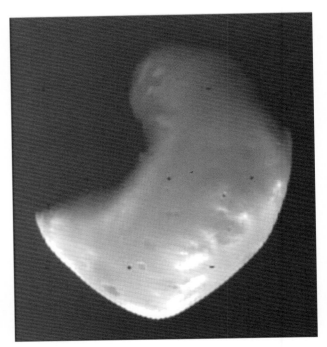

Phobos—This montage of the moon Phobos comprises three separate images taken by Viking Orbiter 1 on October 19, 1978, from a distance of 613–633 km (380–393 mi). The large crater (mostly in shadow) at upper left is known as Stickney—in honor of the wife of Asaph Hall, the discoverer of Phobos. A smaller shaded crater (bottom) is named after Hall himself. Also visible are linear grooves and striations that radiate away from Stickney. (NSSDC, NASA–JPL)

Deimos—A Viking Orbiter 2 image of the outer satellite, Deimos, taken from a distance of 1,400 km (875 mi). Apart from a large crater (left), almost entirely hidden in shade, Deimos appears smooth. However, higher resolution images taken during closer approaches show the surface to be covered with craters, many of which have been partially buried or filled by loose surface material or regolith. Deimos is about 14 km (8.7 mi) from top to bottom in this image. (NSSDC, NASA–JPL)

objects are extremely small—indeed, for many years, Deimos was the smallest satellite known. (The first close-up pictures of Phobos and Deimos were obtained by the Mariner 9 spacecraft in 1972. Since then, they have also been imaged by the Viking orbiters, the Russian Phobos 2, Mars Global Surveyor and ESA's Mars Express.)

Phobos turned out to be a potato-shaped object, measuring 27 × 22 × 18 km (17 × 14 × 11 mi). Its ancient surface displayed a 10-km (6-mi) crater, which was named Stickney in honor of the woman who persuaded

Hall to continue his pursuit of the elusive satellites. Also present are two 5-km (3-mi) wide craters—one of which is named after the discoverer himself.

Radiating away from Stickney are what appear to be chains of small craters and parallel striations that are only 5–10 m (16.5–33 ft) deep but up to 15 km (9 mi) in length. These are almost certainly related to internal fractures caused by the large impact that excavated Stickney. Also visible are linear grooves about 500 m (1,650 ft) wide that may be evidence of lines of weakness produced by tidal forces and impacts.

Deimos also displays a battered, irregular shape, measuring about 15 × 12 × 10 km (9 × 7 × 6 mi). There are some small craters, the two most prominent being named Swift and Voltaire. However, its surface is much smoother than that of Phobos. There are no visible grooves or striations, suggesting that tidal forces and large impacts have played a lesser role in its evolution, and most of the visible craters appear to be partially filled or covered with loose regolith. Close range images also revealed a surface strewn with boulders that have presumably been ejected during collisions with small meteorites.

Each satellite has been affected by tidal forces so that its long axis is aligned with the planet. Like Earth's Moon, both keep the same face toward Mars. The moons are very dark, but with a slight reddish hue, and have densities about twice that of water. Clearly, they were not born in the vicinity of Mars, and most astronomers believe they are captured asteroids. Their dark coloration and spectra indicate a carbon-rich composition similar to C-type asteroids.

Pathfinder and Sojourner

By the 1990s, NASA's space science budgets were once again under considerable pressure. Unable to countenance any multibillion dollar extravaganzas similar to Viking, NASA chief Dan Goldin introduced a "faster, better, cheaper" policy. One of the first projects put forward under the new Discovery program was a technology demonstration mission known as Mars Pathfinder. Although its primary purpose was to test an airbag landing system and a small, autonomous rover, Pathfinder provided the first real opportunity for scientists to study the Martian surface since the early 1980s.

The Mars Pathfinder lander and its piggybacking rover were cocooned inside a tetrahedron-shaped set of airbags. Each of the tetrahedron's four sides comprised six beach-ball-like bags arranged in a triangle. Four minutes after the spacecraft slammed into the upper atmosphere of Mars at

27,200 km/h (17,000 mph), the 24 airbags inflated and the parachute was cut free.

The bags inflated at a height of 300 m (1,000 ft), just eight seconds before Pathfinder hit the surface and bounced at least 15 times, each rebound reaching up to 12 m (40 ft) high. Once the bags came to rest they were deflated so that that petal-shaped solar panels and ramps could be deployed.

The most interesting piece of scientific hardware was the 10-kg (22-lb) Sojourner microrover, which was named in honor of Sojourner Truth, an African-American champion of women's rights and a campaigner against slavery during the American Civil War. (The name was selected from 3,500 entries in a worldwide schools competition organized by the Planetary Society.)

Sojourner would be only the third robotic roving vehicle ever to drive over another world. Little larger than a microwave oven, its upper surface was covered by solar cells which could power the rover for several hours per day, even

Pathfinder Panorama—This image shows part of the 360-degree panoramic view sent back by the Mars Pathfinder lander during sols (Martian days) 8, 9, and 10. At left can be seen a lander petal on top of a deflated airbag, and the ramp which enabled the Sojourner rover to reach the surface on July 5, 1997. On the horizon are the "Twin Peaks," about 1–2 km (0.6–1.2 mi) away. Immediately to the left of the rear ramp is the andesitic rock "Barnacle Bill." Sojourner is using its Alpha Proton X-Ray Spectrometer to study the large volcanic rock nicknamed "Yogi." During its trip to Yogi the wheels dug down several centimeters, exposing some white material. (NASA–JPL)

in the worst dust storms. Its six wheels could independently move up and down, enabling it to drive over a boulder 25 cm (10 in) high.

Since the round-trip time for signals to Mars can take 20–30 minutes, Sojourner could not be driven by controllers on Earth. Instead, the "intelligent" rover was given its destination and tasks for the day, then its computers navigated it around obstacles to reach each target. Motion sensors could detect too much tilt and stop the rover. Top speed was about 0.6 m (2 ft) per minute.

A small array of gallium arsenide cells on top of Sojourner generated electricity from the weak sunlight. The rover operated on about 8 watts of power—about as much as a night light. It carried three cameras—a forward stereo system and a rear color imaging system. There was also a German–US alpha proton X-ray spectrometer which bombarded selected rocks with alpha particle radiation. Once in contact with rocks or soil, it could measure the particles that came back and determine the elements from which they were made.

Pathfinder lifted off from Cape Canaveral, Florida, on December 4, 1996. Once again, engineers at JPL aimed for a landing on Independence Day, and this time they were successful. The spacecraft slammed into the upper atmosphere of Mars on July 4, 1997. Four minutes later, the 24 airbags inflated and the parachute was cut free. After bouncing across the surface for some distance, the spacecraft came to rest on the floor of an ancient northern flood plain, known as Ares Vallis.

The first images from the lander—later named the Carl Sagan Memorial Station—revealed a problem. The air bags had not fully deflated and were preventing deployment of the ramps down which Sojourner would have to travel to reach the surface. Ground controllers had to raise one of the lander's "petals" and operate the airbag retractors for 10 minutes before the ramps could be released.

The following Martian day (Sol 2), Sojourner trundled down a ramp and onto the orange, rock-covered plains. It continued to creep around the landing site for the next 83 Martian days, until a flat battery on the lander stopped communications on September 27, 1997. As the rover

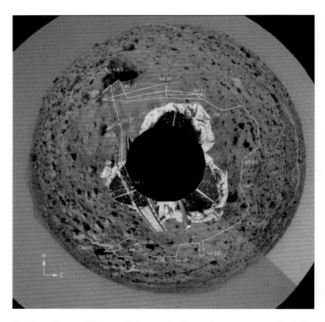

Staying Close to Home—Sojourner's zigzagging traverse around the main lander (renamed the Carl Sagan Memorial Station) is shown in this computer-processed projection. Overlaid onto the projected Martian surface is the path taken during the 83 sols (Martian days) of Pathfinder surface operations. Yogi is at 11 o'clock, with the Rock Garden between 6 and 9 o'clock. (NASA–JPL)

zigzagged about 100 m (330 ft) clockwise around the lander, it sent back 550 images of the landing site.

Almost everywhere its camera pointed, there were rocks of every shape and size. The mission team enjoyed naming some 200 of these boulders, many of them after cartoon characters. Among the rocks examined by Sojourner were "Yogi," "Chimp," and "Barnacle Bill." "You couldn't name anything after a (living) person, a place, a family member or a dog," explained Matthew Golombek, Mars Pathfinder project scientist.

Dozens of rocks were photographed from close range and there were 16 analyses using Sojourner's X-ray spectrometer. Many rocks seemed to be volcanic in origin, though others contained rounded pebbles, suggesting that running water was involved in their formation. The implication was that the climate was much warmer and wetter when these rocks formed.

Although the little rover grabbed the headlines, the lander also provided useful new information. Apart from sending back more than 16,000 images, including some stunning panoramas and views of blue sunsets, it also sent back daily weather reports. Although Pathfinder arrived in the northern summer, temperatures ranged from a minimum of –75°C to a high of –10°C. High ice clouds were often seen in the early morning, but these evaporated during the day.

Air pressure fluctuated considerably during the day, possibly reflecting global changes in temperature, dust, water, and carbon dioxide. Night winds were usually light and blowing from the highlands to the south, swinging around to the northwest during the day. A dozen dust devils—miniature whirlwinds loaded with fine material—were also observed during its 83 days of operation.

Mars Pathfinder–Sojourner gripped the imagination of the public and went on to become by far the largest Internet event in history at the time. A total of 566 million "hits" occurred on the Pathfinder websites during its first month on Mars, with 47 million hits on July 8 alone.

Ice and Dust

Pathfinder was not the only game in town during the latter half of 1996. NASA had also decided to pursue a new, long-term initiative, known as the Mars Surveyor program. The plan was to send an orbiter, known as Mars Global Surveyor, during the 1996 window, followed by two spacecraft during each launch opportunity in 1998, 2001 and 2003. A sample return mission was penciled in for 2005.

The timing proved fortuitous. Scientific and media interest in the Red Planet skyrocketed in August 1996, when a NASA-funded team of scientists held a press conference to announce that they had found convincing evidence for fossilized organisms inside a Martian meteorite known as ALH84001. The team found tiny deposits of carbonate, organic molecules known as polycyclic aromatic hydrocarbons, and iron compounds commonly associated with microscopic organisms. However, it was the electron microscope images of bacterial-like structures that made the front pages.

The results came under a barrage of criticism, starting a debate that has continued to rage ever since. But, true or not, the scientific community had an added incentive to return to Mars. Although Mars Global Surveyor was not designed to look for life, it did carry six experiments that would create the first global maps of topography, magnetic field, and surface minerals. The wide-angle lens on the Mars Orbiter Camera would be able to provide a daily weather map of Mars, while the narrow-angle lens was capable of detecting objects only 2–3 m (7–10 ft) across. (The instruments were mainly reflights of hardware and spares developed for the failed Mars Observer mission of 1992.)

Global Surveyor set off for Mars on November 7, 1996, arriving in a highly elliptical orbit on September 12, 1997. In order to save fuel and weight, the spacecraft was designed to spend the next four months dropping into its mapping orbit by using a technique known as aerobraking. The plan was to use friction with the thin upper atmosphere to slow the spacecraft during each close pass, but a problem with an unlatched solar panel led to a major slowdown. Although some observations were possible from a higher orbit, the primary mapping mission was delayed until March 9, 1999, during the northern summer. (It was originally scheduled for March 1998, during the southern summer.)

From its final nearly-circular, near-polar orbit 380 km (236 mi) above the planet, Global Surveyor's laser altimeter revealed the major variations in Martian topography in remarkable detail. The data showed that Mars has an overall height variation of around 30 km (18 mi)—larger than any of its rocky neighbors. Not unexpectedly, Olympus Mons was found to be the highest point, while the Hellas impact basin in the southern hemisphere was the lowest. Curiously, the Valles Marineris were deepest in the

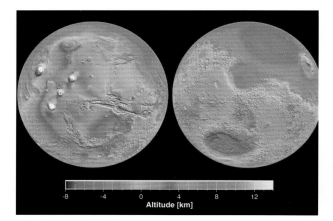

Highs and Lows—These topographic maps were obtained with Mars Global Surveyor's laser altimeter. The spatial resolution is about 15 km (9 mi) at the equator and less at higher latitudes. The vertical accuracy is less than 5 m (17 ft). The southern hemisphere (right) shows the giant Hellas impact basin (purple) surrounded by cratered highlands. The other map includes the Tharsis dome with its huge shield volcanoes (in red and white), and the Valles Marineris which join the northern lowlands at their eastern end. The north polar cap is at a much lower altitude than the southern ice cap. (NASA-GSFC)

center (Coprates Chasma) sloping up toward both the east and west. This cast some doubt over the theory that water had flowed from the canyons onto the northern plains.

The map clearly showed the sudden, steep rise from the northern plains to the cratered southern highlands. This division into two hemispheres accounted for a 6-km (4-mi) difference in altitude between the polar ice caps. The smaller northern cap was shown to be about 1 km (0.6 mi) thick, while the southern cap contained twice as much ice—comparable to the Greenland ice sheet.

Detailed images showed that the ice caps are slowly disappearing. They also revealed a bewildering variety of surface features, including pits, table-shaped mesas, curving stripes eroded by the wind, and "kidney-bean" structures rich in dust. Some impact craters at high latitudes displayed polygonal patterns on their floors, similar to those caused by the freeze–thaw of subsurface ice on Earth.

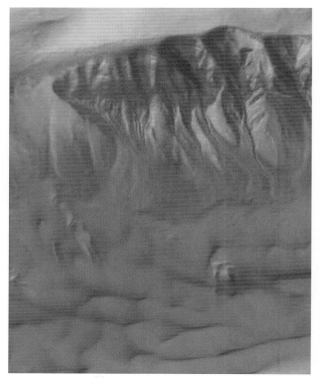

Water-Eroded Gulleys?—No water can survive on the surface of Mars today, but images such as this suggest that rivulets flowed across the surface in geologically recent times. This Mars Global Surveyor image shows gullies a few kilometers long on the wall of an impact crater in the southern highlands. Whitish frost is visible near the top and on the sand dunes below. (Malin Space Science Systems/NASA–JPL)

The search for Mars's missing water took a new twist when detailed pictures of crater walls showed numerous narrow gullies that resembled water-eroded features on Earth. Since they seemed to have formed very recently, perhaps within the past 10,000 years, many scientists argued that subsurface water must have flowed down the slopes and survived long enough to carve the channels. Others searched for alternative explanations, such as eruptions of carbon dioxide gas, arguing that most of the Martian gullies faced away from the warmth of the Sun.

Everywhere the Global Surveyor camera looked, there seemed to be sand dunes or deposits of loose dust. As on Earth, these came in various shapes and sizes—crescent-shaped barchans, and elongated, parallel dunes resembling enlarged ripples on a beach. Also visible were linear walls of rock, called yardangs, that have been sand-blasted over millennia by the prevailing, dust-laden winds.

After observing hundreds of local dust storms, scientists were eventually treated to the biggest global event since the Viking era. What began as a local event around the Hellas Basin in June 2001 stimulated other storms thousands of kilometers away. The entire planet was blanketed for three months, raising the temperature of the upper atmosphere by 44°C. High-resolution pictures from Global Surveyor also showed the dark, twisting paths of countless dust devils criss-crossing the Martian surface. The trails, created by spinning columns of warm, rising air, measured up to 15 m (50 ft) wide and several kilometers in length.

Another surprise was the discovery of a "lumpy" magnetic field, which in places was 10 times stronger than that on Earth. Since Mars no longer has an active dynamo in its core, the localized fields must be left over from the planet's early history. Curiously, the most prominent magnetic sources were aligned in east–west bands of alternating polarity that extended for over 1,000 km (625 mi) north to south like a bar code across the planet's surface. This led some scientists to suggest that they were associated with the creation of new crust—a process similar to sea floor spreading on Earth. Areas without these magnetic "umbrellas" had presumably been demagnetized by huge impacts that took place around 4 billion years ago.

Spirit and Opportunity

After the success of Mars Pathfinder and Global Surveyor, NASA's "faster, cheaper, better" policy suffered two serious setbacks during the next exploration window

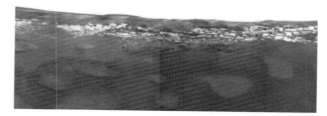

Hematite in Eagle Crater—NASA's Opportunity rover landed in a small hollow, now known as Eagle Crater, on January 24, 2004. Here data from the rover's miniature thermal emission spectrometer have been superimposed on a panoramic view to show the location of crystalline hematite. Red and orange patches indicate high levels of the iron-bearing mineral; blue and green denote low levels. The bedrock outcrop (mainly white) was about 30–45 cm (12–18 in) deep and does not appear to contain much hematite. Also lacking hematite are the rover's airbag bounce marks (foreground). (NASA/JPL/Arizona State University/Cornell)

(1998–1999), when Mars Climate Orbiter and Mars Polar Lander/Deep Space 2 both disappeared without trace on arrival at the Red Planet. Particularly embarrassing was the revelation of the Mars Climate Orbiter Board of Inquiry that engineers at Lockheed Martin and JPL had been involved in a mix-up over metric and English units. The errors in navigation that resulted caused the spacecraft to pass too close to the planet and, presumably, burn up in the Martian atmosphere.

NASA's reaction was to cancel a lander planned for 2001, and press ahead—with due caution—on a dual rover mission to be launched in 2003. The twin Exploration Rovers, named Spirit and Opportunity, were larger, more advanced versions of the Sojourner. In order to simplify design and reduce risk, they were to reuse Pathfinder's airbag system. Unlike the failed Polar Lander, they would be able to send back data during their atmospheric entry and descent.

Although the 2003 lander would not carry any instruments and play a purely passive delivery role, the rovers were equipped with navigation cameras, two high-resolution stereo panoramic cameras, and a spectrometer that could study rocks at infrared wavelengths. The rover's robotic arm carried a microscopic imager, for extreme close-up views of rocks, and two spectrometers that would determine the composition of the surface. Also on the arm was an abrasion tool to remove the weathered outer crust from interesting rocks.

Eager to obtain the maximum scientific return from their robotic explorers, scientists debated long and hard over where to send them. Eventually it was decided to deliver Spirit to Gusev, a 166-km (103-mi) impact crater on the edge of the southern highlands. Orbital images indicated that the crater once contained a huge lake that was fed by a fast-flowing river.

The second rover was also to search for signs of ancient water. Opportunity's target was Meridiani Planum, a broad plain halfway around the planet from Gusev. The region was known to have deposits of gray hematite, an iron oxide mineral that is usually produced in the presence of liquid water.

Spirit lifted off from Cape Canaveral Air Force Station,

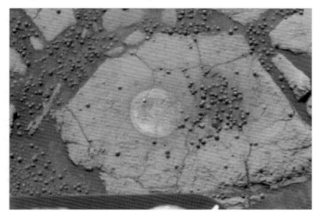

Among the Blueberries—This image taken by Opportunity on the rover's 48th sol is a near-true-color rendering of a rock called "Berry Bowl" in the Eagle Crater outcrop. It shows numerous ball-shaped objects nicknamed "blueberries." Rich in the iron compound hematite, the spherules are thought to have been deposited in water. The circular area on a relatively berry-free spot was brushed clean by the rock abrasion tool to prepare for examination by the rover's spectrometers. (NASA/JPL/Cornell)

Florida, on June 10, 2003, followed by Opportunity on July 8. After a 487-million-km (302-million-mi) voyage, it bounced to a halt on the surface of Mars on January 3, 2004, and wasted no time in sending back images of a flat, relatively featureless plain, strewn with small rocks. The most interesting local landmark appeared to be a shallow depression about 9 m (30 ft) across and about 12 m (40 ft) north of the lander, dubbed "Sleepy Hollow" by mission scientists. Some low hills were just visible some 3 km (2 mi) away.

The mission team determined that the rover came to rest in a region where numerous dust devils had removed surface material and left darker gravel behind. Images from orbiting spacecraft also showed a 200-m (660-ft) diameter impact crater, nicknamed "Bonneville," about 300 m (1,000 ft) northeast of the lander, which might provide a window beneath the surface.

Twelve days of careful preparations culminated in Spirit's first venture onto the Martian surface. Analysis of the soil showed that it was rich in silicon and iron, with significant levels of chlorine and sulfur. One unexpected finding was the detection of olivine, a mineral that is easily weathered. These results led the scientists to conclude that the surface was a fine-grained volcanic basalt.

Meanwhile, Opportunity hit the ground 10,000 km (6,200 mi) away on January 24. After bouncing 26 times, it came to rest inside a 20-m (66-ft) diameter crater. Scientists whooped with delight when the first images revealed a small outcrop of layered bedrock just a short distance away.

The rover's infrared spectrometer unveiled a landscape containing variable amounts of gray hematite, with the highest levels around the outcrop on the flank of the crater. As it began to study the exposed rock, Opportunity discovered thousands of unusual gray spherules scattered across the surface and embedded in the exposed bedrock. Dubbed "blueberries" by the mission team, these were thought to be concretions formed when dissolved minerals solidified around a tiny nucleus.

Over the coming months, both rovers found evidence that confirmed the presence of liquid water at both landing sites in the distant past. This evidence was particularly strong at Meridiani Planum, where Opportunity found

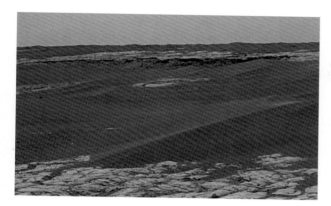

Erebus Rim—Towards the top center of this image is part of the rocky rim of "Erebus Crater." The crater itself is filled with large dunes. This approximately true-color view from the panoramic camera on Opportunity was acquired on the rover's 657th Martian day (November 28, 2005). (NASA/JPL–Caltech/Cornell)

Spirit on Husband Hill—This computer-generated image shows the Spirit rover superimposed onto a mosaic showing the flank of "Husband Hill." The size of the rover was calculated from the width of its tracks in a mosaic taken on the rover's 454th Martian day (April 13, 2005). (NASA/JPL–Caltech/Cornell)

"definitive proof" that the area was "once soaked in liquid water." The data showed that the finely layered rocks in Eagle Crater and elsewhere probably formed as deposits on a salt flat, or playa, which was sometimes covered by shallow water and sometimes dry.

As the months went by, the rovers were sent to explore further afield. Spirit passed by the 200-m (660-ft) diameter Bonneville Crater in mid-March 2004 and set off on a 2.3-km (1.4-mi) trek to the Columbia Hills—named after the shuttle crew that were lost during re-entry in 2003. 156 sols after touchdown, it arrived at the 120-m (400-ft) high uplands, where it found layered rocks containing minerals that were either formed in, or altered by, water. Some of the rocks in the Columbia Hills were found to contain the highest concentrations of sulfates yet measured on Mars.

After a slow, meandering climb, Spirit eventually reached the summit of Husband Hill—the highest part of the uplands—in late September 2005. It was the first ascent of a large hill on another planet by a robotic explorer. From its vantage point, Spirit obtained a sweeping panorama of Gusev Crater that captured dust devils (miniature tornadoes) racing across the plains below.

From that same scenic spot, Spirit looked skyward also snapped images of the two Martian moons, Phobos and Deimos. Other night-time duties included charting meteor showers skirting across the Red Planet's sky.

Meanwhile, Opportunity moved on to the 130-m (430-ft) wide Endurance Crater. Once a safe route was found, it drove inside the crater and spent six months studying the multiple layers of rocks exposed in its cliffs. Opportunity eventually climbed out on December 14, 2004 and set off toward a rugged landscape known as "etched terrain," pausing to take pictures of its crumpled heatshield and the first iron-nickel meteorite ever found on the surface of another planet. On April 26, 2005, the rover dug itself into a small dune when its wheels began slipping. Almost five weeks passed before controllers were able to gingerly manoeuver Opportunity free from its sand trap.

By April 2006, after more than one Martian year of exploration, Opportunity and Spirit had both covered more than 6.8 km (4.2 mi).

Opportunity had worked its way around the western rim of 300-m (984-ft) wide Erebus Crater, en route to the even larger Victoria Crater, though large, wind-blown dunes made navigation tricky. Erebus was surrounded by ancient sedimentary rocks rich in sulfates, but a partial failure in a shoulder joint on the robotic arm restricted measurements of these interesting outcrops.

Spirit descended the Husband Hill to examine "Home Plate," a circular, platform-like structure seen from the summit. Despite the failure of one wheel, it was able to reach a small ridge where it could receive the maximum sunlight during the winter.

The Future is Red

Although the disappearance of both the Climate Orbiter and Polar Lander (along with two dart-shaped surface probes known as Deep Space 2) hit hard at NASA's Mars program, the results from Global Surveyor encouraged the agency to continue its search for Martian water. Despite the cancelation of the 2001 lander, it was decided to go ahead that year with the Mars Odyssey orbiter.

Odyssey carried three scientific instruments to map the chemical and mineralogical makeup of Mars: a thermal emission-imaging system, a gamma-ray spectrometer, and a radiation experiment. Its primary tasks were to find evidence for present near-surface water and to map minerals deposited by past water activity. It would also provide the first measurements of radiation levels at Mars—an important step in preparations to send humans to the Red Planet.

Odyssey launched successfully from Cape Canaveral on April 7, 2001, arriving at Mars on October 23. After using the atmosphere to slow down, the spacecraft entered its mapping orbit in February 2002. Within a matter of weeks, the gamma-ray instruments had detected significant quantities of hydrogen—presumably associated with the presence of permanent water ice—in the south polar region.

By May, it was clear that the soil contained from 35 to 100% of water ice buried beneath a shallow overburden of hydrogen-poor soil. The ice-rich layer was about 60 cm (2 ft) beneath the surface at 60°S, but only about 30 cm (1 ft)

below the surface at 75°S. Odyssey data suggested that 20 to 35% of the subsoil was water ice, filling in most of the pores in a loose jumble of dust and broken rock. Even more water ice seemed to be present in the northern hemisphere, while frozen water made up as much as 10% of the top meter of surface material in some regions close to the equator.

Infrared images showed variations in rock layers similar to those seen in the layered rocks of the Grand Canyon. Mars was confirmed to be a very dusty place, with most of the surface covered by a thin layer of bright, orange–red dust. No evidence of hot springs or geothermal activity was detected on the surface, not even in young volcanic fissures or in exposed layers on cliff walls.

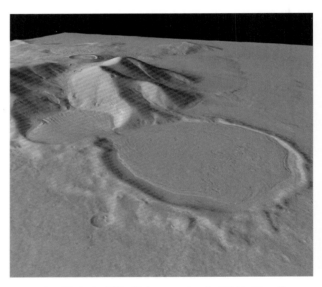

A Martian Glacier?—This 3D image, taken by ESA's Mars Express orbiter, shows an unusual hourglass-shaped structure in Promethei Terra at the eastern rim of the Hellas impact basin. A so-called "block" glacier, an ice stream with a large amount of scree (small rocks of assorted sizes), apparently flowed from the flank of the massif into a 9-km (5.6-mi) bowl-shaped impact crater (left), almost filling it to the rim. The glacier then overflowed into a 17-km (10.6-mi) wide crater, 500 m (1,650 ft) below. (ESA/ DLR/ FU Berlin— G. Neukum)

Lower-Limit of Water Mass Fraction on Mars

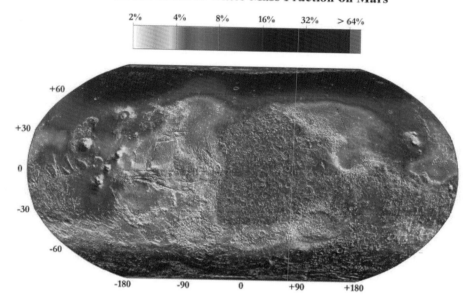

Hidden Water?—This map shows the estimated lower limit of the water content of the upper meter (3 ft) of Martian soil, based on the abundance of hydrogen measured by the neutron spectrometer on Mars Odyssey. The highest water content, from 30% to well over 60%, is in the polar regions. Farther from the poles, significant concentrations are in the uplands east of the Valles Marineris (center), and to the southwest of Olympus Mons (extreme left and extreme right). (NASA/JPL/Los Alamos National Laboratory)

Odyssey's radiation environment experiment showed that the radiation dose at Mars was two to three times greater than that on board the International Space Station, which was protected within Earth's magnetic field.

Knowledge of the planet's inventory of volatiles was further improved when the European Space Agency's Mars Express spacecraft arrived on December 25, 2003. Once its polar orbit was lowered, Europe's first Mars mission began to send back spectacular stereoscopic color pictures of the surface from an altitude of 275 km (170 mi). The first of these was a portion of a 1,700-km (1,060-mi) long and 65-km (40-mi) wide swath across the Valles Marineris.

Mission scientists soon announced the first direct measurements of water ice in the south polar cap. Another result to cause a stir was the "discovery" of methane in the atmosphere—confirmation of previous data obtained by ground-based instruments. Methane levels of about 10 parts per billion were detected, sparking a debate over whether this indicated continuing volcanic activity, some unusual subsurface chemistry, or the presence of life. Hopes of finding subsurface water rose when the antennas of the radar instrument were finally deployed in June 2005.

Bolstered by these successes, ESA began to develop its Aurora program, a road map of robotic exploration that would enable Europe to play a major role in a future international program, culminating in the first human mission to Mars around 2035. ExoMars, the first major mission under the Aurora umbrella, would involve sending a rover dedicated to a search of Martian life in 2011–2013.

Meanwhile, with the Shuttle grounded after the Columbia disaster and assembly of the International Space Station on hold, the United States government also decided it was time for a change of direction. On January 14, 2004, President George W. Bush gave American space policy an ambitious new vision when he told a picked audience of his plan "to take the next steps of space exploration: human missions to Mars and to worlds beyond."

Under the Space Exploration Initiative, future US research on board the International Space Station would be refocused on the long-term effects of space travel on humans, thus preparing the way for crewed expeditions to the Moon and Mars.

In order to send astronauts to Mars within three decades, NASA intends to use the Moon as a stepping stone. Using Earth's neighbor as a testbed for new technologies, human lunar missions could begin as early as 2015, with the goal of living and working there for increasingly extended periods of time.

Simultaneously paving the way for humanity will be a series of increasingly sophisticated robotic missions. First to fly in 2005 was the Mars Reconnaissance Orbiter, equipped with the most powerful telescopic camera yet sent to the Red Planet, together with a radar to search for subsurface water ice. Two years later an improved version of the lander that was canceled in 2001 will finally touch down near the north polar region. Known as Phoenix, the lander will use a robotic arm to scoop up samples and deliver them to tiny ovens so that scientists can study the ices, minerals, and—possibly—organic compounds in the soil.

After the launch of a nuclear-powered Mars Science Laboratory rover in 2009–2011, NASA expects to develop a new series of "testbed" missions that will support future human and robotic Mars exploration, as well as a sample return mission sometime after 2016. Clearly, Mars will continue to dominate the headlines for decades to come.

Trailing an Asteroid—Most asteroids appear as points of light in ground-based telescopes, but during a long-exposure photograph an asteroid's motion results in a long streak. This Hubble Space Telescope image shows a fairly bright asteroid, with a visual magnitude of 18.7, in the constellation Centaurus. The blue trail was caused by a main belt asteroid about 2 km (1.2 mi) in diameter. (R. Evans and K. Stapelfeldt [JPL/NASA])

6 ASTEROIDS: VERMIN OF THE SKIES

The Minor Planets

For most of recorded history there were six recognized planets (including Earth) in the Solar System. No one dreamt that there were any worlds beyond Saturn, but there were occasional suggestions that one or more unseen objects lurked between the Sun's known planetary companions. Johannes Kepler, the seventeenth-century mathematician who formulated the laws of planetary motion, was one who believed that another world must orbit the Sun between Mars and Jupiter.

Then, in 1772, a German astronomer named Johann Bode drew attention to a simple formula that predicted the distances of the planets from the Sun quite precisely (see Chapter 9). (This mathematical curiosity had first been pointed out six years earlier by Johann Titius, something that Bode failed to mention!)

Bode's Law was generally thought to be evidence of God's creative handiwork—but what about the embarrassing 550-million-km (350-million-mi) void between Mars and Jupiter? "Can one believe that the Founder of the Universe left this space empty?" asked Bode.

Nine years later the Law returned to prominence when William Herschel discovered Uranus more or less where the Titius–Bode Law predicted it should have been. This apparent confirmation of the arithmetic progression helped to focus scientific minds.

In 1800, a group of six astronomers, headed by Johann Schröter, gathered in the little German town of Lilienthal with the intention of beginning a systematic search for the missing planet. Their plan was to divide the ecliptic plane, where all the planets were found, into 24 sectors, each of which would be carefully scanned by a different observer.

However, before the group's grand survey could get under way, some unexpected news arrived from Sicily. On January 1, 1801, Giuseppe Piazzi, director of the Palermo observatory, had found a mysterious, star-like object wandering among the stars of Taurus.

Giuseppe Piazzi—Giuseppe Piazzi (1746–1826) was an Italian priest and mathematician. The Director of the Palermo Observatory in Sicily from 1790 until his death, he also established the observatory at Naples in 1817. On January 1, 1801, after 11 years of checking star positions, he came across a faint object in Taurus that did not appear on any chart. Piazzi's newcomer—which he called Ceres—turned out to be the first and largest of the main belt asteroids that orbit between Mars and Jupiter. (Capodimonte Observatory)

At first, Piazzi thought he had found a comet, but the absence of a gaseous coma was troublesome, and he began to have some doubts. Unfortunately, on February 11, Piazzi became seriously ill, and by the time he had recovered, the mysterious wanderer had disappeared in the Sun's glare. With limited information, frustrated astronomers all over Europe began a frantic search for Piazzi's object.

The breakthrough came when a brilliant young mathematician, Carl Gauss, found a way of predicting planetary positions from a limited set of observations. Using his calculations, Franz Xaver von Zach, one of Schröter's "celestial police," rediscovered the faint object on December 31. It was only half a degree from Gauss's prediction.

To everyone's delight, the newcomer was found to follow a nearly circular, planet-like path between Mars and Jupiter. At Piazzi's suggestion, the name Ceres—after the Roman goddess who was the patron of Sicily—was soon adopted.

Its distance from the Sun was 2.77 astronomical units (Sun–Earth distances)—an almost exact fit with the empty 2.8 AU slot in Bode's Law. The one concern was that the 8th magnitude object was obviously very small—more of a minor planet than a fully fledged member of the solar family.

Then, on March 28, 1802, Wilhelm Olbers (another member of the "celestial police") found a second planet at Bode's distance 2.8 AU while he was searching for comets. Pallas, as it was named, seemed to be little more than 160 km (100 mi) across and traveled in a fairly eccentric, inclined orbit, though it did approach Ceres from time to time.

It was already apparent that these minor planets were very different from their larger neighbors. William Herschel's suggestion that they be called "asteroids," because of their star-like appearance and absence of planetary disks in even the largest telescopes, was quickly accepted. Olbers put forward the suggestion that they were remnants of a larger body that had exploded.

Further possible fragments were discovered in 1804 by Karl Harding (an assistant of Schröter) and in 1807 by Olbers. Named Juno and Vesta, they were still smaller than Pallas. Vesta did not even approach the other three asteroids.

The scattered family seemed complete for nearly 40 years until a fifth minor planet (Astraea) was found in 1845, followed by three more in 1847. Since then, at least one asteroid has been discovered every year, with a marked leap in the rate of discovery after the introduction of photographic techniques in the late nineteenth century.

Unwanted asteroid tracks showed up on exposures at the most awkward times, causing one astronomer to dub them "the vermin of the skies." Most scientists lost interest in the thousands of small, seemingly unimportant, chunks of rock that occupied the broad zone between the four terrestrial planets and their giant, gaseous companions. Not until the introduction of new instruments and observing techniques was their curiosity rekindled.

The discovery of objects such as 433 Eros caused a stir, when astronomers realized that asteroids sometimes paid a visit to the inner Solar System, even threatening to collide with our home planet.

Families and Trojans

The orbits of more than 100,000 asteroids have been cataloged since the discovery of Ceres, and the number is rising inexorably, with thousands reported every month. Analysis of the computed orbits shows that the vast majority lie in a belt between 2.2 and 3.3 AU, therefore they are generally closer to Mars than to Jupiter.

About half of all main belt asteroids belong to families that have similar orbital characteristics. Approximately 100 such families have now been recognized, each being named after its largest member. The most populous families—such as the Flora, Eos, Themis, and Koronis groups—may have hundreds of members.

Similarities in their reflected light indicate that each family is made up of fragments of a particular parent body—probably primitive planetoids that were shattered by

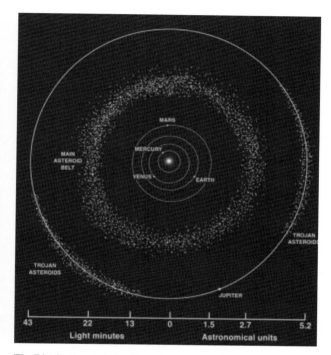

The Distribution of Asteroids—The vast majority of asteroids travel around the Sun in the main belt between the orbits of Mars and Jupiter. The two "clouds" of objects centered 60 degrees ahead and behind Jupiter along its orbit are the Jupiter Trojans. Not shown here are the numerous near-Earth asteroids that approach or cross the Earth's orbit. (NASA)

catastrophic collisions billions of years ago. Certain types of meteorites found on Earth have also been matched to individual large asteroids and their families.

It is interesting to note that the overall composition of the asteroid belt also varies. Stony S-type asteroids are more common near the inner edge, while darker, carbon-rich C-types are more numerous further out. This pattern may reflect differences in the temperature of the solar nebula when the asteroids formed, 4.5 billion years ago.

Certain parts of the asteroid belt—called Kirkwood gaps, after US astronomer and mathematician Daniel

Kirkwood (1814–1895)—are relatively empty, owing to the influence of Jupiter's enormous gravity. The gaps occur where an asteroid's orbital period (year) would be an exact fraction (i.e. 1/2, 1/3, 2/5, etc.) of Jupiter's orbital period.

The most important ratios are 2:1, 3:1, and 5:2. This means that any asteroid that orbits the Sun exactly twice in the time it takes Jupiter to go around once will be removed by the repeated influence of the giant planet on the same sector of the asteroid's orbit. Jupiter's gravity simply keeps the gaps clear by changing the orbits of any asteroids within them.

One interesting exception is the region, well outside the main belt, in which the orbital ratio is 3:2. In this zone, at a distance of about 4 AU, lies the Hilda group of some 700 asteroids. Recent models suggest that Jupiter may have "shepherded" them into this zone more than 4 billion years ago.

The orbits of most asteroids are not very dissimilar to those of the planets. Unlike comets, they all travel around the Sun in the same direction, and generally lie near the plane of the ecliptic, following orbits that are not very

Meteor Crater—Meteor Crater is one of the youngest and best-preserved impact craters on Earth. It was excavated about 50,000 years ago when a 30-m (100-ft) wide, iron-rich meteorite weighing 100,000 tonnes struck the Arizona desert at an estimated 20 km/s (45,000 mph). The resulting explosion left a crater 1.2 km (3,940 ft) wide and 200 m (660 ft) deep. Debris slides and erosion have partially filled the bottom of the crater. It is also known as Barringer Crater, after the mining engineer who first suggested its impact origin. (D. Roddy, US Geological Survey/LPI)

elliptical. However, the bodies are so small that any high-speed collision or outside gravitational influence can easily alter their paths. Many asteroids have been discovered, only to disappear.

As one might expect, there are many exceptions to the general pattern of behavior. Thousands of asteroids have been ejected from the main belt and now inhabit the inner Solar System. There are three groups: the Amors, which travel between the Earth and Mars; the Apollos, which cross Earth's orbit; and the Atens, which stay mainly inside Earth's orbit. Some are large enough to cause considerable destruction and loss of life should they collide with our world.

On the far side of the asteroid belt are two groups that have the same mean distance and orbital period as Jupiter. These Trojan asteroids generally lie either 60 degrees ahead of Jupiter (the Achilles group) or 60 degrees behind (the Patroclus group), though their individual positions can vary considerably. About 1,700 Jupiter Trojans have been found, with around two-thirds of them in the Achilles group. A few Trojans have also been found along the orbits of Neptune and Mars. Rather than being ejected, these asteroids may remain in the positions they have occupied over the lifetime of the Solar System.

Although asteroids and comets are both regarded as leftovers from the birth of the planets, they differ in two important ways. First, comets follow orbits that are highly eccentric (elliptical) and tilted at any angle to the ecliptic—the plane in which the Earth travels around the Sun. By contrast, asteroids have much more circular orbits, typically inclined only about 10 degrees from the ecliptic. The second difference is that comets are largely made of ice, unlike the rocky asteroids. As they move closer to the Sun, the ice vaporizes, creating the comets' distinctive tails.

However, there are a few objects with split personalities that have the characteristics of both comets and asteroids, including a number of asteroid-like bodies that cross the orbit of Jupiter. The first to be discovered was 944 Hidalgo, which follows an elongated path from within the main asteroid belt almost as far as Saturn's orbit.

Perhaps even stranger is 2060 Chiron (not to be confused with Pluto's satellite Charon), discovered in 1977 by American Charles Kowal. Its perihelion lies between the orbits of Jupiter and Saturn, while at aphelion it approaches the orbit of Uranus.

When Chiron last approached the Sun, it brightened rapidly and developed a comet-like coma of gas and dust, with the result that it is now classified as both an asteroid and a comet (95P/Chiron). The path of this weird object means that it is greatly affected by the giant planets, and calculations suggest that in a few million years it will be diverted right out of the Solar System.

Chiron was the first of the Centaurs—icy objects that are classified as asteroids, although they follow comet-like orbits and seem to have originated in the Kuiper belt, beyond Pluto. More than 35 Centaurs have now been discovered.

Ancient Building Blocks

All of the planets in the Solar System probably formed about 4.5 billion years ago from a huge disk of dust and gas orbiting the Sun. This material gradually formed larger and larger clumps as the result of continuous collisions or near collisions.

At some point in this process, Jupiter became so massive that it became the dominant power in the central regions of this solar nebula, capable of scattering large bodies that would otherwise have accreted to form a planet. These bodies may have become satellites of other planets, or they may have been thrown out of the Solar System completely.

Their highly irregular orbits must have perturbed the orbits of smaller bodies, raising their average relative speeds to 5 km/s (3 mi/s). This was too fast for accretion to take place. Instead, asteroids that collided with each other shattered into smaller fragments, as they still do today. Thus the bodies in the asteroid belt have probably been shrinking instead of growing larger since that time. Indeed, it is possible to observe a band of dust associated with collisions in the asteroid belt just before sunrise or after sunset. This is known as the zodiacal light.

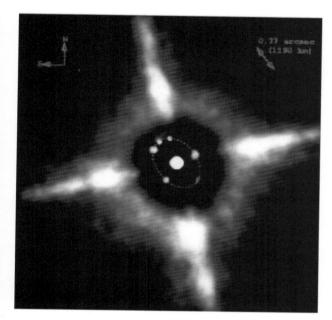

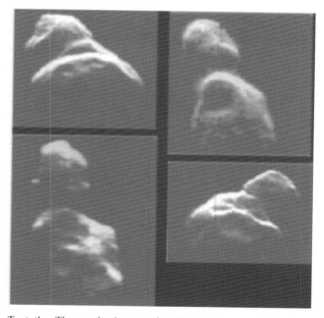

Eugenia and Petit-Prince—This image from the 3.6-m (140-in) Canada–France–Hawaii Telescope, shows a small satellite orbiting main belt asteroid 45 Eugenia. The first asteroid moon imaged from Earth was discovered in infrared images taken in November 1998. It was later given the name Petit-Prince, after the son of Eugenie, Empress of France. Eugenia, the central white dot, is 215 km (134 mi) in diameter, while the moon is about 13 km (8 mi) wide. It follows a nearly circular orbit at a distance of 1,190 km (740 mi), completing one circuit in 4.7 days. The orbit appears oval because it is tilted to the line-of-sight. The large "cross" is an artefact caused by stray light in the telescope. (Laird Close/ESO, Bill Merline/Southwest Research Institute)

Toutatis—These radar images of 4179 Toutatis were obtained on four days in December 1992, when the asteroid passed about 2.5 million km (1.6 million mi) from Earth. They reveal two irregular, cratered objects, about 4 km (2.5 mi) and 2.5 km (1.6 mi) in diameter, that are in contact with each other. The largest crater (image top right) is about 700 m (2,300 ft) across. A 400,000-watt coded radio transmission was beamed at Toutatis from the 70-m (230-ft) antenna at Goldstone, California. The echoes took about 12 seconds to return to Goldstone, where they were decoded and processed into images. (Steve Ostro/NASA–JPL)

Today, there are only a few dozen asteroids with diameters measuring more than 160 km (100 mi). The smaller their size, the more numerous they are. So, although there are probably some 250 asteroids larger than 80 km (50 mi) in diameter, the number of objects larger than 0.8 km (0.5 mi) probably exceeds one million, with billions of rocky shards that are even smaller.

Although they are very numerous, asteroids are not easy to study. Vesta is the only asteroid that ever becomes bright enough to be visible with the naked eye, and it is extremely difficult to directly measure the diameters of even the largest asteroids.

Astronomers have developed various indirect techniques that enable them to distinguish a small, highly reflective asteroid from a large, but very dark one. If the distance is known, the diameter can then be estimated by comparing the observed brightness of the object with its albedo (reflectivity). However, in most cases astronomers use an assumed albedo, leaving considerable room for doubt over their actual sizes.

Measuring the polarization of light provides information about the curvature of the asteroid's surface and thus its size. A more accurate method, known as stellar occultation, measures the sizes of large asteroids by observing the width of their shadows on Earth as they pass in front of stars.

The size, shape, rotation period, and other important characteristics of asteroids can also be measured by bouncing powerful radar beams off their surfaces and processing the echoes that are picked up by large, ground-based antennas. By the end of 2004, radar echoes had been detected from 252 asteroids, including 161 near-Earth asteroids and 91 objects in the main belt.

The largest asteroids, such as Ceres, Pallas, and Vesta, are nearly spherical. Smaller asteroids have highly irregular shapes, evidence of a long history of impacts with fast-moving neighbors.

Radar observations have shown that a surprising number of asteroids resemble two objects that have gently come together and are now orbiting the Sun as a single body. Near-Earth asteroids Toutatis, Castalia, and Hermes are classic examples.

Some sizeable asteroids seem to have very low densities. For example, the Galileo spacecraft found that main belt asteroid Mathilde is not much denser than water, indicating that the battered world is little more than a rocky rubble pile filled with cracks and empty spaces.

Others are more substantial, made up of stony-iron materials that give them a more planet-like density. Some, such as 16 Psyche, seem to be largely composed of the metals nickel and iron.

This inner strength is important, since the rotational periods of asteroids vary from several days to a few minutes. Slowly spinning asteroids have no difficulty holding themselves together, even if they are loosely bound rubble piles. On the other hand, a very rapid rotation indicates that the object must be solid rock, strong enough to overcome the centrifugal forces that are trying to tear it apart.

Many others are binary systems, in which two minor planets orbit each other. Since the gravitational pull of most asteroids is fairly negligible, these moons are more likely to be fragments produced by collisions or tidal disruption during close planetary encounters rather than asteroids that passed too close and were captured.

The first asteroid moon to be discovered was Dactyl, a tiny companion of main belt asteroid Ida, which unexpectedly appeared in images taken by the Galileo spacecraft. Since then, satellites of other main belt and near-Earth asteroids have been found using ground-based telescopes. They typically take a few days to circle each other. About one in six of the near-Earth asteroids larger than 200 m (660 ft) in diameter may be a binary system.

The smallest asteroids can be detected only if they approach close to Earth—often without any advance warning. The closest known approach (apart from a meteoroid that was photographed leaving a smoke trail through the sky over Wyoming in 1972) took place in March 2004, when tiny asteroid 2004 FU162 skimmed past at a distance of only 6,500 km (4,000 mi).

Less than 10 m (33 ft) across, the Earth-grazer posed no real threat, since it would probably have exploded at high altitude if it had been on a collision course with our planet. However, there are thought to be around 1,000 near-Earth asteroids larger than 1 km (0.6 mi) in diameter, each capable of causing huge destruction and loss of life. One of these "missiles" would be expected to strike the Earth on average every 600,000 years.

The Largest Asteroids

Their large distance from Earth and small size means that most asteroids appear as tiny spots or linear streaks in ground-based images. Although it is possible to learn about their composition and period of rotation by studying light curves and surface reflectivity, many questions remain unanswered.

The situation is only a little better for the larger members of the family, such as Ceres. Despite the asteroid's diameter of almost 1,000 km (625 mi), even the Hubble Space Telescope struggles to resolve any details on the surface of Ceres.

Images taken with Hubble's Faint Object Camera in June 1995 revealed an almost spherical shape, measuring 909 km (565 mi) by 975 km (606 mi), but very few features were visible on its bland, uniform surface. The main discovery was a 250-km (155-mi) diameter feature named "Piazzi." Whether this is an impact crater or simply a region of darker material remains uncertain. Although its period of rotation is known to be around 9 hours, the axial tilt and precise location of its poles remain uncertain.

Dark gray Ceres has a density about twice that of water, less than most meteorites, and accounts for about one-third the mass of all the asteroids combined.

The second largest asteroid, Pallas, appears to be elongated, with an average diameter of about 540 km (335 mi). Like Ceres, it orbits in the middle of the main belt and seems to be made of carbon-rich material. However, its orbit is much more elongated and tilted, with an inclination at 35 degrees to the ecliptic. Unfortunately, no Hubble pictures have been taken of Pallas, and its density and composition remain uncertain.

Recent studies suggest that Pallas may be made of similar material to slightly smaller Vesta, even though the latter occupies a region of the main belt that is considerably closer to the Sun.

Vesta itself is a very unusual object. As the brightest main belt asteroid, it is occasionally visible to the naked eye. It has a high density and it is the only large asteroid whose spectrum indicates that it is made of igneous rock that was once molten. Its surface of basaltic rock—solidified lava—is evidence that the asteroid had a hot interior shortly after its formation 4.5 billion years ago. Despite large chunks being

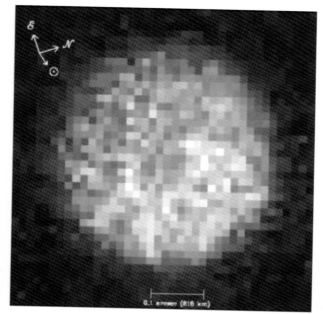

Ceres—Although Ceres is the largest of the asteroids, it is far from easy to observe features on its surface. This Hubble Space Telescope image was taken at ultraviolet wavelengths on June 25, 1995, revealing features approximately 50 km (31 mi) across. Most notable was a dark, 250-km (155-mi) diameter feature (center), for which the name "Piazzi" was proposed. The nature of this feature remains uncertain. Ceres is almost spherical, measuring 909 km (565 mi) by 975 km (606 mi). (Joel Parker [SwRI] et al.)

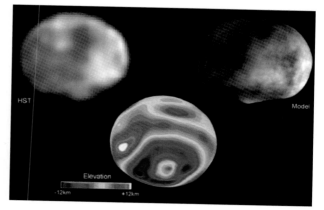

Vesta—Vesta is the largest main belt asteroid yet imaged at high resolution. This Hubble Space Telescope view (left) was obtained in May 1996, when it was 180 million km (110 million mi) from Earth. Its nonspherical shape is clearly visible. The elevation map (center) shows a 460-km (285-mi) diameter impact basin (blue), with a 13-km (8-mi) high central peak near the south pole. A 3-D computer model is shown at right. (Ben Zellner [Georgia Southern University], Peter Thomas [Cornell University], NASA.)

blasted into space by subsequent collisions, the asteroid has remained largely intact ever since.

Knowledge of Vesta advanced by leaps and bounds in 1996, when astronomers used the Hubble Space Telescope to image its surface. The observations were made during Vesta's closest approach to Earth in a decade, when the asteroid was 180 million km (110 million mi) away.

A total of 78 pictures were taken with the Wide Field Planetary Camera 2. The team then created a topographic model of the asteroid's surface by noting where shadows occurred along the limb and at the terminator (day/night boundary), as evidence of irregularities in the surface.

The most notable feature found on the apple-shaped asteroid was a giant crater 460 km (285 mi) across, which is nearly equal to Vesta's 510 km (315 mi) diameter. In the center of the crater is a cone-shaped peak 13 km (8 mi) high that was created when the pulverized rock rebounded back after the impact.

Although the collision gouged out 1% of the asteroid's volume, blasting over two million cubic kilometers (half a million cubic miles) of rock into space, Vesta was strong enough to survive the huge impact. Since the once-molten world is large enough to be differentiated like Earth—with a volcanic crust, core, and mantle—it is possible that the 13-km (8-mi) deep crater may have punched all the way through the crust to expose the asteroid's mantle. The crater is located close to Vesta's south pole. This is probably more than coincidental, since the excavation of so much material from one side of the asteroid would have shifted its rotation axis until it settled with the crater near one pole.

Vesta's violent history may account for the distinctive metallic meteorites that have been collected on Earth. About 6% of the meteorites that fall to Earth appear to be similar to Vesta in composition. If Vesta is the parent body of some smaller asteroids and meteorites, then a gaping wound from a major impact is only to be expected.

How did the distinctive meteorites end up on Earth? There are no strong gravitational forces capable of sending the pieces directly from Vesta toward the inner Solar System. However, its daughter asteroids—literally "chips off the block," with color characteristics similar to Vesta—are near a chaotic zone in the asteroid belt where Jupiter's gravitational tug can redirect fragments into orbits that intersect Earth's orbit.

The number of asteroids increases as size diminishes. There are 22 known main belt asteroids with diameters in the range 410–200 km (255–125 mi). With the exception of Juno, none of the other top 10 largest asteroids is close enough for astronomers to learn much about its surface characteristics. However, some, such as 45 Eugenia, have been found to belong to binary systems, with much smaller moons circling in close proximity.

At present, the biggest asteroid to be imaged up close by a spacecraft is Mathilde, a large rubble pile 66 km (41 mi) wide and very different from its larger cousins. However, the 95-km (60-mi) wide asteroid 21 Lutetia has been targeted for a flypast by ESA's Rosetta spacecraft in July 2010.

The First Flybys

Although asteroids are easy to detect with modern, automated electronic cameras, it is not easy to discover much about their physical attributes. Most asteroids appear as tiny points of light or streaks against a stellar backcloth. The Hubble Space Telescope and modern ground-based telescopes with adaptive optics have struggled to resolve the disks of even the largest main belt asteroids.

In the 1970s, no one was too sure how risky it would be to send a spacecraft into the orbiting debris between Mars and Jupiter. The first reconnaissance was made by Pioneer 10, which entered the main belt on July 15, 1972, and emerged unscathed seven months later. Pioneer's instruments found much less sand- and dust-sized material than expected, although there was an increase in particle population around 400 million km (250 million mi) from the Sun.

Six years later, the twin Voyagers sailed through the main belt without adding to our knowledge of the region,

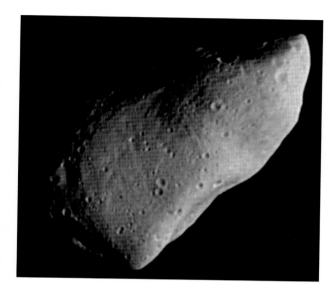

Ida and Dactyl—Main belt asteroid 243 Ida is a member of the Koronis family. This mosaic of Galileo images was taken at a distance of about 10,500 km (6,500 mi), shortly before closest approach on August 28, 1993. Ida measures about 56 × 24 × 21 km (35 × 15 × 13 mi). Brighter, bluish areas around craters (upper left, center, and upper right) suggest a difference in the abundance or composition of iron-bearing minerals. This was also the first image of an asteroid moon, now known as Dactyl. The pair may be fragments of a much larger asteroid that was destroyed long ago. (NASA–JPL)

Gaspra—Main belt asteroid 951 Gaspra is a member of the Flora family. This enhanced color view was taken by the Galileo spacecraft from a range of 5,300 km (3,300 mi), shortly before closest approach on October 29, 1991. Gaspra measures about 18 × 11 × 9 km (12 × 7 × 6 mi). It rotates counterclockwise every seven hours around the north pole, which is located at upper left. The large hollow on the lower right limb is about 6 km (3.7 mi) across. More than 600 craters 100–500 m (330–1650 ft) in diameter are visible. (NASA–JPL)

but a major breakthrough came in 1983 with the Netherlands–UK–US Infrared Astronomy Satellite (IRAS), which detected more than 1,800 asteroids. Among its discoveries was an unusual asteroid, 3200 Phaethon, which seems to be an extinct comet.

Scientists eventually realized that it made sense to take advantage of opportunities provided by any spacecraft that might happen to be crossing the main belt en route to somewhere else. The first chance to see one of these primitive pieces of debris at close range came in October 1991, when NASA's Galileo spacecraft flew past the S-type (stony) asteroid 951 Gaspra on its way to Jupiter.

Galileo's cameras showed that Gaspra resembles a pointed tooth, with flat surfaces suggesting that parts have been sheared off by sizeable impacts. There were also numerous small impact craters and apparent fractures—further evidence of its violent past. The number of small craters compared to larger ones was much greater for Gaspra than for previously studied bodies of comparable size, such as the satellites of Mars. Faint color variations suggested that loose surface material slid downslope under the asteroid's weak gravity.

Twenty-two months later, Galileo returned to the main belt, this time flying past a much larger S-type asteroid named 243 Ida. About twice the size of Gaspra, Ida had a central waist that suggested an amalgamation of two bodies. Although its surface was saturated with craters, one half was more pockmarked than the other.

Studies of their reflected light indicate that both objects are made of metal-rich silicates similar to those that make up a class of meteorites known as ordinary chondrites. However, color variations and a scattering of large boulders suggest that Ida has a thicker coating of surface deposits, and may even resemble a rubble pile throughout.

Galileo's most unexpected discovery was that Ida had a small satellite. Named Dactyl, the moon was about 1.5 km (1 mi) across and seemed to be made of similar rock types to its probable parent. The orbital motion of Dactyl was consistent with Ida having a density 2–3 times that of water.

Some less spectacular flybys of main belt asteroids have taken place during more recent missions. NASA's Deep Space 1 was a low-cost New Millennium mission to try out innovative technology, particularly an ion drive and an autonomous navigation system. The spacecraft flew to within an estimated 26 km (16 mi) of asteroid 9969 Braille on July 28, 1999, but, to the disappointment of scientists, its cameras failed to point toward the dark target and only obtained a few distant views.

From these, it was possible to estimate its size as 2.2 km (1.3 mi) by 1 km (0.6 mi). However, the most significant data came from the spacecraft's infrared sensor, which confirmed that Braille is highly reflective and seems to be one of the few objects that is similar in composition to Vesta.

Three years later, the Stardust spacecraft completed a successful flyby of the asteroid 5535 Annefrank, cruising to within about 3,300 km (2,050 mi) of its rocky surface. The November 2, 2002 encounter offered a unique opportunity to carry out a test run for all the operations planned for Stardust's 2004 encounter with Comet Wild 2.

During the early approach phase, Annefrank proved to be a difficult object to locate and study. Too dark to appear in early scans of the sky by Stardust's camera, the asteroid eventually came into view after the camera was commanded to take longer exposures.

The 70 or so encounter images show Annefrank to be a cratered, potato-shaped object. It was about 8 km (5 mi) in diameter, which was twice as large as expected—an error explained by its very dark surface. Several large craters were seen on the asteroid's limb, but no impacts from nearby dust were recorded on the spacecraft.

Cassini was the most recent mission to take advantage of a passage through the main belt, this time on its way to Saturn. On January 23, 2000, the spacecraft flew past asteroid 2685 Masursky at a distance of 1.6 million km. In an effort to determine the asteroid's basic characteristics, a series of wide-angle and narrow-angle images were taken through various filters between 7 and 5½ hours before closest approach. The long-range views showed that Masursky is roughly 15–20 km (9–12 mi) in diameter, but measurement of its reflected light indicated that it may not be an S-type asteroid like Gaspra, Ida, and Eros, despite its apparent association with the Eunomia family of S-type asteroids.

NEAR Shoemaker

Although some valuable—and intriguing—new insights into asteroids had been provided during spacecraft flybys, scientists were eager for a more in-depth investigation. This required sending a spacecraft to orbit an asteroid. After due consideration, asteroid 433 Eros, the second largest near-Earth asteroid,* was chosen—partly because it was fairly easy to reach and partly because of its scientific promise.

The Near Earth Asteroid Rendezvous (NEAR) spacecraft was designed as one of NASA's "faster, better, cheaper" Discovery-class missions, with a budget cap of $150 million. Its task was to characterize in as much detail as possible the true nature of a large Earth-grazing object by becoming the first satellite ever to orbit an asteroid.

Its target had been under intense scrutiny since it burst upon an unsuspecting scientific community on August 13, 1898, when it was discovered almost simultaneously by the German astronomer, Karl Witt, director of the Urania Observatory in Berlin, and Auguste Charlois of Nice, France. It was named Eros after the Greek god of love, the son of Hermes (Mercury) and Aphrodite (Venus).

Orbital calculations soon determined that Eros was a rogue minor planet, very different from its more predictable

---

* The largest near-Earth asteroid is 1036 Ganymed, with an estimated diameter of 50 km (30 ml).

NEAR Shoemaker—NEAR Shoemaker was the first spacecraft to orbit and land on an asteroid. Launched on February 17, 1996, NEAR sent back detailed pictures of the large main belt asteroid 253 Mathilde in June 1997. After its first rendezvous attempt failed on December 20, 1998, NEAR eventually entered orbit around Eros on February 14, 2000. The mission ended on February 12, 2001 with a soft landing on the asteroid's surface. (NASA/JHU–APL)

Mathilde—253 Mathilde was the first C-type (carbon-rich) asteroid to be imaged at close quarters. This mosaic of four images was obtained by the NEAR spacecraft on June 27, 1997 from a distance of 2,400 km (1,500 mi). The asteroid measures about 59 × 47 km (37 × 29 mi) and the resolution is roughly 300 m (990 ft). Three large craters can be seen, one at lower center, one at the top left viewed edge-on, and one at lower right, also viewed edge-on. The central crater is estimated to be 10 km (6.2 mi) deep. (NASA/JHU–APL)

cousins. Instead of spending its entire life in the main asteroid belt, Eros ventured inside the orbit of Mars and occasionally approached to within 22 million km (13.75 million mi) of Earth. In fact, it was the first near-Earth asteroid to be discovered and the first known representative of the Earth-approaching Amor asteroid group.

Although there is no likelihood of Eros colliding with Earth in the foreseeable future, its periodic close encounters with our planet mean that it is relatively easy to observe. Indeed, astronomers studied the orbit of Eros in order to make an accurate determination of one of the fundamental astronomical distance markers—the value of the astronomical unit. The average distance of the Sun from our planet was determined to be 149.67 million km (93 million mi)—extremely close to today's accepted value.

By the time NEAR set off in pursuit of Eros on February 17, 1996, scientists knew that Eros was irregular in shape and rotated approximately once every five hours, while spectral studies showed it to be an S-type (stony-iron) asteroid.

Sixteen months later, on its first circuit of the Sun, the spacecraft made its first memorable contribution to planetary science as it flew past the main belt asteroid 253 Mathilde at a distance of only 1,212 km (750 mi). In just 25 minutes, it took a series of 534 images that showed a 55-km (34-mi) object marked by enormous scars—evidence of a world that has been brutally battered over billions of years. At least five craters were nearly half the width of the asteroid itself.

Unlike the asteroids previously visited by spacecraft, Mathilde was a very primitive C-type (carbon-rich) object that was as black as charcoal. From the way it altered the spacecraft's trajectory, Mathilde was shown to be only slightly denser than water. The asteroid was little more than a giant, porous rock pile with an unusually long rotation period of 418 hours, about 17 days.

The mission continued to go well, and an Earth flyby on January 23, 1998, put NEAR on course for the encounter with Eros the following year. However, the gremlins struck on December 20, 1998, when a major rendezvous maneuver was aborted by the spacecraft's computer.

With their satellite tumbling and out of touch with the Earth, the NEAR project team at Johns Hopkins University feared the worst. Then came the miracle. Not only were communications re-established, but telemetry showed that the spacecraft remained in good health and had sufficient fuel for another rendezvous attempt next time around.

The scientists and engineers hastily uploaded new commands to enable NEAR to observe Eros as it hurtled past at a distance of 3,827 km (2,378 mi), just three days after the near-disastrous malfunction. Dozens of low-resolution images were captured during a full rotation of the peanut-shaped world. Armed with this new information about the asteroid's size, shape, spin, and gravity, the science team began to prepare for the spacecraft's second coming.

This time, everything went according to plan. On February 14, 2000, NEAR was successfully inserted into orbit around Eros. From its initial altitude of 330 km (205 mi), the spacecraft dropped in careful steps, closer and closer to Eros's pockmarked surface. At first, the camera only captured views of the asteroid's illuminated northern regions, but by June 25, as the Sun began to rise over the south pole, its hidden side began to emerge from the shadows.

On October 26, the spacecraft (now named in honor of US planetary geologist Eugene Shoemaker) swooped to within 5 km (3 mi) of Eros. Even at this altitude, the gravitational field of Eros was so feeble that NEAR's maximum orbital speed was only 23 km/h (14 mph).

However, an even grander finale was to follow. Although NEAR was designed only to orbit Eros, the mission team decided to terminate the mission with the first landing on an asteroid. On February 12, 2001, the tough little spacecraft touched down on Eros at a velocity of about 1.5–1.8 m/s (3.4–4 mph). During its final descent, the spacecraft obtained 69 high-resolution images, the last view showing an area only 6 m (20 ft) across. To everyone's surprise, NEAR sent a signal to show that it had survived the low-velocity, low-gravity impact, but no further scientific data were returned.

A Close Encounter with Eros

After the numerous missions to remote worlds that have marked the Space Age, it is a curious fact that scientists know more about a small chunk of rock named Eros than about almost every other inhabitant of the Solar System. The flood of data returned by NEAR Shoemaker during its year in orbit around Eros transformed the asteroid from a smudge of light in the eyepiece of a terrestrial telescope into a real, three-dimensional world.

Eros was revealed as an elongated, peanut-shaped object, measuring 33 km (20 mi) across from east to west but only 13 km (8 mi) north to south. It spins once every 5 hours 16 minutes around an axis that goes through the little world's narrow waist. The rotation axis itself is tilted at an angle of 89 degrees to its orbit—similar to the planet

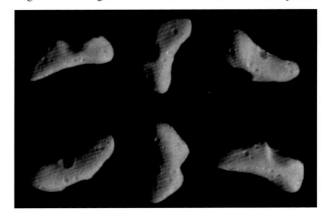

Eros—These NEAR images of 433 Eros were acquired on February 12, 2000, from a range of 1,800 km (1,100 mi). The 5½-hour-long sequence was taken during the final approach. The images show Eros in approximately true color. Its subtle butterscotch hue at visible wavelengths is nearly uniform across the surface. The most notable surface features are a large circular crater, now named Psyche, and a large depression known as the "saddle" on the opposite side of Eros. At 33 × 13 × 13 km (20 × 8 × 8 mi) in size, Eros is the second largest near-Earth asteroid. It spins on its axis once every 5 hours 16 minutes. (NASA/JHU–APL)

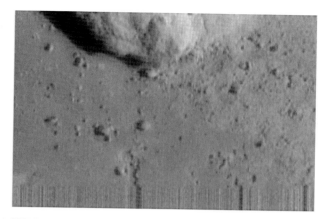

NEAR's Last Image—The last image of Eros received from NEAR Shoemaker was taken from a range of 130 m (420 ft) and shows an area 6 m (20 ft) across. At the top of the image is a large rock, about 4 m (13 ft) across. To the right is a rough area coated with debris, alongside a much smoother area that shows scattered rubble and snake-like patterns of erosion. The streaky lines at the bottom indicate loss of signal as the spacecraft touched down on the asteroid. (NASA/JHU–APL)

Uranus, whose spin axis is also nearly parallel to its orbital plane.

Unlike asteroids such as Mathilde, Eros appears to be relatively unscathed by large impacts, although it has undoubtedly suffered a lot of physical abuse during its long lifetime. Images showed numerous craters in the 0.5–1 km (0.3–0.6 mi) size range, which are evidence of early bombardment and subsequent collisions with other bodies.

However, the asteroid's shape is dominated by two features. On one side of its central waist is a smooth, saddle-shaped depression, which is about 10 km (6 mi) wide and may be a large crater that has been smoothed and degraded over billions of years. On the opposite side of the asteroid is a crater 5.5 km (3.4 mi) in diameter, which is the largest obvious impact structure on Eros.

Are these two indentations somehow related? No one knows, though inspection of the linear grooves and ridges that criss-cross its surface indicates that Eros was already heavily cratered and internally fractured before the saddle's formation.

Not all areas of Eros are equally indented. Some display dense cratering, while nearby regions are relatively smooth. Some particularly unusual craters appear to be square in outline, suggesting that their shape has been guided by extensive, subsurface fractures.

The varied terrain also revealed linear features that mainly trend east–west. These included chains of craters, sinuous, linear depressions up to 2 km (1.2 mi) in length and 200 m (660 ft) wide, and several low ridges. One, which extends around much of the asteroid, is 1–2 km wide and less than 200 m high. Even more striking was a complex ridge that spreads around the northern hemisphere for about 15 km (9 mi) as broken or separate segments.

The asteroid's strange shape also dominated its gravitational field. Calculations of its volume, combined with studies of tiny changes in NEAR's orbital velocity, showed that the asteroid has a density comparable to that of Earth's crust.

Other data indicated that Eros is one of the oldest objects in the Solar System, and probably an ancient fragment from a much larger parent body. Spectral studies confirmed that it is made of material very similar to that found in primitive chondritic meteorites, left-overs from the birth of the planets.

More than 8 million measurements by the spacecraft's laser range finder, which measured Eros's shape with great accuracy, have enabled scientists to determine that Eros is a consolidated object, rather than a loosely bound rubble pile like Mathilde.

Meanwhile, NEAR's X-ray spectrometer, which detects surface fluorescence caused by solar flares, detected low levels of aluminum relative to magnesium and silicon. Scientists conclude that Eros has not experienced the internal melting that occurred in planets such as the Earth and some of its asteroid cousins. Instead of a distinct crust, mantle, and core, it seems to be homogeneous throughout.

Although its surface gravity and escape velocity vary with location, and are higher at the ends of the "peanut" than in the middle, the acceleration of gravity is everywhere thousands of times smaller than on Earth. A person who tips the terrestrial scales at 68 kg (150 lb) would weigh only 16–37 g (0.56–1.3 oz) on Eros—equivalent to a small bag of potato chips. An escape velocity of between 3.1 and 17.2 m/s (10.2 and 56.8 ft/s) means that throwing a ball into orbit would present little difficulty for a visiting astronaut.

To the surprise of the science team, high-resolution images taken from low orbit showed fewer small craters than expected. They had apparently been weathered or buried beneath a layer of loose material that could be several meters deep. Subsequent analysis indicated that quakes caused by meteorite impacts could cause such loose material to slide downslope and gradually obliterate most craters smaller than a football pitch.

At the same time, ejecta blocks—rocks and boulders created by impacts—between 1 and 100 m (3 and 330 ft) across were abundant, although they were not distributed evenly across the asteroid. These boulders appear to be remnants of debris blasted out of impact craters. One possibility is that the cratering rate plummeted a billion or so years ago when Eros was thrown out of the main asteroid belt. After that, there may have been too few impacts to pummel these boulders into smaller pieces.

Sample Return

Although a number of asteroids have been surveyed by remote-sensing techniques, scientists would love to get hold of a sample of material from one of these primitive chunks of debris. To this end, on May 9, 2003, the Japanese space agency launched its Hayabusa (Falcon) spacecraft toward near-Earth asteroid 25143 Itokawa.

Propelled by ion engines, the spacecraft successfully rendezvoused with Itokawa in September 2005. Its first task was to map the surface of the tiny "deformed potato"—which was found to measure only 540 × 310 × 250 m (about 1,800 × 1,000 × 820 ft)—at different wavelengths over a period of several months.

In addition to a multi-band imaging camera, Hayabusa

Hayabusa—Hayabusa (formerly known as MUSES-C) is the first mission designed to land on an asteroid and return a surface sample to Earth for analysis. Launched on May 9, 2003, the Japanese spacecraft used ion engines to arrive at near-Earth asteroid 25143 Itokawa in September 2005. The spacecraft touched down twice, but whether it obtained fragments of surface material is uncertain. Damage during the second attempt means that the spacecraft may not be able to return to Earth. (JAXA)

carried a laser altimeter (LIDAR) to constantly measure the distance between the spacecraft and the asteroid and determine the object's size and shape. A near-infrared spectrometer measured the mineral composition and properties of the surface. The link between asteroids and meteorites found on Earth was also investigated by an X-ray fluorescence spectrometer designed to detect different elements and measure their abundance.

If the sample collection was to succeed, it had to overcome numerous obstacles, including a surface gravity 750,000 times less than that of Earth, a 12-hour rotation period and a rugged, boulder-strewn surface. (Itokawa unexpectedly displayed numerous large boulders and several sizeable, smooth areas.)

Scientists hoped that a unique, close-up look would be provided by a small, hopping robot called Minerva, but the asteroid's gravity was too weak to attract the tiny lander after it was jettisoned, causing it to become a useless mini-satellite of Itokawa.

The efforts to obtain a first sample from an asteroid also ran into trouble. It was hoped to conduct one dress rehearsal and two sampling runs before heading for home. These would involve releasing a softball-sized target marker made of highly reflective material. A flashlight-like device was to illuminate the surface marker and assist in the landing.

During a fleeting touchdown on the surface, Hayabusa would fire a "bullet" into the surface at a speed of 300 m/s (990 ft/s). Any fragments released during the impact should enter a funnel-shaped horn and pass into a sample container that would then be sealed, ready for return to Earth.

However, the brief touchdown and sampling sequence provided engineers with many headaches. The first dry run had to be abandoned. During a second rehearsal, the spacecraft descended to a height of 55 m (180 ft) but lost Minerva. Hayabusa remained on the surface of the "MUSES Sea" for half an hour—much longer than planned—during a sampling attempt on November 20–21, but the sampling gun failed to fire and the spacecraft had to make the first lift-off from an asteroid without its precious rock fragments.

Finally, on November 26, Hayabusa completed a second landing. This time, the mission controllers were confident that two bullets had fired and that sample retrieval had been successful. Unfortunately, the spacecraft then suffered leaks in its chemical thrusters and began to spin out of control, losing contact with the ground. As the seriousness of the situation dawned, the Japanese space agency announced that the return to Earth had been delayed for two years, in order to give engineers more time to gain control. The worst news of all was that the sampling gun may not have fired at all. The only way to be sure is to peer inside the sealed capsule.

Scientists will be hoping that Hayabusa can overcome the odds and safely return to Earth in June 2010, three years later than originally planned. If all goes according to plan, the small re-entry capsule containing the samples will be ejected for a parachute descent near the town of Woomera in Southern Australia. The capsule will be opened in Japan under controlled conditions at a new national facility built for the purpose. Scientists hope that the tiny samples will increase their knowledge and understanding of asteroid surfaces and provide clues to what happened during the earliest epoch of the Solar System's history.

Almost as ambitious is a NASA mission to orbit the large main belt asteroids, Vesta and Ceres. Dawn's goal is to achieve an understanding of the conditions and processes acting in the early history of the Solar System by studying two very different objects that have survived largely unscathed over billions of years.

Scheduled for launch in 2006, Dawn was to use ion propulsion to reach Vesta in 2011. It would then spiral closer to the surface, before making a detailed study of

Dawn—The Dawn asteroid explorer shown with Vesta and Ceres against a background painting of a primordial solar nebula. Dawn will use ion engines during a nine-year mission that will enable it to enter orbit around both of these large primordial worlds. Its measurements will provide detailed insights into the nature of two very different bodies, and enable scientists to compare their characteristics with those of meteorites found on Earth. (UCLA/ William Hartmann)

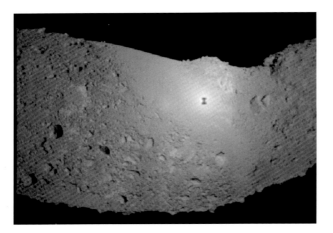

Hayabusa—Itokawa is the smallest asteroid ever seen from close range by a spacecraft. Despite its almost imperceptible gravity, this near-Earth object has a surprisingly diverse surface, with numerous large boulders and several smooth "seas." This image shows the shadow of Hayabusa on the "MUSES Sea" during its descent on November 20, 2005. (ISAS/JAXA)

every aspect of the asteroid's mass, shape, volume spin rate and gravity.

After orbiting Vesta for about eight months, Dawn would head for Ceres, arriving in late 2015 for a similar, in-depth investigation of the queen of the asteroid belt. The schedule is now uncertain, since Dawn was canceled in March 2006 and reinstated a few weeks later. A detailed technical review is currently under way.

A more traditional approach will be undertaken by ESA's Rosetta spacecraft en route to comet Churyumov–Gerasimenko. During two visits to the main asteroid belt, Rosetta will conduct close flybys of two contrasting objects, 2867 Steins and 21 Lutetia.

Steins is relatively small, with a diameter of perhaps 10 km (6 mi). It may be an example of the E-class asteroids—rocks that were once molten and then solidified—that are largely confined to the inner reaches of the main belt. The fleeting encounter with Steins will take place on September 5, 2008, at a distance of just over 1,700 km (1,060 mi). This flypast will take place at a relatively low speed of about 9 km/s (5.6 mi/s).

21 Lutetia is a much bigger object, about 95 km (60 mi) in diameter. The nature of Lutetia is uncertain. It may be an M-class object that was once part of a planetoid's metallic core, or it may be a stony object similar to primitive, unprocessed meteorites known as carbonaceous chondrites. Rosetta will pass within about 3,000 km (1,875 mi) of Lutetia on July 10, 2010, at a speed of 15 km/s (10 mi/s).

Rosetta will obtain spectacular images as it flies by these primordial rocks. Its onboard instruments will provide information on the mass and density of the asteroids, thus telling us more about their composition. It will also search for satellites, measure their subsurface temperature and look for nearby gas and dust.

The Galilean Moons—On January 7, 1610, Galileo turned his 20-power telescope toward Jupiter, and found three small, bright stars in a straight line near the planet—one to the west and two to the east. Observations over the following weeks led Galileo to conclude that there were actually four "stars"—each a satellite orbiting Jupiter. This infrared image from the Hubble Space Telescope shows two of the moons, Io and Ganymede. Also visible are the shadows of these moons and Callisto. (NASA, ESA, Erich Karkoschka/Univ. of Arizona)

7 JUPITER: KING OF THE PLANETS

The Starry Messenger

Since time immemorial, Jupiter has illuminated our nocturnal skies. Ancient civilizations all over the world recognized it as one of the brightest planets or wanderers among the fixed stars. Although it was outshone by Venus, the morning or evening "star," Jupiter had no rival (other than the Moon and the occasional apparition of Venus) after midnight, when it frequently dominated the heavens.

For about six months of each year, Jupiter became an unmistakable presence, shining more brightly and steadily than any twinkling star. Almost inevitably, the planet became associated with the most powerful of all the Greek and Roman gods.

Early astrologers and observers were aware of Jupiter's slow and stately progress through the constellations of the zodiac, completing one cycle every 12 years—longer than any planet apart from Saturn. The logical explanation was that Jupiter must be further away from Earth than Mercury, Venus, or Mars.

Until the mid-sixteenth century, it was almost universally accepted that Jupiter revolved around the Earth, since our world must inevitably be the largest and most important object in creation. The first challenge to this doctrine came from a Polish astronomer, Nicolaus Copernicus, who tried to explain the annual retrograde or reverse movements of Mars, Jupiter, and Saturn against the fixed stars.

In the Copernican system, the Sun was placed at the center and the Earth was relegated to the position of third planet from the Sun. Jupiter was rightly installed toward the edge of the Solar System.

Afraid of ridicule, Copernicus delayed publication of his work until shortly before his death in 1543. For the next 67 years, scholars argued over the merits of the two planetary systems, until an Italian genius named Galileo Galilei found the first irrefutable evidence that Copernicus was right.

In 1609, Galileo had been made aware of a spyglass that had been invented in the Netherlands. Hurrying back to Venice, he soon succeeded in cobbling together a primitive "optick tube" that could magnify objects three times. His paymasters were further impressed with a demonstration of a 10-power instrument that could be used to distinguish sailing ships while they were still far out to sea.

But it was his use of the telescope to study the night sky that confirmed the Copernican revolution in astronomy. Using an instrument that could magnify a magnificent 20 times, Galileo was able to see the mountains and craters of the Moon, and the phases of Venus. He next turned his creation on Jupiter, which was then at opposition and the brightest object in the evening sky.

On January 7, 1610, he observed what seemed to be three fixed stars strung out in a line near the planet. Alerted by this unusual formation, he returned to it the following evening. Instead of being left behind by the planet's motion across the sky, the three little stars were now aligned to the west of Jupiter.

Over the next week he was able to confirm that there were four, rather than three, stars that never left Jupiter. Not only did they appear to be carried along with the planet, but they changed their position with respect to each other and Jupiter. By January 15 he came to the inescapable conclusion that these were not stars, but rather four moons that revolved around Jupiter. By watching their motions

Galileo Galilei—Galileo (1564–1642) was the foremost scientist of his day. At the age of 17, he entered the University of Pisa to study medicine, but he soon became famous for discovering the regular motion of a swinging lamp—the forerunner of the pendulum. He also conducted experiments to study gravity, using balls of different weights and size to disprove Aristotle's "law" that heavy objects fall faster than light objects. In 1609, Galileo constructed his first spyglass. An improved version of this instrument was used to discover the four large moons of Jupiter in 1610. When Galileo argued that this supported the theory of the Earth moving around the Sun, he was taken before the Inquisition, forced to recant, and placed under house arrest. (Rice University)

Two months later, Galileo announced his discovery in a book entitled *Siderius Nuncius* (The Starry Messenger). Although only 550 copies were published, it caused a sensation. For the first time, there was visible proof that everything did not revolve around the Earth. Taking the argument one step further, Galileo suggested that the Earth might not be at the center of the Universe, as Aristotle and Ptolemy had argued. Copernicus had been correct.

Galileo's arguments led him into direct conflict with the Roman Catholic Church, and in 1616 Cardinal Bellarmine prohibited him from denouncing other alternatives in favor of Copernicanism. For the next 16 years, the devout scientist went along with the ban, at least in public.

However, his pointed defence of Copernican theory and ridicule of Aristotle in his book *Dialogue On Two World Systems* led to his arrest and trial before the Inquisition. Galileo was sentenced to house arrest in his villa near Florence, but the breakthrough had been made. Thirteen years after Galileo died in 1642, Huygens discovered a satellite of Saturn. The old ideas were dead and buried.

The naming of the four "Galilean" satellites also turned into a long-running saga. In gratitude to his latest patrons, the Medici family who ruled the state of Florence, Galileo named them the "Medicean Stars." A more attractive alternative was put forward in 1613 by Johannes Kepler, Imperial Astronomer to the Holy Roman Emperor, and supported by a Dutch astronomer, Simon Marius. Kepler favored naming them after maidens who, in Greek legend, had been lovers of Zeus (the Greek name for Jupiter). Thus, the innermost and fastest moving of the moons should be called Io, the second Europa, the third Ganymede, and the outermost Callisto.

Unfortunately, the scientific community preferred to give moons numbers rather than names, starting with the satellite closest to Jupiter, and it was only after the discovery of numerous satellites around Saturn that an alternative system was sought. From the mid-nineteenth century, the names put forward by Kepler came into common use.

over the course of a month, he was able to determine that their orbital periods around Jupiter were 42 hours, 3.5 days, 7 days, and 16 days.

Turbulent Jupiter

Although the introduction of the telescope revolutionized studies of all the planets, perhaps the greatest impact involved Jupiter. By the mid-seventeenth century, improved optics enabled astronomers to measure the size of the disk and to discern features on its visible surface.

It soon became obvious that the ancients who named the planet after the king of the gods had some special insight, for Jupiter is the dominant body in the Solar System, after the Sun. This huge world has more than 2 ½ times the mass of all the other planets combined. 1,300 Earths would fit inside the giant world, but because of its low density, Jupiter is "only" 318 times as massive as Earth.

Although it comes no closer to the Earth than 599 million km (372 million mi), Jupiter is so large that it is a magnificent sight in telescopes of modest aperture. At times of opposition, every 399 days on average, its disk dwarfs every planet apart from Venus.

One of the most obvious characteristics of Jupiter's disk is its unusual shape. The planet appears to have a bulging midriff, so that its equatorial diameter is 9,270 km (5,760 mi) larger than its polar diameter. The reason for this is the rapid rotation of the largely gaseous body, first discovered by Giovanni Domenico Cassini in 1665.

Despite a diameter 11 times that of Earth, Jupiter spins once every 9 hours 55 minutes—faster than any other planet.* This means that Jupiter's equatorial cloud deck is traveling from west to east at more than 45,000 km/h (28,125 mph). As there are almost 2 ½ days on Jupiter for every day on Earth, an alien living on Jupiter's cloud tops would enjoy two or three sunrises over a 24-hour period.

Larger telescopes, improved lenses, and higher magnifications opened the way to studies of Jupiter's surface. It soon became apparent that Jupiter is not a rocky world like the Moon or Mars, but a huge ball of gas with no solid surface. In fact, Jupiter's bulk density was found to be only a little higher than that of water.

As early as 1664, Englishman Robert Hooke sketched a

Jupiter and Earth—Jupiter has more than 2 ½ times the mass of all the other planets combined. 1,300 Earths would fit inside the giant world, but because of its low density, Jupiter is "only" 318 times as massive as Earth. Despite its enormous bulk, Jupiter rotates faster than any other planet. This causes its equatorial region to bulge outward. The clouds are arranged in light and dark bands aligned parallel to the equator. The Great Red Spot (bottom center) is a giant atmospheric storm that was first recorded in 1664. (NASA–JPL)

curious feature that seems to be the first observation of a huge storm in the southern hemisphere. It became known as the Great Red Spot. Although its size has altered and its red coloration has intensified and faded over the centuries, the oval spot—still large enough to swallow several Earths—seems to reign eternal in the giant planet's ever-turbulent atmosphere.

For anyone looking at Jupiter through a small telescope, the most obvious features are the light and dark stripes that

* The equatorial rotation period of about 9 hours 50 minutes is about 5 minutes shorter than the rest of the planet. Radio emissions from Jupiter's interior give a period of 9 hours 55 minutes 30 seconds.

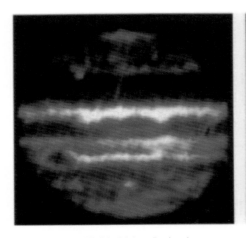 

Jupiter in Infrared and Visible Light—Jupiter has a very warm interior. The Voyager 1 image (right) shows the Great Red Spot amid the zones and belts that lie parallel to the equator. The infrared image (left) was taken by Richard Terrile with the 5-m (200-in) telescope at Mt Palomar on the same day, January 10, 1979. The infrared-bright areas coincide with blue and brown clouds that are lower and warmer. The white and orange–red clouds (including the Great Red Spot) are much higher and colder—so they appear dark in the infrared image. (NASA–JPL)

run parallel to the equator. Like the Great Red Spot, the reddish-brown "belts" and whitish "zones" are permanent features. Nevertheless, they are not unchanging.

Astronomers watch eagerly to record the latest streaks, plumes and spots that evolve in the highly active atmosphere. Some of the spots and plumes survive over a period of several days, months, or even decades. On the other hand, the atmosphere sometimes seems to enter a quiet spell when the belts become fainter and fewer cloud variations are visible.

Above 50 degrees latitude it is a different story. The stripes disappear, to be replaced by reddish-brown clouds that are broken into many small, rotating cells. As is the case near the equator, the different colors indicate different chemical compositions of the cloud particles. The cloud patterns, in turn, reflect different physical conditions—updrafts and downdrafts—in which the clouds form. The light-colored zones are regions of rising gas that spreads out at the cloud tops and sinks back in the neighboring dark belts.

An important breakthrough came in the 1930s, when dark lines in the spectrum of Jupiter were identified as ammonia and methane. Some years later, hydrogen was detected. (Unfortunately, it is extremely difficult to find out what the solids and liquids in the clouds are made of with a spectroscope.) This led the German-born astronomer Rupert Wildt to argue that Jupiter's overall composition must be very similar to that of the Sun.

Although the planet probably had a solid, rocky core, Wildt argued that it was largely made up of hydrogen and helium with a dash of ammonia, methane, and other gases. At great depth, these must be compressed into a thick fluid. The visible atmosphere was only the thin outer layer of an ocean of gases thousands of kilometers deep.

Not until the mid-twentieth century did some hard data on the invisible interior become available. First came the discovery that Jupiter is a strong source of radio waves. The Russian scientist Iosif Shklovsky realized that the signals were generated by energetic electrons trapped in an enormous magnetic field. If our eyes could see Jupiter's magnetosphere, it would appear twice the width of a full Moon in the night sky. One implication was that there must be some kind of "metallic" fluid swirling around inside the planet and acting as a dynamo. A little later, it was found that the bursts of radio static were linked to the motion of the innermost Galilean moon, Io. One explanation was that huge electrical currents were flowing between the planet and Io. Then, in the late 1960s, new infrared detectors flown on a Lear-Jet aircraft discovered that, beneath the −120°C cloud tops, Jupiter was radiating two or three times more heat than it received from the Sun. Was the planet still glowing with heat from its formation 4.5 billion years ago?

The Pioneers

Although some remarkable breakthroughs had been made with ground-based observations, they had often thrown up more questions than answers. The only way to unravel Jupiter's true nature was to send spacecraft to study it at close quarters.

In 1969, NASA approved the first mission to explore beyond the asteroid belt. Known as Pioneer Jupiter, the project involved sending two almost identical spacecraft to fly past the giant planet. However, this posed a number of significant challenges.

In order to reach Jupiter, the spacecraft would have to travel faster than any of their predecessors. This meant that they would have to be as small and as lightweight as possible, and launched by a powerful Atlas–Centaur rocket equipped with an additional upper stage. Even so, they would have to survive a trip of around two years that would enable them to arrive when Jupiter was not too close to the Sun in the sky. For Pioneer 10, this meant a launch window of just a few weeks in the spring of 1972.

Another problem was how to power the spacecraft while avoiding potential damage from Jupiter's radiation belts. With light levels only 4% of those at Earth, normal solar arrays were out of the question. Instead, the Pioneers would become the first civilian spacecraft to use nuclear power.

Pioneer at Jupiter—Pioneers 10 and 11 were the first spacecraft to cross the asteroid belt and explore the giant planets. The nuclear-powered Pioneers continued to send back data for more than 20 years. Pioneer 10 obtained the first close-up images of Jupiter, studied its intense radiation belts and magnetic field, and confirmed that Jupiter is predominantly a liquid planet. Pioneer 11 passed Jupiter on December 2, 1974, then headed for Saturn, 2.4 billion km (1.5 billion mi) distant, where it arrived on September 1, 1979. (NASA–Ames)

Four radioisotope thermoelectric generators would generate 140 watts of power at Jupiter by turning heat from the radioactive decay of plutonium into electricity.

Payload weight was severely restricted. Equipped with minimal computing capability, the Pioneers would have to be "flown from the ground," despite the long delays in communicating with the spacecraft over vast distances. Unlike more sophisticated craft, the spacecraft would rotate continuously around an axis aimed toward Earth. Although this was ideal for observations of energetic particles and magnetic fields, it was a definite handicap for cameras and other instruments that required careful pointing.

In the event, the 258-kg (568-lb) Pioneer 10 was successfully built and assembled in less than three years. Its trailblazing mission began on March 2, 1972, as distant lightning flashes illuminated Cape Kennedy. The Atlas–Centaur vehicle boosted the spacecraft to an unprecedented speed of 51,682 km/h (32,114 mph)—the first time a

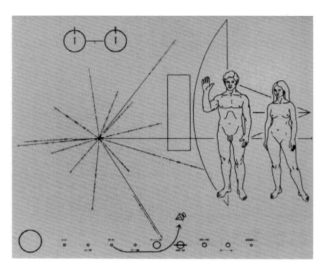

The Pioneer Plaque—Pioneers 10 and 11 carried a calling card from Earth in the form of a 15 × 23 cm (6 × 9 in) gold anodized plaque bolted to the main body. On the plaque an "average" man and woman stand before an outline of the spacecraft. The man's hand is raised in a gesture of goodwill. Other details show humanity's level of scientific understanding. An atom of hydrogen is in the top left corner. The "star" diagram shows the position of the Sun relative to 14 pulsars and the center of the Galaxy. At the bottom is a scale plan of the Solar System, showing Pioneer's trajectory and the relative distances of the planets in binary code. The plaque was designed by Carl Sagan and Frank Drake, and drawn by Linda Salzman Sagan. (NASA–Ames)

spacecraft had been given enough energy to leave the Solar System altogether. The fastest man-made object ever to leave the Earth swept past the Moon's orbit in just 11 hours, crossing the orbit of Mars, about 80 million km (50 million mi) away, in just 12 weeks.

On July 15, 1972, Pioneer 10 entered the potentially hazardous asteroid belt, emerging unscathed seven months later. By early November 1973, with Jupiter fast approaching, the first long-range imaging tests began. Three weeks later, the long-awaited encounter began when the spacecraft crossed the bow shock that marked where the planet's magnetic field slowed the solar wind.

The simple line-scan camera began to build up images as the rotating spacecraft swept across the disk of Jupiter. The pictures were taken in red and blue light so that a "true color" version could be created back on Earth. (The red and blue were used to make a "synthetic" green image in order to produce a three-color view.) By early December, they were showing more detail than the best pictures taken by ground-based telescopes. Near closest approach, the partial frames of Jupiter were three times sharper than anything taken from Earth.

Tension increased as the spacecraft plunged ever more deeply into the radiation belts that surrounded the planet. How would the spacecraft respond to the continuous bombardment from energetic electrons and ions? Several instruments designed to measure charged particles reached saturation point, and a number of false commands were sent to the imaging instrument. A special command sequence sent from the ground successfully countered most of the errors, with the result that only a detailed image of Io and a few partial close-ups of Jupiter were lost.

The calculated risk of sending Pioneer 10 deep into the radiation belts paid off. The intrepid craft eventually swept past Jupiter on December 3, 1973, just 130,354 km (81,000 mi) above the cloud tops. It arrived one minute early—evidence that Jupiter was slightly more massive than expected. Communication then ceased as it moved behind the planet, and scientists held their breath. They need not have worried. Pioneer re-emerged to send back the first images of a crescent Jupiter ever seen by humans.

Meanwhile, its sister ship was already on its way. Launched on April 5, 1973, Pioneer 11 was targeted even closer to Jupiter, in order to receive the greatest possible gravity assist for a follow-on flight to Saturn. Continually reprogrammed by commands from the ground, the spacecraft performed flawlessly as it penetrated the radiation belts, this time flying from south to north in order to minimize exposure to the damaging particles. More worrying for the scientists was a threatened strike in Australia that could have resulted in a loss of 6–8 hours of data every day.

It eventually passed by Jupiter only 43,000 km (26,725

mi) above the cloud tops on December 2, 1974. With its speed almost doubled by Jupiter's gravitational grasp, the second phase of Pioneer's adventure began as it looped high above the ecliptic and headed toward the sixth planet from the Sun.

Pioneer Puzzles

During their high-speed penetration of the hitherto unexplored Jovian system, the Pioneers sent back a mass of intriguing data that left scientists eagerly awaiting the launches of the more advanced Voyagers. As is often the case, some age-old questions were solved but a number of new puzzles appeared.

Most public attention was given to the images that painted Jupiter in pastel shades interrupted only by the orange oval of the Great Red Spot and occasional transits by satellites and their dark shadows. Not only was it possible to see features as small as 500 km (310 mi) across, but the cameras were able to observe the huge planet from viewpoints never available on Earth.

Close-range pictures revealed cloud patterns that suggested rising and descending currents in the atmosphere. This interpretation was supported by heat maps, which showed the colder, lighter zones rising many kilometers above the dark belts.

An unexpected variety of cloud shapes near the boundaries between the light zones and dark belts suggested that conveyor belts of air were moving at different speeds, but the detailed "snapshots" provided by the Pioneers were too few in number to create a movie sequence that would reveal these motions.

Pioneer 11 produced some of the most intriguing views of Jupiter as it swept over the planet's northern polar region. For the first time, astronomers could clearly see a mottled pattern of individual cells of rising air instead of the linear belts and zones that dominated closer to the planet's equator. The cloud tops were found to be substantially lower at the poles than near the equator, but they were

covered by a thicker, transparent atmosphere. Blue sky at the poles was attributed to scattering of sunlight by gases—just as on Earth. Unfortunately, the Pioneers could not clear up the mystery of the colorful clouds. Scientists speculated that they were probably caused by the action of solar ultraviolet radiation on chemical compounds such as ammonia, carbon, sulfur, or phosphorus. The deeper color of the Great Red Spot might be the result of chemical reactions on gas rising far above its surroundings.

It was clear that Jupiter has no solid surface beneath the clouds. This put an end to the theory that the Great Red

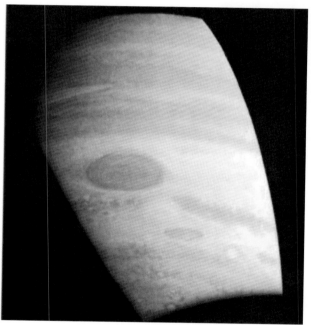

The Great Red Spot—One of the most exciting aspects of the Pioneer missions was the unique close-up views of Jupiter's cloud features. This Pioneer 11 image straddles most of the southern hemisphere. The picture, taken 545,000 km (339,000 mi) above the cloud tops, shows structures 240 km (148 mi) across. Even so, the resolution was not sufficient to show the turbulence that surrounds the mysterious storm system—there is no clear evidence of cloud motion or winds in or around the Spot. (NASA–Ames)

Heading North—This picture, taken just after closest approach on December 2, 1974, was described as perhaps the most scientifically important image of Jupiter obtained by Pioneer 11. It covers an area from the equator (lower right) almost to the north pole. Particularly noticeable is the change from a banded cloud structure at low latitudes to circular cells created by rising gas. At the time, Pioneer 11 was heading north, high above the ecliptic plane, at the beginning of its voyage to Saturn. (NASA–Ames)

Spot was created by interactions with an underlying mountain. Although the exact reason for its existence remained uncertain, the Spot was undoubtedly an atmospheric phenomenon.

Unfortunately, Pioneer 10's planned close-up pictures of Io were lost through radiation interference, and the remaining moons were only glimpsed at long range. The best views showed contrasting light and dark spots on Ganymede, and a pale, icy Europa. However, careful tracking of the spacecraft showed that Jupiter was about 1% heavier than previously thought, while the masses of several Galilean moons had to be revised.

The Pioneers confirmed that Jupiter possessed an internal heat source that enabled the planet to emit almost twice as much energy as it received from the Sun. The temperature of the cloud tops was around –123°C, and the poles were, surprisingly, as warm as the equator.

Several Pioneer instruments studied the atmosphere above the upper ammonia ice clouds and found an initial drop in temperature followed by a rise as pressure increased with depth.

Measurements of Jupiter composition showed that the planet is probably 85–90% hydrogen, with most of the remainder in the form of helium. In other words, the giant planet is very similar to the Sun in its internal makeup. By combining the various measurements, scientists were able to improve their understanding of the hidden interior. Calculations suggested that the pressure reaches some 3 to 5 million Earth atmospheres and a temperature of 11,000°C—twice as hot as the surface of the Sun—at a depth of 24,000 km (15,000 mi).

Crushed by the overlying gas, the hydrogen turns into a metallic fluid that readily conducts both heat and electricity. At the very center of the gas giant there might be a rocky, iron-rich core slightly bigger than the Earth.

The Pioneers were also able to give the first detailed information on the enormous magnetic field that was generated by motions in the metallic hydrogen. This magnetic bubble in space was found to be blunt on the sunward side, where the solar wind slammed headlong into it. Pioneer 10 first entered the magnetic field about 6.4 million km (4 million mi) from the planet—much further out than scientists had expected.

The spacecraft entered and left the bubble on a number of occasions as gusts in the solar wind caused it to pulsate in and out. In contrast, on the leeward side, the bubble was shaped into a long, tapering tail that stretched for millions of kilometers into space.

In many ways, the field resembled that of the Earth, although it was some 2,000 times stronger. The magnetic field's axis was tilted 11 degrees with respect to the north–south rotation axis. Spacecraft instruments confirmed theories that the magnetic field is the opposite way round to the Earth's, with magnetic north near the south pole. The magnetic axis was also offset by about 10,000 km (6,250 mi) from the center of the planet.

Because of the tilted magnetic field, the radiation belts are also tilted, wobbling up and down as Jupiter spins. Most of the trapped particles were found to originate from the planet and its satellites, particularly Io. The intensity of the radiation exceeded anything ever measured in nature—comparable to the levels measured after the explosion of a large nuclear bomb in Earth's upper atmosphere.

Voyages of a Lifetime

In 1969, the same year that the Pioneer Jupiter project received official approval, NASA began to prepare for a far more ambitious follow-up. Known as the Grand Tour, the plan was to send four identical robotic probes equipped with state-of-the-art components on 12-year journeys to the outer Solar System. One pair of spacecraft would visit Jupiter, Uranus, and Neptune, while the others would complete the reconnaissance by exploring Jupiter, Saturn, and Pluto.

The idea was to take advantage of an alignment of all the giant gas planets that occurs only once in every 176 years. This alignment would enable the spacecraft to use Jupiter's gravity to boost their energy and modify their trajectories so that they were redirected toward each outer planet in turn. Unfortunately, political and budgetary considerations intervened, and in 1972 the $750-million Grand Tour was scrapped. Managers at NASA's Jet Propulsion Lab came back with a scaled-down mission based on the Mariner technology that had successfully sent exploratory robots to Venus and Mars. Initially called Mars–Jupiter–Saturn 1977, the two-pronged mission was renamed Voyager by a democratic vote of the team just before launch.

Officially, the revamped mission was designed specifically for encounters at Jupiter and Saturn, but the flight plans had specially been chosen to allow the second spacecraft to complete its own mini-Grand Tour. If all

went well, Voyager 2 would be able to use a gravity assist by Saturn to send it toward Uranus and Neptune.

Like their predecessors, the Voyagers required a high-speed launch that would eventually enable them to escape the Solar System altogether. Unable to rely on solar arrays, the spacecraft would also use radioactive power sources that could generate 450 watts at launch and keep them alive for decades. However, in contrast to the relatively small and simple Pioneers, the 815-kg (1,800-lb) Voyagers were extremely complex and capable. Each spacecraft was equipped to carry out 11 science experiments that would investigate every aspect of the planetary systems they would visit. Five of the remote-sensing instruments, including two TV cameras, were placed on a scan platform that could swivel in almost every direction and accurately point toward the targeted moons and surface features.

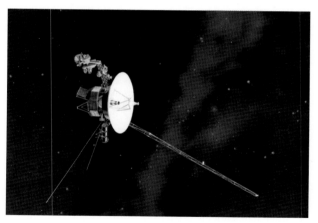

Voyager—The twin Voyagers revolutionized our knowledge of the outer Solar System. Voyager 1 explored Jupiter and Saturn before heading out of the Solar System. Now more than 14 billion km (almost 9 billion mi) from the Sun, it is the most distant human-made object. Voyager 2 flew past all four gas giants and is also heading toward interstellar space. Voyager's main features were its large, dish-shaped high-gain antenna, the long magnetometer boom (right), the radioactive power source (bottom), and the scan platform for the camera and other instruments that required accurate pointing (top). (NASA–JPL)

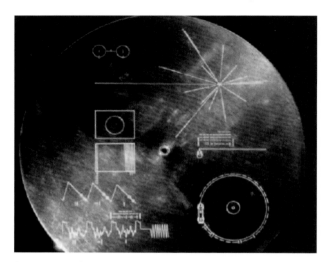

Sounds of the Seventies—The Voyagers carried "Sounds of Earth" records, complete with instructions on how to play them and where they had come from. The explanatory diagram appears on both the inner and outer surfaces of the cover, since the outer diagram will be gradually eroded by the space environment. The 30-cm (12-in) gold-plated copper disks contain greetings in 60 languages, samples of music from different cultures and eras, and natural and human-made sounds from Earth. They also contain electronic information that can be converted into diagrams and photographs. (NASA)

Redundancy was an important feature, so each Voyager carried two radio receivers and four transmitters. Although the transmitter power was only 23 watts—less than the power of a small light bulb—the craft could transmit data at the then enormous rate of 115,200 bits/s from a distance of 1 billion km (625 million mi). An onboard digital tape recorder had enough capacity to store almost 100 images for later transmission back to Earth.

Perhaps the most important advance was the addition of a computer "brain" that would control the thrusters, scan platform, and instruments. Software updates loaded into this electronic intelligence system would be vital to ensure that the spacecraft survived their long, hazardous odyssey to interstellar space.

In an unusual quirk of numbering, Voyager 2 lifted off first from Cape Canaveral, Florida, on August 20, 1977. The powerful Titan–Centaur launch vehicle blasted Voyager past the Moon in a mere 10 hours, but the early part of the mission was plagued with technical glitches, including an erroneous signal that the scan platform had not fully extended.

Worse was to come in November, when one of the radio receivers began to lose power. The following April, the primary radio receiver suddenly failed and shocked engineers discovered that the backup was also faulty. Less than halfway to Jupiter, they had to find a way of communicating with a partially deaf spacecraft. Fortunately, it was possible to upload a sequence of new commands, including a backup automated mission that could be executed at Jupiter and Saturn if the secondary receiver was totally lost.

Voyager 1's launch was delayed a few days in an effort to avoid some of the teething problems of its sister ship. Lift-off finally took place on September 5, 1977, and the early part of the mission went more or less according to plan—including a historic image looking back at the Earth and Moon. Then, on February 23, 1978, the scan platform ground to a halt during a sequence of maneuvers. Although the platform later responded to commands in the normal way, it was treated with kid gloves for the remainder of the mission.

Following a faster trajectory, Voyager 1 overtook its twin in the asteroid belt, and by January 1979 it was close enough to send back images of cloud features that could be assembled into time lapse movies. Voyager was traveling so fast that it swept through the inner Jovian system in less than 48 hours, with closest approach to the planet on March 5, 1979. Among the highlights of the encounter were the discoveries of a thin ring around Jupiter and widespread volcanic eruptions on Io.

Four months later, Voyager 2 arrived. This time, the inward leg involved close flybys of Callisto, Ganymede, and Europa before the spacecraft swept past Jupiter on July 9 at a range of 654,000 km (400,000 mi)—more than twice as far away as its predecessor. Once again, the brief encounter was a spectacular success. Saturn and opportunities for further discoveries now beckoned.

Perpetual Motion

After the *hors-d'oeuvres* served up by the pair of Pioneers, scientists were eagerly looking forward to the main course—made up of two Jupiter flybys by the much more capable Voyagers. They were not disappointed. Excitement mounted as the colorful cloud formations of Jupiter loomed into range of Voyager 1's cameras, revealing a scene of incessant motion and swirling turbulence. There was so much going on that it seemed impossible to keep track of everything and then come up with a sensible explanation. And yet, the scientists slowly began to create order out of chaos.

One of the first surprises was that Jupiter was presenting a very different, more active face than the one seen by the Pioneers. In the mid-1970s, the Great Red Spot was fairly uniform in color and surrounded by a huge bank of white cloud. Pale brown bands circled the northern hemisphere.

By 1979, the south temperate region to the west of the Great Red Spot had evolved into a complex mass of plumes. Further, more subtle changes were seen when Voyager 2 arrived four months later. By then, a broad, white band below the Great Red Spot had slimmed down considerably, the swirling mass of cloud to the west had calmed down a little, and the Spot itself seemed to be reverting to the appearance observed by the Pioneers.

Time lapse movies that compressed 10 Jupiter days (about 100 hours) into a matter of minutes showed planet-girdling jet streams that swept the clouds along at different speeds. By studying the movement of the more permanent cloud features, it was possible to pick out the prevailing winds at different latitudes.

Voyager images showed broad bands of cloud in the equatorial zone and the north temperate belt that were sweeping eastward at more than 100 m/s (225 mph) relative to the planet's rotation. Other clouds, notably at around 18 degrees north and south, traveled almost as quickly in the opposite direction. Indeed, it soon became apparent that, as

Changes Galore—Jupiter's cloud patterns changed significantly between the two Voyager flybys. Most of the changes were due to the prevailing winds moving features at different speeds in relation to each other. In these two views, taken in January (left) and May 1979, the large white oval storms south of the Great Red Spot shifted 90 degrees to the east. There were also major changes in the shapes of the plumes and eddy patterns to the west (left) of the Spot. (NASA–JPL)

Stormy Weather—Voyager 1 captured this image of the Great Red Spot on March 1, 1979. Three Earths could fit inside this huge anticyclonic (high-pressure) storm. Details within the Spot are clearly visible, along with a layer of thin ammonia ice cloud that covers its outer reaches. Like an island in a stream, the Spot acts as a barrier, causing considerable turbulence in the surrounding clouds, particularly on its windward (west) side. Also visible is a white spot, another anticyclonic storm that raced eastward past its larger neighbor, propelled by a faster moving jet stream. Features as small as 600 km (375 mi) can be seen. (NASA–JPL)

on Earth, the northern and southern hemispheres were more or less mirror images of each other as far as the broad pattern of clouds and winds was concerned.

Superimposed on these parallel stripes of light and dark material was a plethora of spots, barges, swirls, and curls. Along the northern edge of the equator was a series of cloud plumes that were probably linked to rising air and appeared to be evenly spaced around the planet.

At the opposite color extreme were occasional brown ovals, which appeared to be windows into the warmer atmospheric depths. These were fairly common in the northern hemisphere and have been found to have a typical lifetime of one to two years.

Much more conspicuous was the scattering of white spots, including three white ovals, each about the size of Earth, which raced around Jupiter to the south of the Great

Red Spot. First seen in ground-based telescopes back in 1939, the ovals were anticyclonic storms that rotated in an anticlockwise direction.

Similar, but smaller, bright spots displayed the same characteristics of high-pressure systems, spinning anticlockwise in the southern hemisphere and clockwise to the north of the equator. The more modest wind speeds at high latitudes meant that the white spots furthest from the equator were more circular and less stretched out than their near-equatorial counterparts.

Of course, the star of the show was the Great Red Spot, a huge anticyclonic storm that resembled an orange–red whirlpool in the middle of a fast-flowing stream. Voyager images revealed a six-day, counterclockwise rotation, with winds blowing east to west on its northern flank and in the opposite direction on the southern side. Within the Spot itself, there seemed to be very little horizontal motion. Temperature measurements confirmed that the Spot was a plume of rising gas that towered far above its surroundings. The fates of clouds that entered its domain were seen to vary. One bright feature was seen to circle the Spot for 60 days without any appreciable change. Some clouds swept around the northern edge and kept on going. Some were captured and consumed. Yet others would split in two when they reached the western or eastern end of the giant oval, and then travel in opposite directions.

Voyager 1 saw a Red Spot partially covered to the north by a veil of high ice cloud—probably ammonia. By the time of the second Voyager encounter, a more marked ribbon of white cloud had developed around its northern edge, blocking the motion of small spots coming from the east and forcing them to turn tail and head back to their source.

Although the Jovian weather dominated much of the Voyager media coverage, the giant planet also served up other surprises. Voyager 1 detected very strong ultraviolet emissions from auroras over the planet's north and south poles. The temperature of the atmosphere in these regions was at least 700°C. A little later, its wide-angle camera scanned the planet's night side and detected the largest aurora ever observed—nearly 29,000 km (18,000 mi) long. Also visible in the same image were other Jovian "fireworks"—19 bright patches created by lightning superbolts several thousand kilometers below the aurora. Several meteor trails even intruded on the nocturnal scene.

Incandescent Io

Prior to the late 1970s, Io, the innermost of the four Galilean moons, was a world of mystery. The satellite was known to be reddish in color, and a sparse atmosphere had been detected by the Pioneer 10 spacecraft in 1973—an observation that was hard to explain since Io is only a little larger than Earth's Moon.

Then in 1974 a yellow glow was detected around the satellite, and found to be due to scattering of sunlight by sodium on its surface. Even so, most astronomers expected Io to have an ancient cratered surface, so the scientific world was astounded when Voyager close-up pictures showed a lurid, orange–red, but crater-free landscape. This implied a young surface that had been modified within the past one million years.

Voyager 1's highest resolution photos showed radiating flow patterns around dark circular depressions which resembled calderas or large volcanic craters. One hemisphere alone revealed more than 100 such features over 25 km (16 mi) across. The visual evidence raised the possibility of Io being one of a rare breed—a world with active volcanoes. Within a few days, the mystery was solved by accident.

On March 8, 1979, three days after closest approach to Jupiter, Voyager 1 was looking back at the receding planet and its moons in order to snap a number of navigation images that would be used to calculate its course. One long-exposure image of a crescent Io, taken from a distance of 4.5 million km (2.8 million mi), showed the satellite against the field of background stars. Later that day, an optical navigation engineer named Linda Morabito began to analyze the picture on her computer. To her surprise, there appeared to be a crescent-shaped feature extending hundreds of kilometers beyond the edge of Io. Since the

The "Pizza" Moon—Volcanic Io is seen here in a mosaic of Voyager 1 images taken from a range of 400,000 km (250,000 mi) on March 5, 1979. Two of the most active volcanoes are visible: Pele (center), with its horseshoe-shaped deposits, and Loki (upper left). Galileo images later showed that Io is mainly yellow or light greenish, rather than orange–red. Accurate color images were not possible from the Voyager data because there was no coverage in red light. Dark areas are very hot lava flows, often on caldera floors. (NASA–JPL)

orange moon had no appreciable atmosphere, what could it be? Morabito and her excited colleagues set out to eliminate all possibilities, such as another moon in the line of sight, and concluded that it must be the result of an ongoing volcanic eruption of incredible violence.

On the Monday morning, the Voyager science team set to work to search for evidence of ongoing eruptions in other pictures. By mid-morning, Joseph Veverka and Robert Strom had discovered several additional volcanic plumes.

Meanwhile, on March 11, John Pearl and Rudy Hanel had independently found evidence of strong heat emissions from different places on the satellite. The most prominent was a source nearly 200°C hotter than its surroundings. Pearl brought the new results to the imaging team, and, indeed, the hot spot was located near one of the volcanic plumes.

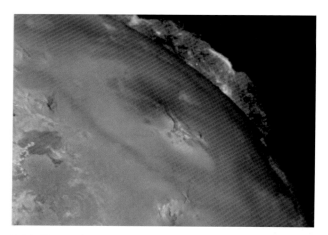

Pele's Plume—Computer processing of Voyager 1 images produced this remarkable view of an umbrella-like volcanic plume rising 300 km (190 mi) above the surface of Io. The eruption originated in Pele, one of Io's largest volcanic centers. Pele's vent is the dark spot just north of the triangular-shaped plateau (right center). The plume fallout covers an area the size of Alaska. When Voyager 2 arrived four months later, Pele was inactive. (NASA–USGS)

A month later, continuing analysis of infrared spectra yielded identification of sulfur dioxide gas over this same erupting volcano. At last a source had been located for the enigmatic sulfur and oxygen ions trapped in Jupiter's magnetic field. Eight active eruptions were eventually located in the Voyager 1 pictures of Io. (Six out of seven volcanoes sighted by Voyager 2 were still active four months after the Voyager 1 flyby.)

With hindsight, the volcanoes explained many hitherto puzzling observations. A few months earlier there had been a report of a sudden infrared brightening of Io. Meanwhile, the Voyager ultraviolet experiment had been unable to track down the source of intense sulfur emissions or to explain the changes in the gas clouds around Io since the Pioneer 10 and 11 flybys

Continuous, widespread volcanic activity also explained the young, crater-free surface and bizarre features seen in the Voyager images. Io was apparently the most geologically active body ever encountered in the Solar System. The

discovery was not a total surprise. One week prior to Voyager 1's encounter with Jupiter, Stanton Peale published a paper suggesting that tidal heating might produce volcanism on Io. As the moon passed between gigantic Jupiter and its large satellite neighbors, Europa and Ganymede, it was squeezed and unsqueezed during each orbit. This flexing pumps energy into the satellite, which heats its interior.

One model suggested that the satellite had a subcrust of solid silicate at a depth of perhaps 20 km (12.5 mi). Above this would be an ocean of molten sulfur overlain by a solid crust of sulfur mixed with solid and liquid sulfur dioxide. Heating of the sulfur dioxide would turn it into a gas that could explode on the surface, blasting fine particles at least 100 km (62 mi) into space. A fine snow of solidified sulfur and sulfur dioxide would then drift back down onto the surface, forming the colorful rings that surrounded the active vents.

Calculations suggested that almost all of the 100,000 tonnes of material erupted each second returns to the surface as snow. The remainder is captured by the Jovian magnetic field or populates the thin atmosphere that surrounds Io.

The liquid sulfur also escapes from calderas to flow over the surface. The longer flows are a red–yellow color, whereas sulfur at higher temperatures (above 250°C) becomes black and viscous and remains close to the calderas.

Appropriately, the volcanic features were given mythological names related to fire and volcanic legends. Thus the largest volcano, recognizable from its "hoof print" deposits, was called Pele after the Hawaiian volcano goddess.

Europa

Prior to the Voyager flybys, no one expected Europa, the smallest of the Galilean moons, to provide any major surprises. Its density, similar to that of Io, suggested a largely rocky world with a fair scattering of ice. Ground-based observations showed it as a brilliant white world

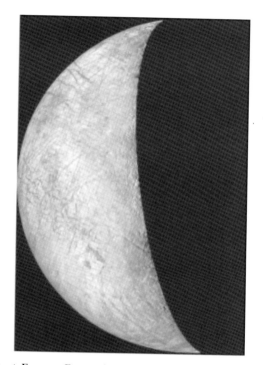

Crescent Europa—Europa is the smallest and smoothest of the Galilean satellites. This Voyager 2 mosaic is centered at about 300° E. The bright areas are probably fresh ice deposits, whereas the darkened areas may be areas with "dirt" or a more patchy distribution of ice. Long, linear structures cross the surface in various directions. Some of these are over 3,000 km (1,860 mi) long and 2–3 km (1–2 mi) wide. They may be fractures or faults that have disrupted the surface. (NASA–JPL)

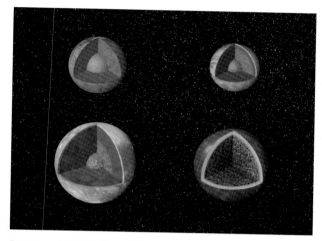

Interiors of the Galilean Moons—Cutaway views of Ganymede (lower left), Callisto (lower right), Io (upper left), and Europa (upper right). Information returned by the Voyager missions indicated that Io is dry and rocky; Europa might have a thick crust of ice over a rocky interior; Ganymede and Callisto were probably half rock and half ice. More recent Galileo data suggest that most of them have metallic cores of iron and nickel (shown in gray) and rocky mantles (brown). Io's rock or silicate shell extends to the surface, while the rocky interiors of Ganymede and Europa are surrounded by possible water oceans or layers of ice (blue and white). Only Callisto is shown as a fairly uniform mixture of ice and rock, though it may have a more complex core. (NASA–JPL)

covered in ice. In the early Voyager pictures, it maintained a bland, snowball appearance, in contrast to the spottiness of Ganymede and the orange color of Io.

During a distant flyby at a range of 734,000 km (456,000 mi), Voyager 1 could see no craters or mountains. Instead, it observed a curious spider's web of criss-crossing linear features, some up to 3,000 km (1,860 mi) in length. Was this resemblance to a cracked egg shell the result of internal forces or some other global tectonic process?

In July 1979, Voyager 2 sent back spectacular images from four times closer. Although the detailed coverage was limited, the pictures could still resolve features as small as 4 km (2.5 mi). They confirmed that Europa was covered in dark streaks measuring up to 70 km (44 mi) in width and thousands of kilometers long. Although most of these linear features were straight, others were curved or irregular in shape. The dark bands resembled great faults or fractures in the ice crust that have been filled with slush from below. (In fact, they were not really dark; the contrast with the nearby white ice was only about 10%.) However, there was no evidence of valleys or linear depressions. Amazed scientists said it was as if they had been "made with a felt-tipped pen."

The surface over which they passed was also found to be "as smooth as a billiard ball." The main exceptions seemed to be regions of mottled terrain that could be small interlocked hills and depressions. Their origin was uncertain.

Only three identifiable impact craters were seen, each about 20 km (12.5 mi) across. Once again, astronomers seemed to be gazing upon a remarkably young surface, perhaps only a few hundred million years old—a complete contrast to the cratered landscape they had originally predicted. One suggestion was that the craters disappeared through gradual "relaxation" of the ice crust. More active obliteration by some form of flooding or ice volcanism was also a possibility.

Even more enigmatic were the smaller, lighter streaks that had never been seen on any other world. Generally only about 10 km wide, they formed scallops or curved cusps that repeated on a scale of a few to hundreds of kilometers. In images taken with the sunlight grazing the surface, they could be seen to rise a few hundred meters above their surroundings.

The general impression from the Voyager pictures was of pieces of icy crust fitting together in an irregular fashion—rather like the ice sheet that covers Earth's Arctic Ocean. No evidence of lateral movement of the pieces was visible. However, one theory suggested that the lighter streaks could have been created by modest folding where neighboring blocks of ice collided. The dark streaks might be related to tension that pulled the blocks apart, allowing "contaminated" slush to seep up from below and fill in the gaps.

It was generally agreed that the driving force that formed these surface features was related to tidal forces or heat-driven convection beneath the solid crust. Perhaps, as tidal forces slowed the satellite's spin, its equatorial bulge was reduced, enabling radial tides to form concentric patterns of cracks that pointed directly toward or away from Jupiter.

Scientists could only speculate about the nature of the satellite's interior. Since its density is only a little lower than that of Io, the bulk of the satellite was thought to be silicates (rock), overlain by a 100-km (62-mi) layer of water or soft slushy ice, and topped by a thin ice crust, also perhaps no more than 100 km thick.

Once again, tidal heating involving squeezing between Jupiter and the other Galilean moons could be invoked to explain the satellite's warm interior—though to a lesser extent than for Io. In addition, internal heating by natural radioactive elements almost certainly contributed. However, the details of how the satellite was resurfaced and how its patterns of unusual streaks were preserved remained a mystery.

Ganymede and Callisto

As the outermost Galileans and the largest members of the quartet, Ganymede and Callisto were expected to be very similar. Both were known to have low densities, indicating a bulk composition of about half ice and half rock, and both were far enough away from Jupiter to avoid the intense bombardment by charged particles trapped in the planet's radiation belts. In the event, they proved to be just as unique as their smaller cousins.

It was already known that Ganymede outsized the planets Mercury and Pluto, but it was thought to be second to Saturn's moon Titan in the satellite rankings. After the Voyager flybys, Ganymede's diameter was increased to 5,270 km (3,275 mi). This upward revision meant that Ganymede superseded Titan as the largest satellite in the Solar System.

Callisto proved to be a little smaller, with a diameter of 4,840 km (3,007 mi)—roughly identical in size to Mercury. However, it had the lowest bulk density of all the Galileans. Callisto was also found to be the darkest of the four Galilean satellites, although its ancient icy crust was still twice as reflective as our Moon.

Predictions that both satellites would display evidence of a long history of bombardment proved correct, but there were important differences. Ganymede showed tremendous variety in its landforms, ranging from dark regions that

Gigantic Ganymede—This Voyager 2 photo of Ganymede was taken on July 7, 1979, from a range of 1.2 million km (750,000 mi). It shows a dark circular feature about 4,000 km (2,500 mi) in diameter with narrow, closely-spaced pale bands traversing its surface. The bright spots are relatively recent impact craters, while paler circular areas mark older impact sites. The light branching bands are ridged and grooved terrain that has been resurfaced relatively recently. The nature of the brightish region covering the northern part of the dark circular feature is uncertain, but it may be some type of frost. (NASA–JPL)

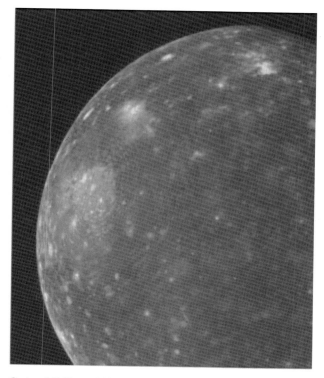

Cratered Callisto—This Voyager 1 mosaic of Callisto was obtained on March 6, 1979, a distance of about 400,000 km (250,000 mi). Callisto was found to be the most heavily cratered of the Galilean satellites, suggesting that it has changed little over billions of years. The most prominent impact feature is the multi-ringed Valhalla (left). Unlike craters on the rocky planets, Valhalla is quite flat, with no encircling mountains or central peak. The bright central region is around 600 km (375 mi) across, but the surrounding rings extend to a distance of 1,500 km (940 mi) from the ancient impact center. (NASA–JPL)

displayed numerous ancient craters to broad grooves made up of parallel lines of mountains and valleys.

Although Ganymede showed a whole range of crater forms, from vague ghost-like images to fresh, young features surrounded by white rays of clean ice, there was no direct visual evidence of huge impacts.

A huge circular region of dark, ancient surface was seen, however, on the side facing away from Jupiter. Measuring 4,000 km (2,500 mi) across, this huge feature, now known as Galileo Regio, preserved numerous ring structures that were associated with a tremendous impact billions of years ago. The actual crater in the center of this "ghost" basin had been filled in and disappeared long ago.

Of most interest to geologists were the lighter, grooved bands that criss-crossed Ganymede. The parallel mountain ridges they contained were typically 10–15 km (6–9 mi) across, about 1 km (3,280 ft) high and hundreds of kilometers long.

In many places, the grooves divided up the darker

terrain into isolated polygons, as if they had been created by slushy material pushing up from below along fault lines. This process must have been repeated many times, for grooves could be seen to cross and lie on top of each other, so that some of them covered up underlying craters. In some places, the grooved terrain was offset, evidence of lateral movement along faults.

Although crater counts supported the idea that these grooves were younger than their surroundings, none of them seemed to be less than 3 billion years old. Whatever activity Ganymede exhibited in its youth had long since died out.

Callisto was found to be the most crater-saturated body known. Curiously, there was a distinct lack of craters larger than 150 km (93 mi) across. Instead, the entire surface seemed to be uniformly covered with 50–100 km (31–62 mi) impact structures, many with bright rims indicative of exposed water ice. In some places, they were so closely packed that there appeared to be no room for any more. Scientists concluded that Callisto had changed very little over billions of years. Unlike its three fellow Galileans, the moon had apparently been geologically inert almost since the time of its formation.

The craters themselves were very different in shape from those on the Moon. Instead of deep, saucer-shaped basins, the craters on Callisto were very shallow. Scientists suggested that the ice had slowly flowed into the hollows over billions of years, gradually smoothing and obliterating their original, rugged appearance.

The monotonous landscape of medium-sized craters was relieved only by a few giant, circular impact regions. The largest, known as Valhalla (after the home of the Norse gods), was a bright, flat region about 600 km (375 mi) across, presumably all that was left of the original impact basin. It was surrounded by numerous concentric rings that extended outward about 1,500 km (940 mi) from the impact center.

The rings—very different from those on rocky Mercury—were thought to have been caused by melt and flow of the surface followed by freezing. This would explain the absence of a large basin or fringing mountain range like those found on Mercury or the Moon.

A Ring and New Moons

By the time the Voyagers arrived at Jupiter, rings were known to exist around both Saturn and Uranus, but a Jovian equivalent was not generally predicted. Nevertheless, the imaging team decided to program the spacecraft to take a single image as Voyager closed on the planet—just in case. The search paid off, although the ring proved to be extremely elusive.

Voyager 1's sole image of the ring was snapped on March 4, 1979, a little under 17 hours before closest approach to Jupiter, from a distance of about 1.2 million km (750,000 mi). The observation involved pointing the camera off to the right, into the "empty" space beyond the planet's equator.

Sure enough, smeared across the 11-minute exposure was a blurred streak surrounded by numerous, hairpin-like images of background stars. The imaging team had been lucky. Voyager had captured the very tip of the ring. A little further to the right and the camera would merely have taken a picture of the stars in the Beehive star cluster. Within a few days, infrared images taken by astronomers in Hawaii succeeded in confirming its existence. Scientists hastily revised the Voyager 2 encounter in order to find out more about their fortunate find.

The second Voyager was able to reveal much more about the ring system. Looking down on it from above the ring plane, the spacecraft was able to show that it was very different from the broad disk that surrounded Saturn. Other images taken from behind Jupiter showed the backlit rings standing out in stark contrast to the black body of the planet. The observations showed some structure in the rings, which extended all the way from Jupiter's cloud tops to a distance of 53,000 km (33,125 mi). The main rings were quite narrow, comprising a fairly diffuse segment about 5,000 km (3,125 mi) wide, and a bright outer ribbon about 800 km (500 mi) across.

The rings were very thin—perhaps no more than 1 km (0.6 mi) deep—and very transparent. The dark particles of unknown composition ranged from microscopic dust to chunks several meters across, but were widely dispersed. (In fact, Pioneer 11 had passed through the ring in 1974, with no obvious adverse effects.) Like tiny moonlets, they followed individual paths around the planet every 5 to 7 hours.

Since particles seemed to slowly migrate from the ring to Jupiter's surface, they must be replenished somehow. One suggestion was that nearby Adrastea and Metis, discovered by Voyager 2, could be sources of material.

Prior to 1979, there were 13 known Jovian satellites. The largest was Amalthea, the only moon known to orbit closer to Jupiter than Io. Discovered by Edward Emerson Barnard in 1892, the fifth Jovian moon traveled around the planet once every 12 hours at a distance of 181,000 km (113,125 mi) from its center.

Although the Voyager encounters were dominated by the four Galilean giants, the opportunity to study Amalthea

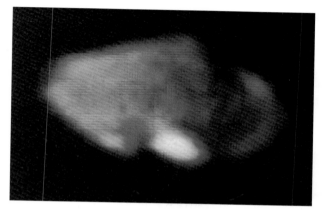

Red Amalthea—Amalthea was thought to be the innermost of Jupiter's satellites before the Voyagers found Adrastea and Metis. The elongated moon measures 270 × 170 × 150 km (170 × 107 × 95 mi). Like most satellites of Jupiter, Amalthea is in a synchronous rotation, with the long axis always pointing toward Jupiter. It orbits only 181,300 km (112,650 mi) from the center of the planet—equal to 2.5 Jupiter radii. At this distance, the satellite is bombarded by intense radiation and high-velocity micrometeorites. The dark-red surface was attributed to sulfur and other material arriving from Io. (NASA–JPL)

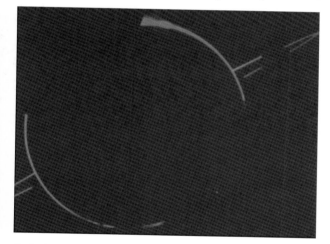

Jupiter's Ring—This image of Jupiter's newly discovered ring was taken by Voyager 2 as it looked back toward the Sun (hidden behind the giant planet) from a distance of 1.5 million km (900,000 mi). The gaps in the ring are the result of combining several different pictures. The ring is mainly made of extremely fine dust particles derived from two tiny satellites near its outer edge. The bright part of the ring is about 5,200 km (3,200 mi) wide. (NASA–JPL)

was too good to miss. Voyager 1 images revealed Amalthea to be an elongated, reddish object that measured about 270 × 160 km (170 × 100 mi). One hemisphere was dominated by a large, bright feature—presumably an impact crater.

The other eight satellites orbited far beyond the Galileans, and far beyond the Voyagers' cameras. They were all faint and small, probably no more than a few tens of kilometers across, and they all followed quite elongated paths, with the four outer satellites (Sinope, Pasiphae, Carme and Ananke) traveling in a retrograde direction, opposite to that of the inner satellites (and the spin of Jupiter).

Jupiter's retinue grew by three as a result of the Voyager flybys. 1979 J1 was discovered by David Jewitt and Ed Danielson of Caltech in October 1979, during routine examination of Voyager 2 ring images. Unlike the other minor moons, it orbited very close to Jupiter, only 58,000 km (36,250 mi) above the planet's cloud tops. This placed it just outside the main ring and well inside the orbit of Amalthea. Following with tradition, it was named Adrastea, after a nymph who nursed the infant Zeus.

The fifteenth and sixteenth satellites of Jupiter were discovered in 1980 by Steven Synott of JPL. They, too, were found close to the giant planet. 1979 J2 (named after the Greek nymph, Thebe) was located between Amalthea and Io. The discovery image, taken during the Voyager 1 flyby about one year earlier, showed a 75-km (47-mi) spot alongside a dark, smudged shadow.

1979 J3 (named Metis, after the first wife of Zeus) was found to travel only 1,000 km (620 mi) inside the orbit of Adrastea, skimming once around the outer edge of Jupiter's main ring in a little over seven hours.

Galileo the Spacecraft

Three hundred and eighty years after Galileo Galilei discovered the four planet-sized satellites of Jupiter, a spacecraft named in his honor left Earth to explore the giant planet's realm. The Italian pioneer would have marveled at the technological progress that had taken place since the invention of the telescope.

NASA's Galileo spacecraft was designed to rewrite the history books by becoming the first man-made object to enter orbit around a giant planet and the first to release a probe into a gas giant's atmosphere.

The 2½-ton Orbiter would build on the Voyagers' brief reconnaissance of the Jupiter system. Its 10 scientific instruments were designed to carry out a multi-year survey of the planet, its moons, magnetic field, and particle environment.

One important innovation was the Orbiter's "dual spin" design. The main body, which carried the high-gain antenna and various instrument booms, would rotate about three times a minute in order to sample the surrounding environment. The aft section of the spacecraft, where the Probe was located, would remain "fixed" in order to provide a stable platform for the remote-sensing instruments.

The Launch of Galileo—This artist's impression shows the deployment of the Galileo spacecraft from the Shuttle Atlantis on October 18, 1989. Once released from the cargo bay, a two-stage Inertial Upper Stage boosted the spacecraft out of Earth orbit. Over the next six years, Galileo used gravity assists from Venus and Earth to gain enough momentum to reach Jupiter. It made two visits to the asteroid belt, completing close flybys of Gaspra and Ida. In July 1994 Galileo obtained the only direct images of the impacts when the fragments of comet Shoemaker–Levy 9 plunged into Jupiter's night side over a six-day period. (NASA)

The scan platform on this aft section carried a TV camera (which was actually a Voyager spare equipped with new, solid-state electronic sensors), two spectrometers to measure the composition of clouds and satellite surfaces, and an instrument that detected reflected light in order to measure temperatures and surface properties.

Like the Voyagers, Galileo was powered by two radio-isotope thermal generators that produced electricity from the heat generated during the decay of plutonium dioxide. Together, they could produce about 500 watts, although this would gradually drop as time passed.

Since huge amounts of data, including many thousands of images, were expected to be returned during the mission, Galileo was provided with a solid-state tape recorder with a capacity of 900 megabits.

Galileo in Orbit Around Jupiter—On December 7, 1995, Galileo became the first spacecraft to enter orbit around Jupiter. Unfortunately, the mission was handicapped by the failure of the umbrella-like high-gain antenna (left) to unfold properly. Nevertheless, despite prolonged exposure to intense radiation, the orbiter continued to operate until September 21, 2003. It was deliberately destroyed in the planet's dense atmosphere, in order to avoid possible contamination of Europa. (NASA–JPL)

Science data were to be transmitted from a dish-shaped high-gain antenna at a rate of 135 kilobits/s, with two slower, low-gain antennas that would be used during the early stages of the mission and to send telemetry. However, the umbrella-like main antenna was to be launched in the furled position, and only opened out once the Orbiter was well on its way to Jupiter. This new design proved to be the Achilles' heel of the sophisticated spacecraft.

Galileo was designed to be the first US planetary mission to be launched from the Space Shuttle, but this decision led to major problems for the mission team. Originally scheduled for 1982, the launch was put back four years by delays in development of the Shuttle and the powerful Centaur upper stage that would boost Galileo on a direct path to Jupiter.

Then came the Challenger disaster of January 1986, which led to a cancelation of the Centaur stage on the grounds of crew safety. Grounded for three years, Galileo was placed in storage until the Shuttle became operational once more and a smaller, safer, upper stage was prepared.

The much-delayed mission eventually got under way on October 18, 1989. After deployment from the cargo bay of Shuttle Atlantis, the Inertial Upper Stage fired to accelerate the spacecraft out of Earth orbit. Over the next six years, Galileo used gravity assists from Venus and Earth to gain enough momentum to reach Jupiter.

Six months later, ground controllers sent the long-awaited signal to unfurl the 4.8-m (16-ft) main antenna, but to their dismay the dish only partially opened and then jammed. All efforts to shake it free failed. From now on, Galileo would have to rely on its low-gain antennas, which could only transmit data at a rate of 10 bits/s. Galileo's promised flood of images was inevitably reduced to a trickle, despite efforts to enhance the Deep Space Network and completely replace the flight software.

Further disaster threatened when the onboard tape recorder stuck only eight weeks before its scheduled arrival. Fortunately, although the opportunity to image Io from close range was lost as a result of the malfunction, the mechanical blip proved to be temporary.

After releasing the small, piggybacking Probe into the planet's swirling cloud tops, Galileo finally braked into orbit around the planet on December 7, 1995. Despite the continual threat from technical glitches caused by intense radiation, it went on to complete one of the most productive missions in the history of Solar System exploration.

Bouncing around the Jovian system like a celestial billiard ball, the craft was able to utilize the gravity of Jupiter's large moons to change course. When Galileo's prime mission came to an end in December 1997, two mission extensions were granted. After circling the Solar System's largest planet 35 times, its 14-year mission of exploration was brought to a spectacular close when it plunged into Jupiter's atmosphere on September 21, 2003. This dramatic end was chosen to ensure that Galileo would not collide with Europa and contaminate the pristine environment of this potential alien ecosystem.

From launch to impact, the stalwart spacecraft had traveled more than 4.6 billion km (almost 2.9 billion mi), completed 35 satellite flybys, and survived more than four times the cumulative dose of harmful radiation it was designed to withstand. Despite its communication handicap, it returned over 30 gigabytes of data, including 14,000 pictures.

Probing the Depths

Although the Galileo Orbiter carried an advanced suite of instruments designed to measure every aspect of Jupiter and its environment, scientists were also eager to obtain the first "ground truth" data from within the planet's turbulent atmosphere. Galileo was accordingly provided with a small Probe, similar in concept to those flown on the 1977 Pioneer Venus mission.

The 339-kg (747-lb) Probe was composed of two major segments. A *deceleration* module, which made up approximately half of the overall weight, was required to enable the Probe to survive a fiery ballistic entry into Jupiter's upper atmosphere. Once it was released from its protective cocoon, a *descent* module would transmit data back to the Orbiter for at least one hour as it carried out a relatively gentle parachute descent.

For almost seven years, the inert Probe simply piggybacked on the nose of the Orbiter as it journeyed across the Solar System. Finally, on July 13, 1995, it was set free. Spinning at 10.5 rpm to provide rotational stability, the Probe was despatched along a precise path that would cause it to plunge toward the distant planet's equatorial region.

Over the next five months, the Probe picked up speed as it was pulled inexorably inward by Jupiter's overwhelming gravitational grasp. When the moment of truth arrived on December 7, the Probe slammed into the sparse outer atmosphere at 170,700 km/h (106,000 mph)—equivalent to traveling once around the Earth in 15 minutes. The entry site was pinpointed as 7.5°N and 4.46°W.

An enormous release of energy caused by supersonic compression and frictional heating created a 15,500°C cloud of incandescent plasma ahead of the Probe. Erosion of the forward heatshield reduced its mass from about 152 kg (335 lb) to about 65 kg (143 lb), but the deceleration module did its job. Within a matter of minutes, the spacecraft had slowed to a relatively sedate 430 km/h (250 mph).

The next step was to fire a mortar that deployed a pilot parachute. The aft cover of the deceleration module was then released, allowing the main parachute to be deployed and the forward heat shield to be jettisoned. Over the next 57.6 minutes, the Probe's seven experiments* transmitted data to the Orbiter, 215,000 km (134,000 mi) overhead. By the time rising temperatures caused the communication system to fail, the Probe had fallen about 200 km (125 mi) below the visible cloud tops.

The historic measurements sent back during the suicide mission into the giant planet's atmosphere provided fascinating new insights and posed many problems for the scientists. However, the first discovery came from the Energetic Particle Instrument, which started to take measurements of radiation levels in the previously unexplored inner regions of Jupiter's magnetosphere three hours before atmospheric entry.

The instrument discovered a new radiation belt between Jupiter's ring and the upper atmosphere that was approximately 10 times stronger than Earth's Van Allen radiation belts, and included high-energy helium ions of unknown origin.

As the Probe plunged into Jupiter's atmosphere, the Atmosphere Structure Instrument measured variations in the temperature, pressure, and density of Jupiter's atmosphere until it succumbed at a pressure of about 24 bars—24 times the air pressure at sea level on Earth, equal to the pressure at a depth of 230 m (750 ft) in the ocean—and a temperature of 152°C.

The density and temperature figures were significantly higher than expected at upper levels, although they matched

---

* There was no camera on board.

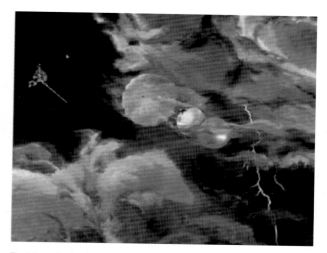

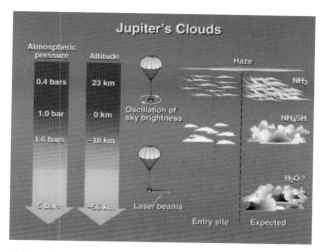

Probing Jupiter's Depths—The Galileo Probe slammed into Jupiter's atmosphere on December 7, 1995. Traveling at a speed of 170,700 km/h (106,000 mph) it had to survive temperatures twice as hot as the Sun's surface and deceleration forces as great as 230g (230 times the acceleration of gravity at Earth's surface). For almost one hour, it sent back information about the surrounding atmosphere until the communication system failed due to the high temperatures about 200 km (125 mi) below the visible cloud tops. (NASA–Ames)

Galileo Probe Results—The Galileo Probe sent back some surprising data about Jupiter's clouds. One experiment detected a high-level cloud layer—probably ammonia ice (NH_3) particles— near the 0.6-bar pressure level (1 bar is the air pressure at sea level on Earth). Another experiment found one well-defined cloud layer above the 1.6-bar pressure level as well as faint traces of cloud particles at many altitudes. No sign of the anticipated water cloud was detected. (NASA–Ames)

theoretical predictions deeper down. The results appeared to confirm that Jupiter has a significant internal heat source.

It was anticipated that the Probe would pass through three layers of cloud at different depths, but, to the scientists' surprise, no thick layers of dense cloud were found. Instead, only very small clouds and haze were detected along the entire descent trajectory. A single, well-defined cloud structure, probably made of ammonium hydrosulfide ice particles, was found above the 1.6-bar pressure level.

One surprising discovery was the dryness of the atmosphere and the absence of water cloud at lower levels. However, infrared observations obtained with ground-based telescopes showed that the Probe—against all the odds—somehow picked out one of the few "hot spots" on Jupiter where the air is clear and dry.

Helium abundance was 24%, similar to that found in the Sun. On the other hand, carbon (mainly in the form of methane gas), nitrogen (in the form of ammonia), and sulfur (in the form of hydrogen sulfide gas) were present in higher concentrations than on the Sun.

It was no surprise that wind speeds were very high, reaching around 700 km/h (435 mph) beneath the clouds and showing little variation with depth.

No optical lightning flashes were observed close to the Probe, although some 50,000 discharges were observed at radio frequencies. The nearest thunderstorm was far away (at least 12,000 km or 7,500 mi). The lightning was much stronger than on Earth, but the radio signals suggested that electrical storm activity was actually 3–10 times less common than on Earth.

Jupiter Unveiled

Although most attention was focused on its menagerie of moons, magnificent Jupiter inevitably came under scrutiny by the Galileo Orbiter. In fact, Galileo's first contribution to improving our understanding of the giant planet's atmosphere came in 1994, when it was still far from its destination.

The spacecraft's instruments sent back the only direct observations of the explosive impacts involving Jupiter and some 20 fragments of comet Shoemaker–Levy 9. Huge plumes of material erupted high above the main cloud deck in the middle of the southern hemisphere, leaving dark "bruises" in their wake. The brownish scars imaged by the Hubble Space Telescope eventually merged into a new dark belt which persisted for the next 18 months.

During the orbital mission, Galileo's gaze occasionally turned toward Jupiter to keep tabs on its dynamic weather. Images taken with different filters were able to distinguish between clouds at various altitudes, confirming that the reddish belts were regions of sinking gas separated by bright zones where the atmosphere was rising. For the first time, temperature maps were also able to show features smaller than 2,000 km (1,240 mi) across, revealing waves of warm air that appeared to have no counterpart in the visible cloud structure.

After the Galileo Probe coincidentally plunged into a rare hot spot, scientists made a point of using the near-infrared spectrometer to study these warm, arid regions. Orbiter images of one hot spot showed winds converging from all directions, like water pouring down a large sink hole in the ground.

In September 1998, another instrument recorded what may be the darkest Jovian spot ever seen, at around 15°N. Despite its ominous appearance in visible light, infrared measurements showed that the 7,500-km (4,700-mi) wide blemish was simply a huge, tunnel-shaped clearing through the main cloud decks that enabled heat to escape from the deep interior.

In contrast, the rising gas in the Great Red Spot was shown to take the form of a broad dome whose center towered some 10 km (6 mi) above the surrounding ammonia clouds and tilted slightly toward the east. Not surprisingly the Spot (and the similar, but slightly smaller white oval storms) was found to be very cold, although warmer clouds around it indicated areas of sinking air. Instead of forming a broad blanket, the internal clouds showed a pattern of curved "arms" separated by clearer air, making the overall structure resemble a leaning spiral staircase.

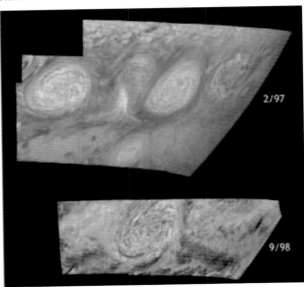

Merging Storms—Storms on Jupiter tend to have extremely long lives, but Galileo was able to see the "before" and "after" of a rare merger involving three white ovals that had circled the southern hemisphere since 1939. By February 19, 1997, two of them were closing on each other, squeezing an intervening low-pressure area into a pear shape (top). A year later, when Galileo was behind the Sun and unable to communicate with Earth, the two high-pressure areas merged. When Galileo's camera returned to the scene on September 25, 1998, the sole survivor was revealed (bottom). Light blue clouds are high and thin, reddish clouds are deep, and white clouds are high and thick. (Caltech)

Although the atmosphere was known to contain ammonia gas, it proved extremely difficult to find a pure cloud of young, fresh ammonia ice. Such clouds were usually obscured by unknown substances that attached themselves to the ammonia ice particles within a few hours of their formation.

Scientists were, therefore, jubilant when, during its first orbit of Jupiter, Galileo detected the first discrete cloud of ammonia ice ever seen on the planet. The spot, perhaps 15 km (9 mi) thick, was named the Turbulent Wake Anomaly because it lay in the roiling air downstream from the Great Red Spot. This region was characterized by powerful wind currents moving in opposite directions that pulled up ammonia gas from below. The ammonia then cooled rapidly to form ice in the 120°C atmosphere.

One of the most important observations involved the amalgamation of two enormous white ovals that had been circling Jupiter for almost 60 years. Unfortunately, Galileo was not able to communicate with Earth during the actual merger in February 1998, but the "before" and "after" images showed how the Earth-sized anticyclonic storms had squeezed a region of intervening low pressure into a pear shape, before destroying it altogether. The resultant hybrid that appeared at 33°S seemed only slightly larger than its twin parts, indicating either a loss of some material or an increase in height.

Numerous thunderheads up to 50 km (30 mi) tall were identified just north of the Spot, where columns of warm air were thrusting up towering thunderheads. Many others were seen on the planet's night side, showing up as broad, bright splotches illuminated from below by lightning flashes within Jupiter's water cloud, 50 to 75 km (30 to 45 mi) below the ammonia cloud. The strikes were hundreds of times brighter than lightning on Earth.

One image sequence showed a convective thunderstorm, located 10,000 km (6,218 mi) northwest of the Great Red Spot. The dense white cloud measured 1,000 km (620 mi) across and rose 25 km (15 mi) above most of the surrounding clouds. The cloud base nearby was shown to be at least 50 km (30 mi) below the general cloud level, so deep that it could only be made of condensed water.

The Turbulent Northern Hemisphere—This false-color mosaic shows Jupiter's clouds between 10 and 50°N. The scene is created by alternating eastward and westward winds, together with regions of rising and falling air. Mingling with the parallel belts and zones are large white ovals, bright and dark spots, plumes, interacting vortices, and areas of chaotic turbulence. Features appear foreshortened toward the north (top). The images were taken on April 3, 1997 at a range of 1.4 million km (875,000 mi). The two white, interacting storms (center) each measured 3,500 km (2,190 mi) north–south. (Caltech)

Unfortunately, the chemical ingredients that gave the Spot its orange–red color—and resulted in the planet's overall gaudy appearance—remained a mystery. Was it due to phosphorus, or the presence of organic compounds resulting from the interaction between intense radiation, methane, and ammonia?

Galileo views of the night side also revealed the Jovian auroras (equivalent to Earth's Northern and Southern Lights) in great detail. The bright, oval arcs over the planet's magnetic poles are created by electrically charged particles striking the upper atmosphere from above. Galileo's cameras showed the northern auroral ring rotating with the planet. The illuminated ribbon was hundreds of kilometers wide and raised about 250 km (156 mi) above the planet.

Turmoil on Io

Galileo's first color image of Io was taken on June 25, 1996, at a distance of 2.2 million km (1.4 million mi). Subsequent long-range views confirmed that major changes had taken place on the volcanic moon since the Voyager flybys 17 years earlier. Over that period, about a dozen areas at least as large as the US state of Connecticut had been resurfaced.

However, it was not until October 1999 that scientists were prepared to risk the spacecraft by exposing it to the intense radiation in the vicinity of Io. Clearly visible were the volcanic plumes blasted outward from violent eruptions, and ever-varying patterns of colorful sulfurous deposits around major centers of activity. High-resolution images, 5–500 m (16–1,600 ft) per pixel, provided spectacular views of lava in many forms—lakes, flows, fire fountains, and curtains.

Near true-color images replaced the tomato ketchup coloration made famous in the Voyager images, proving that most of the sulfur-dioxide-spattered surface is yellow or light greenish. (Accurate natural color renditions were not possible from the Voyager images taken during the 1979 flybys because there was no coverage in the red.)

The few red areas may be associated with very recent volcanic "ash" deposits erupted in huge, umbrella-shaped plumes. An apparent fading over time indicated that this fresh material may be unstable. Dark patches represented very hot lava flows, hidden beneath a thin crust of solidified

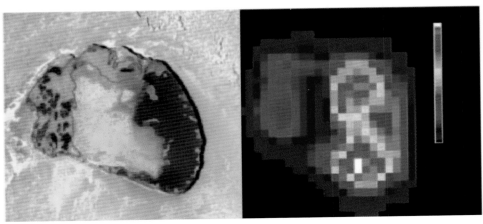

Io Hot Spots—Two views of Tupan Patera, an active volcano on Io. The visible light image (left) shows a depression, about 75 km (47 mi) across, surrounded by cliffs about 900 m (3,000 ft) tall. The floor of Tupan is covered with a complex pattern of dark black, green, red, and yellow materials. A thermal image obtained on October 16, 2001 (right), shows temperature variations within the caldera. The hottest areas (red) are probably lava lakes. The central region may be an island or a topographically high region. Parts of it are cold enough for sulfur dioxide to condense. Tupan was named after the Brazilian native god of thunder. (NASA–JPL)

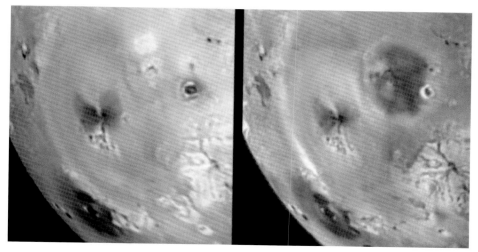

Io Metamorphosis—These Galileo images show dramatic changes on Io as the result of volcanic activity between April 4, 1997 (left) and September 19, 1997 (right). Most noticeable is a new dark spot, 400 km (249 mi) in diameter, which appeared around the Pillan Patera volcanic center. The pinkish deposits around Pele, southwest of Pillan, are also different, perhaps due to interaction between the two large plumes. (NASA–JPL)

rock and usually filling caldera floors. Bright white materials corresponded to sulfur dioxide frost, while bright yellow materials represented recent lava flows surrounding centers such as Ra Patera.

Galileo images revealed no fewer than 74 new hotspots, bringing the overall total to 120. This compares with 600 active volcanoes on the much larger Earth. However, instead of dome-shaped shields or pointed cones, most of Io's volcanoes were mere holes in the ground, huge calderas (depressions) filled with crusted-over lakes of molten lava.

The most famous of these is Loki, which boasts a lava-filled crater some 200 km (125 mi) across. Galileo images showed a gigantic iceberg of solidified sulfur dioxide floating on a dark crust that covered the viscous lake of molten rock. Along one shoreline was a ribbon of glowing lava that had broken through the overlying crust.

Curiously, despite the extensive volcanic activity, the giant plumes seen by Voyager were largely absent for most of the Galileo era—with the notable exception of Pele. However, four plumes erupted in the north polar region during the final 13 months of the mission.

Early observations by Galileo indicated that plumes associated with a volcanic vent known as Prometheus had shifted some 70 km to the west over 17 years. The mystery of the moving plumes was later solved when it was found that they were associated with a shifting lava flow. When the fresh lava broke through the overlying crust at a new location, it interacted with sulfur dioxide ice to create a huge, umbrella-like cloud.

On one occasion, the Orbiter made an accidental excursion through a swarm of microscopic particles blasted from a volcanic vent. The plume from Pillan Patera—a neighbor of giant Pele—soared to a height of 500 km (310 mi) and turned out to be the most powerful ever seen.

The first direct images of large-scale volcanism were captured on February 22, 2000, when Galileo's cameras imaged a 60-km (37-mi) curtain of fresh, hot silicate lava at the edge of a 200-km (125-mi) wide crater called Tvashtar Patera.

Voyager data had indicated the presence of low-temperature, sulfur-based volcanism, but Galileo's infrared images painted an entirely different picture. The temperature of the molten rock reached a sizzling 1,700°C, making it the hottest lava observed anywhere in the Solar System. This high-temperature fluid is able to travel long distances over the generally flat surface. The longest lava flow recorded by Galileo was at Amirani and stretched 300 km (187 mi) across the surrounding plains.

Galileo also imaged huge mountains (some more than twice the height of Everest) and plateaus, further evidence of Io's tectonically active crust. Nearly half of the mountains were located alongside the volcanic craters, implying a link between the two.

Instead of Earth's sideways-moving crustal plates, Io seemed to be dominated by vertical movements. One suggestion was that the process that drives mountain-building—perhaps the tilting of blocks of crust—makes it easier for magma to get to the surface.

In this scenario, liquid rock rises from the deep interior and spreads out over the surface as lava flows. Older lavas are continuously buried and compressed until they break, thrusting tall mountains upward and opening up new pathways for the lava to follow.

One example of this tectonic connection was Tohil Mons, a crescent-shaped mountain 6 km (4 mi) high, that bordered a caldera. The rugged ridge had apparently been uplifted along large faults in the crust. Although there was evidence of major landslips on the steep mountain slopes, the debris seemed to have disappeared in a lake of fresh lava.

Galileo confirmed that Io has a dense core of iron and nickel, but it also showed that the moon does not generate a magnetic field. At the mercy of Jupiter's radiation belts and intense magnetic field, Io feeds charged particles from its thin upper atmosphere toward Jupiter along a narrow, doughnut-shaped "flux tube," known as the Io torus. As the moon swims through the planet's magnetosphere, a current of 5 million amps flows along this invisible conduit.

Europa's Hidden Ocean

Galileo did not disappoint in revealing fiery Io's extreme reaction to tidal pumping, but scientists were also eager to examine its more subtle effects on icy Europa. The Voyagers had provided tantalizing glimpses of a smooth white ball criss-crossed by a tangle of intersecting dark fracture lines. The absence of large impact craters suggested a young surface that was continually being resurfaced. Was Europa covered by an ocean of water hidden beneath the Jovian equivalent of the Earth's Arctic ice floes?

One important clue to answering the puzzle came from measurements of Europa's gravity field during several close flybys by Galileo. These showed that the satellite was probably coated with a layer of water 100–200 km (62–125 mi) thick, but was this in the form of ice, slush, or liquid?

Meanwhile, spectroscopic measurements showed evidence of "salts"—perhaps magnesium sulfate—on top of cleaner ice. These were particularly common on Europa's trailing hemisphere, and probably originated in salty water that seeped to the surface and then froze.

Close-up images taken by Galileo revealed a mass of strange iceforms that provided intriguing circumstantial evidence for a subsurface liquid layer. The interwoven pattern of fractures, ridges, grooves and spots all pointed to a world that was continually being modified by internal forces.

Many of the freshest fractures seemed to have been formed when tidal pumping squeezed the surface until it cracked. Analysis indicated that the stress pattern had swept around the satellite over time. Perhaps Europa's surface had rotated faster than the interior.

Close-ups of the gray grooves showed paired ridges separated by a narrow valley. It seemed likely that liquid water or warm ice had risen to the surface along deep fractures. Where the rising material forced the surrounding ice upward, the surface may have been warped, creating a double ridge on either flank. The widest grooves appeared to be multiple versions of these parallel bands, caused by repeated upwellings.

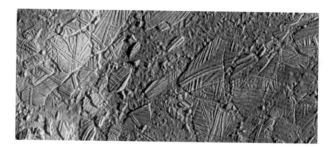

Ice Rafts on Europa—This view of Europa's Conamara region shows a disrupted ice crust that resembles ice floes in the Arctic Ocean on Earth. The white and blue colors outline areas that have been blanketed by fine ice particles ejected during the formation of the 26-km (16-mi) crater Pwyll, some 1,000 km (625 mi) to the south. A few impact craters, less than 500 m (1,650 ft) in diameter, can be seen. The reddish brown area has been coated with minerals spread by water vapor escaping from below the crust. The image covers an area approximately 70 × 30 km (44 × 19 mi) and has a spatial resolution of 54 m (177 ft). (NASA–JPL)

More difficult to explain were scattered circular or elliptical features, typically 50–100 km (30–60 mi) across, that created a mysterious mottled appearance. (Scientists called them "lenticulae," Latin for freckles.) Many of the markings were found to be domes, but others were small hollows; some were smooth, others were rough.

Scientists believe that this menagerie of markings can be accounted for by large masses of rising "magma." The surfaces of the domes seem to be made of old ice that has simply been pushed up from below. Their formation may be compared to the behavior of a lava lamp, with blobs of warm ice moving up through the colder, near-surface ice. Rough terrain resulted if the blobs disrupted the surface. Smooth, dark patches were created if meltwater spread onto the surface and then quickly froze.

Some large, dark patches form chaotic terrain, as in the Conamara region of Europa where individual blocks of ridged ice left over from older plains were found to be embedded in younger ice. Some of the blocks appeared to be tilted or partially rotated as if they were icebergs floating in a frozen ocean. Once again, the story is of a warm slush or liquid welling up and modifying the existing surface.

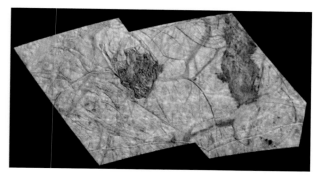

Spots and Ridges—Thera and Thrace are two dark, reddish "chaos" regions that may be created by complete melt-through of Europa's icy shell or upwelling of warm ice from below. Thera (left) measures about 70 × 85 km (43 × 53 mi) and appears to lie slightly below the surrounding plains. The curved fractures along its boundaries suggest that collapse may have been involved in its formation. Thrace (right) is longer, shows a hummocky texture, and appears to stand at or slightly above the bright plains. Running along its southern flank is a gray band known as Libya Linea. (PIRL/Univ. of Arizona)

Galileo also confirmed that Europa is almost devoid of large or small impact craters. Even the two largest impact features (Callanish and Tyre) appeared as smooth patches surrounded by multiple rings and numerous secondary impact craters, rather than dramatic basins flanked by towering mountains. The rings followed fractures created in the ice when the central crater collapsed.

Of 28 sizeable impact structures on Europa imaged by the Galileo and Voyager spacecraft, at least six are large enough have central peaks. The largest crater of this type, known as Pwyll, measures 26 km (16 mi) across. Although its floor is level with the surroundings, the complex central peak reaches a height of 600 m (1,980 ft), well above the crater's rim. Apart from a dark central region, Pwyll is surrounded by brilliant white rays of ejected debris that extend outward for hundreds of kilometers.

Such structures place important constraints on the nature of Europa's crust. Clearly, the impacts that created them did not completely melt through the surface ice layer or vaporize the moon's surface. If a layer of warm, convecting ice existed immediately beneath Pwyll's peak, the formation would have disappeared very quickly.

The clinching evidence came from an unexpected source. As it flew close to Europa, Galileo discovered marked deviations in the magnetic environment. The only conclusion was that Europa generates a magnetic field. However, the rapid and systematic variations could not be explained by motions within a liquid iron core. The most reasonable explanation seemed to be an electrically conducting fluid such as salty water. If the saltiness was equal to that of Earth's oceans, then a depth of at least 10 km (6 mi) would be required.

Scientists are still debating the thickness of the icy crust; estimates range between 10 and 40 km (6 and 25 mi). However, comparisons of Voyager and Galileo images have shown no obvious changes in surface appearance, suggesting that current geological activity is very modest and slow.

Meanwhile, exobiologists speculate about the existence of alien life in an ocean that has a greater volume than all the oceans on Earth.

Ganymede and Callisto Revisited

One of Galileo's most significant discoveries was made during its very first orbit around Jupiter. About half an hour before closest approach to Ganymede, the spacecraft began to pick up complex radio signals and the magnetometer readings shot up five-fold. Detailed observations on subsequent flybys confirmed suspicions that Ganymede must have a magnetic field—the first ever associated with a planetary moon. In fact, Ganymede's field was found to be stronger than that of Mercury.

At the same time, precise tracking of the spacecraft's trajectory allowed scientists to probe Ganymede's gravity field and learn more about its interior. Analysis showed that the satellite probably has a thick ice shell above a deep rock- ice mantle and a dense, iron-rich core.

At first, it was believed that currents in the molten inner core were the most likely source for the moon's magnetism,

although models suggest that this liquid core should have cooled within one billion years, bringing the internal dynamo to a grinding halt. By 2000, more detailed analysis suggested that Ganymede's magnetism probably came from a stable layer of liquid water between two layers of ice, about 150–200 km (90–120 mi) underground.

Up above, Galileo images of immense faults and fractures painted a picture of a youthful moon blessed with a hot interior and mobile crust. The high-resolution views revealed Ganymede's major grooves to be an intricate web of criss-crossing striations peppered with small impact craters. One relatively smooth, bright band, known as Arbela Sulcus, was seen to cross over an older, more cratered, landscape—evidence of a complete separation of Ganymede's icy crust, just like the bands on Europa.

Stereo images indicated that the warmer ice a few kilometers down had been stretched and pulled apart, causing a mass of parallel faults to develop at the surface. This, in turn, created numerous small rifts and ridges where the icy crust slipped.

Although impact craters were seen to be much more numerous than on Europa, the larger structures (termed "palimpsests," after reused parchment from which previous writing has been partially erased) were very similar, often taking the form of low, bright patches with very little surface relief. It seems that ancient impacts punched through Ganymede's crust, enabling warm ice to fill the initial basins.

One of the first areas imaged by Galileo was the dark, northern "eye" known as Galileo Regio. The stunning views revealed a patchwork made up of bright uplands—ridges, hollows, and isolated hills—separated by darker craters and furrows. After losing most of their ice, slopes facing the Sun appeared to be coated with a dark, rocky, clay-like residue that slowly crept downhill into the hollows and valleys. Scientists speculated that walking on Galileo Regio might be comparable to crunching across frozen mud.

One of the greatest surprises came after Galileo's death plunge into Jupiter's atmosphere. Scientists discovered the gravity signatures of irregular lumps beneath Ganymede's

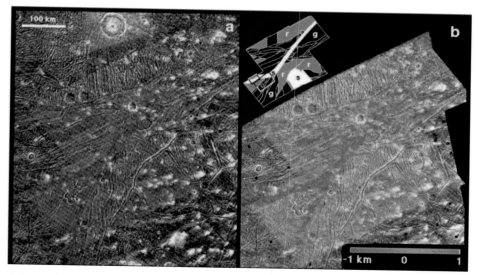

Sippar Sulcus on Ganymede—These views show different terrains within an area called Sippar Sulcus in Ganymede's southern hemisphere. (a) is a Galileo mosaic superimposed on less detailed Voyager images. A swath of grooved terrain runs from upper right to center left. Irregularly shaped hollows are thought to be calderas, perhaps caused by subsurface collapse. The numerous bright patches are secondary impacts associated with a large crater, Osiris (out of view to the right). (b) is a digital elevation model of the same scene. Highest areas are purple and red. The inset is a geological map highlighting areas of grooved terrain (g, black), reticulate terrain (r, gray), smooth terrain (s, white), and calderas (hatched). (LPI/NASA–JPL)

icy surface. Since there was no surface evidence of their existence, scientists could only speculate that they comprised large concentrations of rock within or beneath the icy crust.

Ganymede and Callisto probably formed under very similar conditions, so it is not surprising that they have very similar sizes and densities. Furthermore, about half of Ganymede's surface is heavily cratered, just like Callisto.

Voyager images indicated that pockmarked Callisto had hardly changed over billions of years, but this bleak portrait had to be modified after Galileo's close-up pictures failed to find the expected number of small impact features. Callisto seems to be undergoing some type of slow, glacial resurfacing which continually obliterates the craters. Indeed, the new views showed that the entire satellite seems to be coated with a blanket of fine, dark debris of uncertain

origin. Once again, downslope "soil" creep appears to be all-pervasive.

In other detailed images, Galileo found unusual, kilometer-sized "pits" and small craters that were not entirely circular. One possible explanation for the pits was that they were caused by slumping after most of the ice was vaporized by the feeble Sun. Bright frosts on Sun-facing slopes were also a challenge to explain.

Callisto's internal structure also provided quite a shock. Although the interior was probably undifferentiated, with no metallic core, the satellite's magnetic signature was found to indicate the presence of a saltwater ocean—perhaps only 10 km (6 mi) deep—beneath an icy shell, 200 km (125 mi) thick. It was the third subsurface ocean to be found on a Galilean moon.

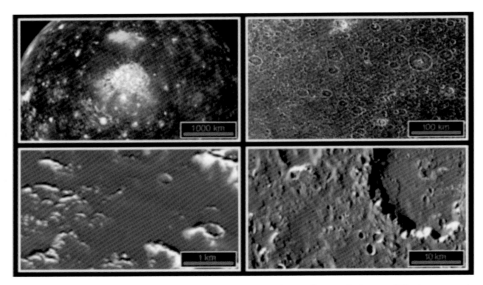

Callisto Quartet—These four Galileo views of Callisto show how increasing resolution modifies interpretation of the surface. In the global view (top left), many small bright spots are seen around Valhalla, the largest impact structure on Callisto. The regional view (top right) reveals them to be large craters with bright rims. The local view (bottom right) brings out smaller craters and shows bright material on some crater rims amid a smooth, dark layer of material that covers much of the surface. In the close-up frame (bottom left), where spatial resolution is 30 m (100 ft), the surface is surprisingly smooth between the craters. (DLR)

Callisto's magnetic field, like Europa's, was associated with electrical currents that flowed near the satellite's surface and fluctuated in time with Jupiter's rotation. However, in the absence of tidal flexing, the sole heat source available to maintain the ocean was radioactive elements in the interior.

Moons and Rings

One of the highlights of the Voyager missions was the discovery of a faint, dark ring system. The Voyagers observed three main components: a bright main ring, an inner halo, and a possible faint outer ring.

Galileo not only confirmed the existence of the elusive outer region, but revealed that, in fact, the so-called "gossamer" ring comprises two distinct parts. Most of the material that makes up the rings was shown to originate from the four inner moons, Adrastea, Metis, Amalthea, and Thebe.

The denser main ring, which lies inside the orbits of Adrastea and Metis, is fed by debris blasted from their surfaces during collisions with small meteoroids (fragments of asteroids and comets). The main ring itself is very thin, since both satellites orbit close to the planet's equatorial plane.*

---

* The orbits of Adrastea and Metis are only about 1,000 km (625 mi) apart.

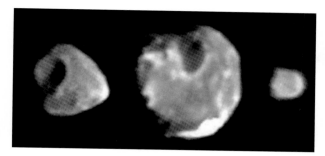

Thebe, Amalthea, and Metis—These Galileo images, taken in January 2000, are the highest resolution images ever obtained of the inner satellites Thebe, Amalthea, and Metis (left to right). The moons are shown to scale. The prominent impact crater on Thebe is about 40 km (25 mi) across and is known as Zethus. The large white region near the south pole of Amalthea sits inside a large crater named Gaea. It marks the location of the brightest patch of surface material seen anywhere on these three moons. (NASA–JPL)

Thebe and Amalthea, the next two satellites out from Jupiter, supply the fine dust that forms the deeper "gossamer" rings. The inner part of the gossamer rings, fed by Amalthea, is rather more dense and embedded within the sparse outer segment supplied by Thebe. The gossamer rings have much greater depth than the main ring since the two source satellites orbit Jupiter on inclined paths which enables them to spread the material into a broad band.

The spacecraft also detected sparse clouds of dust around each of the four large Galilean moons. There was even a hint of a dust ring that trailed all the way round the orbit of Ganymede.

Galileo was able to detect a fair amount of surface detail on the four inner moons. They all turned out to be potato-shaped objects with a history of impact cratering. All were tidally locked so that the same side of the satellite always points toward Jupiter (in the same way that the same side of our Moon always points toward Earth).

Since these satellites are very small, their surface gravities are very low: a person weighing 68 kg (150 lb) on Earth would weigh about 0.45 kg (1 lb) on Amalthea, and about 28 g (1 oz) on Adrastea.

Battered Thebe had a maximum diameter of around

Gossamer Rings—Galileo discovered three main components of Jupiter's rings. (The inner halo is not shown.) The outer edge of the main ring (red) is patrolled by the satellites Metis (m) and Adrastea (a). It is formed mostly from debris created when small meteoroids collide with the two innermost moons. One faint "gossamer" ring (orange) ends abruptly at the orbital distance of Amalthea (A). An even fainter ring (green) extends out to the orbit of Thebe (T). The gossamer rings are much broader because Thebe and Amalthea spread out the dusty impact debris as they travel above and below Jupiter's equator. (NASA/Cornell University)

116 km (72 mi), and revealed three or four very large impact craters. Metis was notably elongated, with a maximum width of 60 km (37 mi). Smallest of all was Adrastea, a mere 20 km (12 mi) across.

Amalthea, the largest of the inner quartet, measured about 270 km (168 miles) in length and half that in width. Galileo imaged two large craters and many smaller neighbors. Pan, the larger crater, was located in the "northern hemisphere." It measured 90 km (55 mi) long and at least 8 km (5 mi) deep. The other large crater, Gaea, was 75 km (45 mi) long, even deeper than Pan, and straddled the south pole of the moon. Also visible was a bright patch inside Gaea and a white streak that was 50 km (30 mi) long and possibly the crest of a ridge.

The moon's internal make up proved to be something of a surprise. During Galileo's final flyby in early November 2002, scientists studied Amalthea's gravitational effect and found that the red-tinted satellite has a very low density, indicating that it is full of holes.

Although its overall density is close to that of water ice, 'fluffy' Amalthea is probably a loosely packed pile of rubble—mostly rock with only a little ice—rather than a solid chunk of ice. As is the case with a number of other irregularly shaped moons and asteroids (e.g. Mathilde), the empty gaps take up more of the overall volume than the solid, boulder-sized chunks.

Amalthea's irregular shape and low density suggest that the moon has been broken into many loose-fitting pieces that are held together solely by the pull of each other's gravity.

Another serendipitous discovery that resulted from the November 2002 encounter was the discovery of seven to nine small pieces of debris close to Amalthea. (Two might possibly have been duplicate sightings.) The discovery was made by Galileo's star scanner, a telescope used to determine the spacecraft's orientation by sighting stars.

The objects appeared as bright flashes and scientists declared that they could be anywhere from gravel to stadium-size. Their origin was uncertain. They could have been gravitationally captured into an orbit near Amalthea or they might be fragments that had broken away from the moon during past collisions.

Multiple Viewpoints

Although most of what we know about the realm of giant Jupiter has come from the in-depth exploration by the Voyagers and Galileo, scientists have also been able to call upon a wide range of instruments and spacecraft to further enhance their understanding.

Regular monitoring by ground-based telescopes (including the increasingly capable amateur instruments equipped with CCD detectors) and the Hubble Space Telescope has meant that almost every cloud formation and storm has come under intense scrutiny. However, one of the most significant advances has been the discovery of dozens of small, irregular moons since 1997. Most of these were found by Scott Sheppard and David Jewitt (University of Hawaii) and a Canadian team of Brett Gladman, Lynne Allen and J. J. Kavelaars, with the aid of sensitive CCDs that could search for these captured objects in the sky around Jupiter. By mid-2005, the count had risen to 63.

One of the most surprising findings came from observations made in the summer of 1999, when Jupiter's north pole was most visible from Earth. Images taken by the Infrared Telescope Facility and Hubble revealed a hexagonal-shaped vortex that extends like a tunnel deep into the atmosphere. (The Voyagers found a similar cloud structure at Saturn's north pole.) The lower temperature of the air within the vortex created an eastward wind that isolated it from the rest of the atmosphere. The system rotated slowly eastward at 1.2 degrees of longitude per day, and was capped by a layer of thin haze.

The Hubble Space Telescope has frequently turned its attention to monitoring changes in cloud formations, particularly the Great Red Spot. This proved particularly valuable when pieces of comet Shoemaker–Levy 9 punched into Jupiter's atmosphere in July 1994, leaving a series of dark "bruises" that survived for many months (see Chapter 12).

Hubble observations of the planet's ever-changing auroras have also allowed scientists to study the interaction

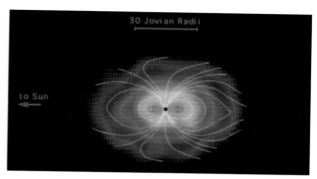

Jupiter's Magnetic Bubble—Jupiter's magnetosphere takes the form of a long tail on Jupiter's night side that stretches all the way to the orbit of Saturn. This image, taken with Cassini's ion and neutral camera, shows the normally invisible swarm of charged particles that whirl around Jupiter. Also added are: Jupiter (black circle), magnetic field lines (white), and an outline of the Io torus—a doughnut-shaped ring of charged particles that spread around the moon's orbit. The picture was taken shortly after Cassini's closest approach to Jupiter, from a range of about 10 million km (6 million mi) on December 30, 2000. (NASA/JPL/JHU Applied Physics Laboratory)

Auroral Rings—These Hubble Space Telescope views, taken on September 20, 1997, show Jupiter's auroras glowing in ultraviolet light. Apart from the oval auroral "curtains" extending several hundred kilometers above Jupiter's limb, the inset views show a bright streak of light around each outer rim. These curved streaks mark the footprints of energetic particles flowing from Io and striking the upper atmosphere. (John Clarke, University of Michigan/NASA)

between Jupiter's magnetic field, the solar wind, and particles spewing from Io's volcanoes. The first direct images of the aurora taken in ultraviolet light were obtained by Hubble's Faint Object Camera in 1992, showing warped oval rings at the north and south poles (offset from Jupiter's spin axis by 10–15 degrees). Nearby was the auroral "footprint" created by 1 million amps of electrical current flowing between Jupiter and the volcanic moon Io.

If you were flying under Io's footprint at Jupiter's cloud tops, the aurora would fill the entire sky. You would see an explosion as the gases 400 km (250 mi) above rapidly became as hot as the surface of the Sun. The aurora would

speed overhead from east to west at 5 km/s (10,000 mph) as Jupiter's rotation moved it rapidly beneath Io.

Further insights into the dynamic environment of charged particles and magnetic fields came with the Cassini spacecraft's flyby of Jupiter in December 2000. Scientists were able to take advantage of a double-take by Cassini, which was outside the magnetosphere monitoring the solar wind, and Galileo, which was much closer to the planet and inside the magnetosphere. Other simultaneous observations were made with the Chandra X-ray Observatory and ground-based radio telescopes.

The first "stereo" observations showed how Jupiter's magnetic bubble inflated and deflated, changing shape as it responded to the pressure of the solar wind. The joint data showed that shock waves carried from the Sun by the solar wind stimulated radio emissions from deep within Jupiter's magnetosphere and brightened auroras at Jupiter's poles. Electron density and electric currents in the magnetosphere clearly increased when it was compacted by the shock waves.

Cassini also carried a type of magnetosphere-imaging instrument never before flown on an interplanetary spacecraft. The instrument not only showed some structural detail of Jupiter's huge magnetosphere, it also detected a cloud of neutral atoms escaping from the planet's magnetic domain.

The joint campaign also searched for particles in the intense radiation belts close to Jupiter and detected extremely energetic electrons traveling near the speed of light close to Jupiter.

Jupiter's immense gravity has so far been exploited by five spacecraft needing an extra boost en route to the outer Solar System. On each occasion, scientists have taken the opportunity to learn more about the largest planet and its environment. This sequence is set to continue in coming years, with flybys involving missions such as Pluto New Horizons.

Perhaps the most unusual gravity assist was given to Ulysses on February 8, 1992, when the ESA–NASA nuclear-powered spacecraft swept over Jupiter's north pole and used the planet's gravity to plunge southward. In this way, Ulysses was able to journey where no spacecraft had gone before—over the poles of the Sun. Twelve years later, it returned to Jupiter's vicinity, this time remaining outside the Jovian magnetosphere.

Although Ulysses carried no camera, it did carry instruments to study the solar wind and a dust detector identical to that on Galileo. This enabled Ulysses to detect 11 streams of smoke-sized dust that were traveling away from Jupiter at speeds of up to 20 km/s (45,000 mph). It also dived through Io's doughnut-shaped torus and sent back new information on the radiation belts and magnetic field. The density of material in the torus was less than expected, suggesting that Io's volcanoes were temporarily less active.

Apart from occasional fleeting flybys, no further missions are expected to visit giant Jupiter for at least 10 years. Both NASA and ESA are considering preliminary concepts to send orbiters to explore the planet, but, meanwhile, Jovian observations will largely be restricted to ground-based instruments and Earth-orbiting observatories, such as the Hubble Space Telescope.

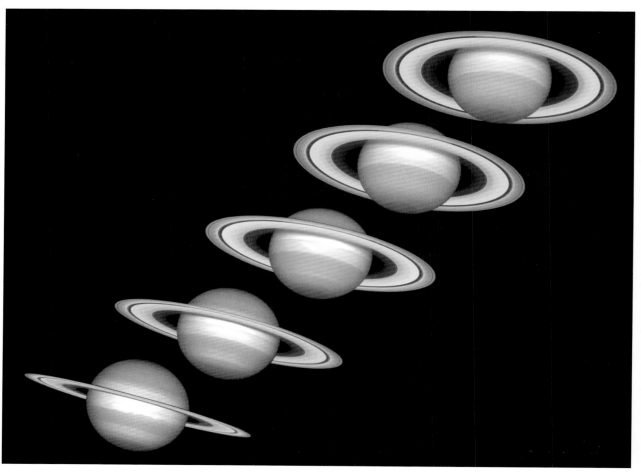

Ringing the Changes—In the *Systema Saturnium* of 1659, Christiaan Huygens showed how a tilted ring explains the changing appearance of Saturn. As Saturn moves around the Sun, different ring phases are visible from Earth. These Hubble Space Telescope images, taken between 1996 (bottom) and 2000 (top), show Saturn's rings opening up as the southern hemisphere moves toward its summer. (NASA and the Hubble Heritage Team, STScI/AURA)

8 SATURN: LIGHTWEIGHT LORD OF THE RINGS

The Planet with Handles

Lurking at the outer edge of the Solar System known to the ancients was the planet Saturn. Its slow motion through the constellations of the zodiac and modest brightness clearly indicated that the pale yellow "wandering star" was further away than any of its fellows.

The oldest written records to mention the planet date back to the Assyrians, around 700 BC. A few centuries later, the ancient Greeks named it in honor of Kronos, their god of agriculture and the father of Zeus. (The Roman equivalent was Saturnus.) This turned out to be very appropriate, since Saturn was later found to be a giant, second only to Jupiter among the Sun's family.

Without the benefit of optical aids, early astronomers were restricted to studying its movement against the "fixed" stars. Every 13 months, or so, they would see Saturn brighten as it reached opposition—when it was directly opposite the Sun in our sky and at its closest to Earth. With modern measurements of the Solar System, we now know that—even at opposition—Saturn remains 1.2 billion km (750 million mi), or eight Sun–Earth distances, away.

For the next millennium, our knowledge of Saturn changed very little. Then, came the invention of the telescope, and, almost overnight, Saturn became a worthy object of study. As was the case with many of the other planets, it was the Italian genius, Galileo Galilei, who first took advantage of the new instrument to observe Saturn.

In 1609, using a home-made telescope that could magnify images by a mere 20 times, Galileo noticed that Saturn seemed to be sandwiched between two smaller planets. To his amazement, when he returned to the scene in 1611, the two smaller planets had vanished and Saturn was now in splendid isolation. He eventually suggested that Saturn must have had arms or "cup handles" that mysteriously grew and disappeared periodically.

Although others tried to solve the mystery, confusion reigned for more than 40 years, until a Dutch astronomer named Christiaan Huygens found the solution with the aid of an improved telescope that could magnify 50 times, and a great leap of imagination.

In a brief paper entitled *De Saturni luna observatio nova* (New observation of a moon of Saturn), Huygens announced that he had discovered a satellite (now known as Titan). He also noted that he had found an explanation for Saturn's *ansae* or "handles." Curiously, he decided to disguise the answer in the form of an anagram, and invited others who thought they knew the answer to the mystery to come forward.*

Three years later, with no one any closer to unraveling the solution, Huygens revealed the answer in his book, *Systema Saturnium* (The Saturn System). The mystery apparition was a thin, solid ring that circled above Saturn's equator. Since the planet was inclined by about 30 degrees to the plane of its orbit, the ring, too, was tilted by the same amount.

---

* The solution to the anagram was: *Annulo cingitur, tenui, plano, nusquam cohaerente Dissertatio de natura Saturna, ad eclipticam inclinato* (It is surrounded by a thin flat ring, nowhere touching, and inclined to the ecliptic).

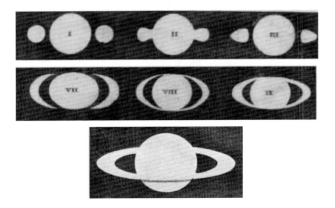

Ring Drawings—The six drawings of Saturn (top and center) show the changing appearance of the "appendages" that so puzzled early telescopic observers, with Galileo's drawing at top left (I). The existence of a ring system was first noted as a complex anagram by Christiaan Huygens in 1656. Not until the publication of his book, *Systema Saturnium*, in 1659, was the ring openly explained for all to see (bottom).

This fixed inclination meant that the ring's appearance changed as Saturn orbited the Sun once every 29½ years. When the ring appeared edge-on to the Earth, during Saturn's spring and autumn, it seemed to virtually disappear. Then, as the seasons changed to summer or winter, the planet's poles would be tipped toward the Sun and Earth. At such times the ring appeared in all its glory. Then the entire cycle would operate in reverse until the disk was viewed edge-on once more.

As is often the case with revolutionary ideas, Huygens' ring theory initially received a critical reception. A decade passed before the existence of a ring was widely accepted. But was it a solid ring as Huygens proposed?

The next important breakthrough was provided by Giovanni Domenico Cassini, a renowned Italian astronomer who, at the invitation of King Louis XIV, had taken up the position of director of the Paris Observatory. By the time he became a naturalized Frenchman in 1673, Cassini had already discovered two more satellites of Saturn—eventually named Iapetus and Rhea.

However, his key contribution to improving our understanding of the Saturnian system came in 1675, when he observed the first evidence of structure in the supposedly solid ring. While studying the wide open ring, he noticed a continuous dark region, about two-thirds of the way out from Saturn, which he correctly surmised to be a gap in the supposedly solid structure. The "empty" gap—now known as the Cassini Division—clearly subdivided the ring into an outer A ring and an inner B ring.

Contradicting almost all of his contemporaries, Cassini also stated his belief that the ring was actually composed of a large number of small satellites orbiting Saturn. However, another 82 years were to pass before the truth of his theory was finally accepted by the scientific world.

The Titans

When Christiaan Huygens initially observed Saturn, his attention was drawn away from studies of the planet's odd shape by a bright star which lined up with Saturn and its "handles." After some weeks of observing this "star" as it moved around Saturn, he came to the inescapable conclusion that it must be a moon in orbit around the planet. It was only the sixth planetary satellite ever found.

The discovery of Saturn's first satellite was announced in 1656. Three years later, Huygens provided further details about the newcomer in the *Systema Saturnium*. Just as the four moons of Jupiter had persuaded Galileo to reject the old ideas, so Huygens saw Saturn's satellite as further proof that the Copernican system of a heliocentric (Sun-centered) Universe was correct.

His painstaking observations also enabled him to calculate that the satellite's orbital period around Saturn was 15 days, 23 hours, 13 minutes (not far from the modern value of 15 days, 22 hours, 41 minutes). Although notably fainter than Jupiter's four large companions, when its great distance was accounted for, Titan was obviously of comparable size.

Huygens' discovery of giant Titan was followed in quick

Christiaan Huygens—Dutchman Christiaan Huygens (1629–1695) developed advanced pendulum clocks, invented improved eyepieces, and constructed "air" telescopes up to 75 m (250 ft) in length. With these he discovered Jupiter's equatorial bulge, the Martian polar caps, and a dark surface feature on Mars, later named Syrtis Major. He discovered Saturn's satellite, Titan, on March 25, 1655, and was the first to recognize that Saturn was surrounded by a ring. He was a member of the Royal Society in Britain, and became a founding member of the Academie Royale des Sciences in Paris, where he worked with Giovanni Domenico Cassini. Fearing persecution as a protestant, he returned to the Netherlands in 1686. (ESA)

succession by further satellite observations. In 1671, Giovanni Domenico Cassini noted the existence of a strange satellite (now known as Iapetus) that appeared bright on one side of Saturn and much dimmer on the other side. Using impeccable logic, the great observer deduced that the moon must be two-faced, with a lighter, trailing (rearward-facing) hemisphere abutting a much darker leading side.

A year later, the Director of the Paris Observatory (now known to his French colleagues as Jean Dominique Cassini) found a moon—later called Rhea—inside the orbit of Titan. For some time his attention turned to other matters, notably the nature of the ring system, but two more satellites (Dione and Tethys) were detected in 1684. The ringed planet already had more known moons than Jupiter, its larger, more illustrious neighbor.

Once again, the question of what to call the moons surfaced. Following Galileo's precedent, a grateful Cassini wanted to name them after his patron, Louis XIV, but this was hardly a practical proposition bearing in mind the opposition of rival nations. Most astronomers continued to refer to them by number.

However, when two more satellites of Saturn (Enceladus and Mimas) were found by William Herschel in August and September 1789, confusion reigned. The naming of Saturn's moons was becoming a priority. In 1847, Sir John Herschel, William's son, suggested that they should be named after mythological figures associated with Saturn. When, the following year, William Lassell and William Bond independently discovered Saturn's eighth satellite (Hyperion), it was generally agreed to accept Herschel's system. The satellites were named after the Titans, the mythological family of Greek gods led by Kronos (Saturn).

The ninth satellite, Phoebe, was found in 1898 by William Pickering during observations at the Harvard Observatory's outpost in Peru. Phoebe was the black sheep of the family, since it followed a retrograde (backward) path at an average distance of 13 million km (8.1 million mi) from the planet.

While Saturn's family grew, astronomers continued to debate the nature of the rings. In 1785, Pierre Simon, Marquis de Laplace, mathematically demonstrated the instability of two broad, solid rings orbiting Saturn. His alternative theory favored numerous solid rings separated by small gaps.

Although William Herschel was able to show that the rings could be no more than 500 km (310 mi) thick, Laplace's suggestion remained in favor until American professionals William and George Bond, and English

Saturn's Largest Moons—This montage shows the relative sizes of the nine satellites of Saturn known before the Space Age (left to right: Mimas, Enceladus, Tethys, Dione, Rhea, Titan, Hyperion, Iapetus, and Phoebe). They are also shown in order of distance from Saturn, with the closest (Mimas) to the left. Most of them are between 400 and 1,500 km (250 and 935 mi) across and have very low bulk densities, evidence that they are mostly composed of water and other ices. Orange, haze-covered Titan, similar in size and density to Jupiter's largest satellites, Ganymede and Callisto, is the sole exception. (Paul Schenk, Lunar and Planetary Institute)

amateur, the Reverend William Rutter Dawes, independently discovered a third, inner ring in 1850. The faint C or "Crepe" ring extended inward from the B ring about half way to Saturn's cloud tops. The planet could clearly be seen through the dusky, diaphanous feature.

Meanwhile, calculations by French mathematician Edouard Roche had shown that a moon would be torn asunder by tidal forces if it dared to approach too close to Saturn. The distance at which this would occur—the so-called Roche Limit—more or less corresponded to the position of the rings. It was looking more and more as if the rings were made of millions of fragments from a disrupted moon.

Another breakthrough came in 1857, when a young Scottish physicist named James Clerk Maxwell mathematically proved that a very thin, solid ring was unstable and would be destroyed by gravitational forces. His prize-winning paper noted that the only possible explanation was that the rings were composed of small particles in separate orbits around the planet. The inner ring particles must travel faster than those further out.

Finally, in 1895, Maxwell's theory was proved by James Keeler at the Allegheny Observatory in Pittsburgh. Keeler used a spectroscope to show that the rings were, indeed, rotating around Saturn in the manner described by Maxwell. (Keeler and German astronomer Johann Encke

also observed a sizeable gap, now known as the Encke Division, near the outer edge of the A ring. Nearby is another narrow division, now named after Keeler, even though he probably never saw it!)

Observations of the edge-on rings during the twentieth century made it clear that the disk was no more than 15 km (9 mi) thick. Other gaps were also found, although they were so narrow that they were impossible to capture on long photographic exposures. Astronomers began to suspect that these mysterious empty regions were swept clear by Saturn's satellites.

The Sixth Planet

Although Saturn is the second largest planet in the Solar System, at first glance it seems to bear little resemblance to its even bigger cousin, Jupiter. Instead of turbulent, colorful ribbons of cloud and huge storm systems, Saturn appears as a fairly bland, tangerine disk—a muted version of magnificent Jupiter, marked only by pale bands and occasional spots. As anyone looking at the giant world through a telescope for the first time will attest, the abiding glory and mystery of its rings inescapably draws the eye away from the hub of the system.

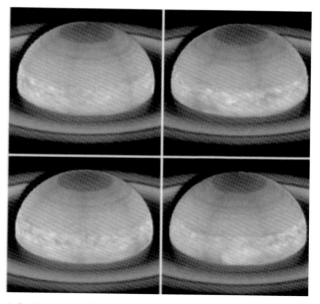

Saturn and Earth—Saturn is the second largest planet in the Solar System, with an equatorial diameter of 120,000 km (74,500 mi). If the planet and its main ring system (rings A to G) were placed between Earth and Moon, there would not be much room to spare. The Earth–Moon distance is 384,400 km (238,900 mi) while the overall width of the main ring system is 374,000 km (233,750 mi). 764 Earths would fit inside Saturn, but as the gas giant's bulk density is well below that of water, it weighs in at only 95 Earth masses. (NASA)

A Sudden Storm—Saturn's fairly bland appearance is periodically altered by bright, equatorial storms that are generated by large-scale upwelling of warmer gas. These Hubble Space Telescope images of a major disturbance were captured during Saturn's northern summer. They were taken on November 17, 1990, two months after it was first noticed by amateur astronomers. The storm eventually extended all the way around the planet. Such storms are rare: the previous major event along Saturn's equator occurred in 1933. (STScI/NASA)

Nevertheless, ground-based observers were able to determine the basic facts about the sixth planet from the Sun. Until the discovery of Uranus in 1781, it was regarded as the outpost of the Solar System, with an average distance from the Sun of about 1.4 billion km (886 million mi)—equivalent to almost 10 AU. Following a great circle around our star, Saturn swept along at a modest 10 km/s (21,600 mph), taking 29½ years to complete each lap.

Despite its relative remoteness, Saturn revealed a sizeable disk—clear evidence that it was a gigantic world. In fact, with a diameter more than nine times that of our world, more than 760 Earths would sit inside its gaseous sphere.

In terms of size, the ring system is even more impressive. Saturn and its main rings would barely fit into the void between Earth and the Moon. A tape measure spread across the entire system would stretch for 374,000 km (233,750 mi).

Saturn turned out to be rather less impressive in terms of mass, only 95 times heavier than Earth. Clearly, Saturn is not a very substantial body, with an average density 0.7 times that of water—in an ocean universe, Saturn would find no difficulty in floating.

This combination of great size and low mass means that a human standing on the visible surface of Saturn would actually weigh less than he or she does on Earth—a purely hypothetical case, of course!

One clue that helps to explain the curiously low density is the non-spherical shape of the planet. Telescopic observers could not fail to notice that Saturn looks as if it has been squashed: the poles are markedly flattened while the equatorial region bulges outward. In fact, the diameter across the equator is almost 13,000 km (8,125 mi) greater than the polar diameter—a difference of 11%. Although many planets, including Earth, display a bulging midriff to some extent, such oblateness is much more pronounced in the case of Saturn.

This unusual shape is obviously related to the planet's rapid rotation, which has been recognized for more than two centuries, and its gaseous composition. By observing occasional long-lived cloud features, William Herschel calculated that a day on Saturn lasts 10 hours 16 minutes—only two minutes longer than the figure generally accepted today. This is a little longer than Jupiter's rotation period but noticeably shorter than the other seven planets of our Solar System.*

Until the Space Age, astronomers agreed that Saturn generally resembled a more somnolent version of Jupiter, presumably because it received less heat from the distant Sun. However, notable atmospheric disturbances were recorded in the equatorial region in 1876, 1903, 1933 and 1960—suggesting a periodic upwelling of warm air roughly every 30 years. The 1933 event grabbed the headlines when British amateur Will Hay, a well-known stage and screen comedian, noted a conspicuous white spot that quickly grew and spread toward the east. However, within a few months, the enormous outbreak faded and died. This association of storms with mid-summer in the northern hemisphere was confirmed when another equatorial eruption appeared in 1990. Over the next few weeks it spread all the way around the planet before eventually dissipating.

---

* The period of rotation was shown to about 20 minutes longer at higher latitudes the result of different wind speeds at cloud level. Similar variations were found on Jupiter.

The exact nature of Saturn remained uncertain for a long time, though it was thought to be largely gaseous with a fluid interior. A major breakthrough came with the introduction of the spectroscope, which led to the 1932 discovery of the gases methane and ammonia—simple compounds linking hydrogen with carbon and nitrogen. This led Yale University's Rupert Wildt to suggest that the youthful Saturn had formed by drawing in a huge amount of hydrogen, helium, and other gases from the primitive solar nebula. He envisaged an Earth-like, rocky core surrounded by a thick layer of water, ammonia, and methane ices, topped by a hydrogen-rich atmosphere. Subsequent theories suggested that the hydrogen atmosphere was compressed so much that it changed into metallic hydrogen at a depth of about 20,000 km (12,500 mi).

By the late 1970s, astronomers were also fairly sure that Saturn had its own internal heat source and a sizeable magnetic field. Meanwhile, infrared and radio observations indicated that the rings were extremely cold, around −190°C, and made of water ice particles no more than 1 m (3 ft) in diameter.

Despite these advances, it was clear that there was a limit to how much could be learned using ground-based instruments. A robotic explorer would have to be sent to seek out the answers.

Pioneer Paves the Way

By the summer of 1979, two state-of-the-art Voyager spacecraft were well on their way toward Saturn, but another, less sophisticated explorer was about to give scientists a taste of delights to come.

After its encounter with Jupiter, Pioneer 11 followed a looping trajectory that carried it high above the ecliptic plane. Five years later, after it had crossed from one side of the Solar System to the other, NASA's nuclear-powered spacecraft was ready to undertake humanity's first close reconnaissance of Saturn.

Pioneer 11 at Saturn—This was the best wide-angle view of Saturn returned by Pioneer 11. It was taken from a distance of 2.5 million km (1.5 million mi) on August 29, 1979, 58 hours before closest approach. The banded structure of the planet's clouds is clearly visible, as is the shadow of the rings. The rings were imaged from the shaded north side—an aspect never seen from Earth. Sunlight passing through the rings allowed an assessment of ring density and size of material to be made. The dense B ring appears dark, because it blocks sunlight. The sparser C and A rings appear much brighter, as does the Cassini Division. (NASA–Ames)

Although the Jupiter flyby had boosted Pioneer in the general direction of Saturn, there had been plenty of time to fine tune the trajectory. Several options were considered, including a dive directly through the Cassini Division, and a slightly less hazardous plunge through the D ring, midway between the C ring and the cloud tops. The other main alternative was a safer, more routine passage well outside the visible rings.

Despite the science team's preference for an "all or nothing" sweep past Saturn at the closest possible distance, Tom Young, NASA's Director of Planetary Programs, decided to play safe and go for the outer option. So it was that Pioneer 11 flew past Saturn on September 1, 1979, sweeping beyond the visible edge of the A ring before passing 21,000 km (13,000 mi) above the cloud tops and

then crossing the ring plane for a second time. By modern standards, the fuzzy images sent back were hardly spectacular, but they did enable scientists to learn a great deal about the strange Saturnian system.

Although the TV images had a higher resolution than any pictures taken from Earth, scientists were disappointed by the absence of detail on the bland, tangerine sphere. The most fascinating feature visible on the pale clouds was the black shadow of the rings, occasionally separated by bright slots where sunlight pierced the narrow gaps.

One of the most confusing aspects of the images was the unusual lighting conditions. At the time, the rings were almost edge-on to the Sun, with their southern side barely illuminated. However, Pioneer swooped down from above the ecliptic, and its TV camera saw the rings' shaded, northern face. This meant that Pioneer's pictures showed major variations in ring brightness as sunlight struggled to penetrate the flattened disk. Whereas the dense, closely packed B ring blocked most of the sunlight and appeared as a broad, dark band, the major gaps and the translucent C ring were brightly illuminated.

Pioneer's TV camera showed clear evidence of structure in the main rings, as well as a dimmer region in the center of the bright Cassini Division. This was eventually found to be due to the presence of particles in the middle of the "empty" gap. Even more fascinating was the discovery of a thin ring (now known as the F ring) about 3,600 km (2,250 mi) beyond the outer edge of the A ring. Unfortunately, few details could be resolved by Pioneer's camera, and the gap between the A and F rings became known as the Pioneer Division.

The F ring discovery image also revealed another surprise—the presence of a satellite between the new ring and the orbit of Mimas. The moon, designated 1979 S1, was seen in the same general location as two small satellites observed outside the A ring when the system was edge-on in 1966. But which one was it? No one could tell.

Further confusion followed a few minutes after Pioneer crossed the ring plane, when the spacecraft's instruments detected sudden, dramatic changes in the magnetic field and particle population—clear evidence that it had almost

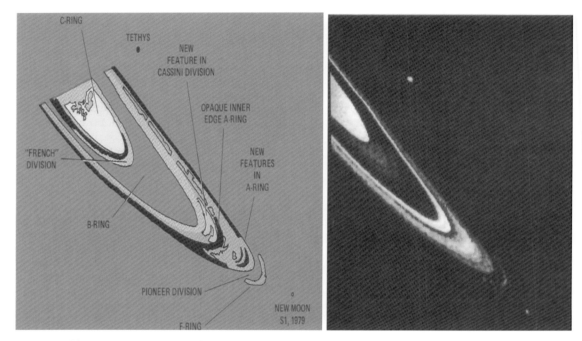

A New Ring and a New Satellite—The computer-processed picture (right) from Pioneer 11 was taken on August 31, 1979, from a distance of 943,000 km (585,950 mi). The newly discovered F ring is visible about 3,600 km (2,250 mi) beyond the outer edge of the A ring. Also visible at lower right was a "new" satellite, provisionally known as 1979 S1. It was later determined that it was a co-orbital moon, possibly one that was discovered by Audouin Dollfus during ground-based observations in 1966. This unusual view was taken below the rings, on their shaded side. Other discoveries included the so-called French Division, whose existence had long been questioned, and variations within the A ring. (NASA–Ames)

collided with a sizeable satellite. This chance encounter enabled scientists to estimate that the object was about 200 km (125 mi) across and orbiting just below the ring plane at a distance of approximately 152,000 km (95,000 mi) from the center of Saturn. Its orbital period would be about 17 hours.

Curiously, this was the same distance as 1979 S1. Further calculations showed that this moon would have been in just the right place to be intercepted by Pioneer 11. 1979 S1 and 1979 S2 were one and the same.*

Unfortunately, only tantalizing, fuzzy images were returned of Saturn's growing family of moons. Little was learned apart from the fact that its planet-sized moon, Titan,

* Scientists came to the conclusion that 1979 S1 was actually Janus, a moon that had erroneously been placed in a slightly different orbit when it was briefly spotted in 1966. Even more confusion arose in 1980, when the rings were edge on. Ground-based observations showed two satellites in the same orbit (Janus and Epimetheus) and another moon (Helene) co-orbiting with Dione. These were the first co-orbital satellites ever seen.

was too cold for life. Clouds of charged particles were also detected along the orbits of Enceladus and Tethys.

Pioneer also broke new ground by charting Saturn's magnetic field and particle environment. The magnetic bubble extended out to the orbit of Titan in the sunward direction, but formed a huge, extended "tail" away from the Sun. The magnetic field and radiation belts dwarfed those of Earth but were much weaker than at Jupiter. Indeed, the field strength at the cloud tops was actually weaker than we experience on Earth because the metallic hydrogen source region was so deep in the interior.

Unlike the other planets, Saturn's magnetic poles were found to almost coincide with the geographical poles—though, as on Jupiter, a magnetic compass from Earth would point south rather than north.

The Voyager Encounters

Even as the plucky Pioneer 11 completed humanity's initial passage through the Saturnian system, two advanced Voyager spacecraft had already passed Jupiter en route to the "lord of the rings." If Pioneer was the cut-price Model T Ford, the $250-million Voyagers were the Rolls Royces of NASA's space fleet. Equipped with state-of-the-art instruments, they were designed to investigate every aspect of the worlds that dominate the outer Solar System.

One obvious advantage of flying two spacecraft was the redundancy provided if one ship succumbed to technical failure or the harsh environment of space. It was also impossible to fully explore the mysteries of Saturn's rings and orange Titan with one flypast. Furthermore, the second spacecraft could certainly be reprogrammed to take advantage of any discoveries by its forerunner.

The mission team also had another agenda. Although the pair of nuclear-powered spacecraft were officially targeted only to explore the environs of the two inner gas giants, scientists still hoped to complete a "grand tour" of the outer Solar System by utilizing the gravity of Saturn.

The key to the Voyager 1 encounter was tantalizing

Voyager 1 at Saturn—This Voyager 1 image was taken on November 16, 1980, four days after closest approach to Saturn, from a distance of 5.3 million km (3.3 million mi). The view of a crescent Saturn can never be obtained from Earth. Voyager 1 was targeted to fly only 7,000 km (4,375 mi) from Titan's center, 18 hours before arrival at the planet. From the time of the Titan flyby until five hours after closest approach on November 12, it was looking at the dark, shaded side of the rings. Saturn's gravity then bent its path so that it headed north to observe their sunlit side. (NASA–JPL)

Titan. In order to observe this huge moon, while obtaining a good view of the rings and images of some smaller satellites, the spacecraft was targeted to arrive from above the ring plane. It would then swing over Saturn's southern hemisphere at a height of 124,200 km (77,000 mi), and return northward using the planet's gravity. If all went well, Voyager 2 would be free to pursue the grand tour of Uranus and Neptune.

On November 11, 1980, Voyager 1 skimmed past Titan on its inward leg just 4,520 km (2,800 mi) above the orange clouds. It then sent back close-ups of Tethys before plunging beneath the rings. Detailed pictures of Mimas, Enceladus, and Dione followed before Voyager recrossed the ring plane and obtained close-range images of Rhea.

On the outward leg, as seen from the Earth, the spacecraft disappeared behind the planet, reappeared briefly as it flew behind the gap between Saturn and the C ring, then passed behind the main ring system. When it emerged, the spacecraft was heading north and beginning its long trek toward interstellar space.

Voyager 2 at Saturn—The encounters with Saturn by Voyagers 1 and 2 were very different. Voyager 2 approached from above the rings, so that it looked down on the sunlit side. It crossed the ring plane only once, soon after its closest approach to Saturn on August 25, 1981, then continued south to look back at the shaded, southern side of the rings. The faint cloud bands are enhanced in this false-color image, taken on July 12, 1981, when Voyager 2 was 43 million km (27 million mi) from Saturn. Also visible are some bright spots in the northern hemisphere. (NASA–JPL)

discovered G ring. The theory that some of the narrow rings were confined by nearby companions was confirmed by the discovery of a tiny "shepherd" on each side of the F ring and another just outside the A ring.

The greatest disappointment was the inability of Voyager's cameras to pierce the smog that shrouded Titan. However, important information was returned about the temperature, pressure, and composition of its dense atmosphere and the nature of the high-altitude haze layers. Scientists were intrigued to find that Titan had a nitrogen-rich atmosphere in which complex organic compounds were created.

Voyager 1's successful survey of Titan meant that its successor was free to explore Uranus and Neptune. First, however, it had to negotiate the Saturn flyby. Voyager 2 arrived more than nine months after its sister craft, suffering from the loss of its primary radio receiver and a faulty backup system. Nevertheless, the flyby was in many ways more successful.

Blessed with a more sensitive camera and an approach trajectory well above the ring plane, Voyager 2 was ideally placed to focus on the spectacular rings. The northern face of the planet's beautiful encircling disk was also receiving more direct sunlight and so appeared much brighter.

From early June to late September 1980, Voyager 2 sent back more than 18,500 images of Saturn, its rings and satellites. Closest approach occurred on August 25, just 101,000 km (63,000 mi) above the cloud tops.

This time, Titan was just a side show, with Hyperion, Enceladus, Tethys, and Iapetus getting considerably more attention. Voyager 2 crossed the ring plane only once, providing the opportunity to look back at the shaded side of the disk as it exited the Saturn system en route to Uranus. Unfortunately, two-thirds of the Enceladus pictures were lost through jamming of the scan platform that carried the cameras—a problem that arose when the spacecraft was behind Saturn, only 45 minutes after it passed below the rings. Fortunately, the platform was back to normal nine days later, when Voyager made a distant pass of Phoebe, the outermost satellite then known.

This time, Voyager resolved thousands, rather than

Although the flight from Titan to Rhea lasted little more than 24 hours, Voyager's instruments were continually monitoring the Saturn system for almost four months in total. The remarkably clear, high-resolution images showed hitherto unsuspected features that gave scientists many headaches in the months to come. The rings were found to be incredibly complex, with the number of ringlets running into the hundreds. Some of these were seen to be elliptical, while the F ring seemed to comprise several braided strands.

Another puzzle involved shadowy spokes in the B ring that survived despite different rotation rates within the ring material. The existence of the D and E rings was confirmed, and the alphabetic nomenclature was extended by the newly hundreds, of ringlets, though a programmed search for small satellites embedded within them was unsuccessful. Saturn's bland clouds came alive with images of brown and white spots, and undulating, ribbon-like jet streams. The spectacular successes of the twin Voyagers left delirious scientists buried beneath a backlog of data and looking forward to an even more comprehensive exploration by some future Saturn orbiter.

Saturn Unveiled

Although the bland, butterscotch planet appeared a very different proposition to psychedelic Jupiter, scientists were curious to discover whether the two giants of the Solar System were really so different deep down. Observations from Earth of occasional long-lived cloud features had enabled astronomers to recognize differing rotation rates and wind speeds at various latitudes. The Voyager cameras provided a much more comprehensive picture of the atmospheric circulation.

As on other planets, the winds were arranged in a symmetrical manner either side of the equator, though the overall pattern was more Earth-like and less complex than on Jupiter. Only above 40 degrees latitude did Saturn's atmospheric circulation mimic that of its larger neighbor.

On Saturn most winds flowed eastward, in the same direction as the planet's rotation.* Most notable was a tremendous equatorial jet located between 35°N and 35°S, where westerly winds hurtled around the planet at up to 1,800 km/h (1,125 mph). This was more than four times the velocity of Jupiter's equatorial jet stream and much faster than any wind measured elsewhere in the Solar System. On the other hand, easterly winds were weak and relatively rare, with only four such currents being identified through Voyager's images.

Despite the planet's large axial tilt, the Voyagers noticed no major seasonal differences between the northern and southern hemispheres.

Beneath the obscuring haze, Saturn's atmosphere was

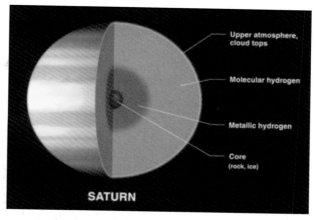

SATURN

The Interior of Saturn—As with Jupiter, Saturn is composed mainly of hydrogen and helium. However, the bulk density is about 0.7 that of water—by far the lowest density of all the planets. Saturn would actually float if it could be placed in a large enough ocean. Beneath the upper ammonia cloud there are thought to be clouds of ammonium hydrogen sulfide and water ice. Below this is a layer of liquid hydrogen, which changes to liquid metallic hydrogen as temperature and pressure increase. The core is thought to be composed of rock and ice. (LPI)

found to be surprisingly turbulent. Where wind speeds were relatively modest, away from the equator, warm rising gas from the interior created sizeable storm systems, with numerous white or brown spots and peculiarly shaped convective clouds that were probably associated with gigantic thunderstorms.

As on Earth, the smaller eddies lasted only a few days. However, some of the larger features found by Voyager 1 were recovered in Voyager 2 images, demonstrating that they could survive for nine months or more.

The most impressive storms on Saturn were oval spots that resembled smaller versions of Jupiter's great white ovals. Like their Jovian counterparts, they were anti-cyclonic in nature, rotating in a clockwise direction in the

---

* Meteorologists refer to these as "westerlies" because they come from the west.

Swirls and Spots—Use of camera filters and computer processing allows the high-level haze to be pierced. This false-color image of the northern hemisphere was taken by Voyager 1 on November 5, 1980, at a range of 9 million km (5.5 million mi). Small-scale convective cloud features can be seen in the dark belt (center); an isolated convective cloud with a dark ring is seen in the lighter zone (right); and an undulating wave is visible in the pale-blue zone. The smallest features in this photograph are 175 km (108 mi) across. (NASA–JPL)

As would be expected with the rings casting huge shadows over the planet, some differences in atmospheric temperature were detected, although seasonal effects were quite small. Calculations suggested that the immense atmosphere might be so large that it could only offer a sluggish response to summer heating or winter cooling.

The absorption of sunlight by hazes meant that temperatures reached –130°C at high altitudes, compared with –140°C at the main cloud deck. The clouds of ammonia ice crystals seemed to extend over an altitude range of about 100 km (62 mi). Deeper layers of cloud presumably existed, but they could not be seen. No hot spots pierced the all-embracing blanket.

In addition to the dominant components, hydrogen and helium, the Voyagers detected ammonia and methane, together with trace quantities of other hydrocarbons—phosphene, ethane, acetylene, methylacetylene, and propane—in the upper atmosphere.

It was also confirmed that, despite its smaller size, Saturn was a stronger heat source than Jupiter. This unlikely situation led theorists to suggest that some process other than simple emission of heat left over from its formation was at work.

The favored suggestion was the separation of helium and hydrogen deep beneath the surface. The excess heat would then be created as helium droplets rained down to deeper levels. This theory was supported by evidence that helium only accounted for 11% of the mass of the upper atmosphere, well below the predicted value of 20%.

Auroral emissions were detected as electrons spiralled down magnetic lines of force and collided with the upper atmosphere, though reflected light from the bright rings prevented them from being captured by Voyager's cameras. Saturn's auroral zone was about 9,000 km (5,625 mi) across and centered on the north and south poles.

Radio signals typical of lightning were also detected, though visual confirmation was again absent. One unique aspect of this electrical activity was that it occurred in the vicinity of the B ring, where discharges were 10,000 times more intense than those we experience on Earth. How these were generated remains a mystery. One unusual suggestion

northern hemisphere. Although nothing comparable to Jupiter's Great Red Spot was seen, Voyager 1 did discover a much smaller red storm, about 16,000 km (10,000 mi) across, at latitude 55°S.

Curiously, the pattern of belts and zones seemed to have little connection with the pattern of winds. Scientists suggested that the major seasonal variations caused by Saturn's pronounced axial tilt and the umbrella effect of the huge ring system might account for this apparent anomaly. Bands of cloud running parallel to the equator were surprisingly common, with 24 counted in the southern hemisphere alone. A number of undulating, ribbon-like streamers were also seen within unstable, high-speed jet streams.

was that a tiny iron moon in the rings somehow acted as a dynamo.

Spokes and Multiple Rings

The rings inevitably (and justifiably) were the stars of the Voyagers' Saturn encounters. Particularly impressive was the manner in which order was imposed upon the countless billions of icy particles in orbit around the planet. Whereas Voyager 1 imaged around 100 ringlets, the second spacecraft discovered many thousands.

By carefully measuring the variations in light from a star as it passed behind the rings, Voyager 2's photopolarimeter was able to map out fine-scale structure down to a few hundred meters. Scientists concluded that each ringlet was typically less than 100 km (62 mi) wide, with a more likely figure in the range 1–10 km (0.6–6 mi). However, most of the ringlets were not completely distinct but simply faded into each other.

Although individual particles were too small to be seen, the Voyagers showed that they ranged in size from particles of smoke to house-sized boulders. Different regions of the rings were dominated by particles of different sizes. For example, the F ring was mostly dust, whereas the ringlets of the Cassini Division contained sizeable chunks tens of meters in diameter.

Subsequent analysis of the particles' behavior showed that they jostle each other very gently as they career around Saturn at speeds of about 3 km/s (75,000 mph). Like tiny snowballs, the frequency of these collisions determines their state of compaction. The sheer density of the particles makes them move in a similar way to a fluid.

Colorful Rings—Saturn's ring system comprises four main rings separated by gaps (the largest is the Cassini Division—seen here as dark blue–black). The ring particles range from centimeters to tens of meters in size and are mainly made of ice. This enhanced color view was taken by Voyager 2 on August 17, 1981, from a distance of 8.9 million km (5.6 million mi). It shows possible variations in chemical composition between the C ring (blue, at top right) and the largely empty Cassini Division, the inner B ring and its outer region (where the spokes form), and the outer A ring. Spatial resolution is about 20 km (12.5 mi). (NASA–JPL)

Ring Spokes—This enhanced color image shows the outer half of Saturn's main rings. The dark Cassini Division, which separates the outer A ring from the B ring, contains several faint rings. The most prominent structure in the A ring is the dark Encke Division, about one-third of the way from the ring's outer edge. The B ring, more densely packed with material than the A ring, appears brighter, and several faint "spokes" are visible as dark streaks across it. At lower right is the outer part of the much fainter C ring. (NASA–JPL)

Spiral density waves, like those in a spiral galaxy such as our Milky Way, were seen to form at so-called resonant locations, where particles overtaking on the inside received repeated tugs from satellites outside the main system. After two years of careful analysis, scientists also recognized spiral "bending waves" or warps in the ring plane.

The behavior of the spiral patterns provided important clues to the varying nature of the ring regions. These were supported by studies of how radio signals and light were scattered by the rings. As a result, scientists were able to estimate that the mass of the entire ring system was comparable to that of 400-km (250-mi) diameter Mimas.

The densest (and most massive) region was the B ring, the only section able to block light and radio waves. On the other hand, the narrow F ring was comparable in mass to a moonlet only 1 km (0.6 mi) across. This agreed with the theory that the rings were created by the break-up of one or more small satellites in Saturn's not-too-distant past.

Since their creation, the rings seem to have been fairly stable. Although they are composed predominantly of water ice, minor color variations were noted. The C ring and the ringlets within the Cassini Division were similar in color, but three ringlets nestling within the C ring bore a close resemblance to the hue of the dense B ring. Whether these represented real variations in composition remained uncertain.

The major boundaries between the four main rings were clearly visible as sudden, drastic changes in particle density. The sharp outer edge of the A ring was maintained by a nearby satellite, Atlas, while the outer edge of the B ring (also the inner boundary of the Cassini Division) was located where particles orbited Saturn twice for every orbit of the moon Mimas. There were no apparent reasons for the other boundaries.

Within the four main rings were a few marked gaps that were largely empty of particles. One division in the C ring contained an eccentric (noncircular) ring that varied in width from about 90 km (56 mi) at its furthest from Saturn to only 35 km (21 mi) at its closest point. The A ring's Keeler Gap contained at least two kinked, discontinuous ringlets.

Ringlets such as this were thought to be created by the gravitational influence of small satellites that were either embedded in the rings or acting as shepherds. Partial confirmation of this theory came in 1990, when analysis of Voyager images by Mark Showalter revealed a 20-km (12.5-mi) satellite, now known as Pan, patrolling inside the Encke Division.

The largest gap, the Cassini Division, 5,000 km (3,100 mi) wide, was found to contain at least 100 ringlets. Some of these were noncircular, as were the inner edge of the Cassini Division and a number of others in the C ring. A thin, dense ring near the inner boundary of the Cassini Division was thought to be shaped and confined by one or more unseen satellites.

Particularly amazing was the F ring, revealed by Voyager to comprise three separate, intertwined strands, each approximately 20 km (12.5 mi) wide. Although the braided bands were knotted and kinked, they retained their shape for at least 15 orbits. Scientists concluded that the unprecedented contortions must be due to complex gravitational interactions with the two shepherding satellites. However, magnetic forces acting upon the tiny particles were not entirely ruled out.

Magnetism was also called upon to explain the presence of mysterious, wedge-shaped "spokes" that marred the bright B ring. Voyager movies showed that these huge radial features, often tens of thousands of kilometers across, could appear in a matter of minutes and persist for several hours before fading away.

One theory suggested that their formation could be triggered by small comet fragments, no bigger than a meter across, hitting the rings at high speed. Traveling at perhaps 30 km/s (67,500 mph), the meteoroids would explode, sending a cloud of electrically charged gas across the rings. Microscopic grains knocked off the ring particles could then interact with the planet's magnetic field to create the huge, wispy spokes.

Tantalizing Titan

Long before Voyager 1 approached the sixth planet from the Sun, ground-based observations had shown that Saturn's largest satellite, Titan, was unique among planetary moons. Although it was thought to be the largest satellite in the Solar System, scientists were intrigued by spectroscopic evidence of the gases methane and hydrogen.

Estimates for the surface pressure ranged from 20 millibars to 20 bars, i.e., anything from one-fiftieth of Earth's surface air pressure to 20 times that amount. Speculation was rife that, beneath its clouds, Titan might have a nitrogen-rich atmosphere and an icy surface coated with tarry organics or oceans of liquid methane.

Expectations were high that the Voyagers would be able to reveal the true nature of this strange world, but, to the disappointment of everyone except the atmospheric scientists, the satellite refused to open its orange veil for the cameras. Voyager 1 found that Titan was smothered by an all-embracing blanket of hydrocarbons, relieved only by a dark polar hood and high level hazes.

The images also showed that the haze was much brighter above Titan's southern hemisphere—which was enjoying late summer—than above its northern, winter hemisphere.* Earth-based observations confirmed that this difference in brightness changes with Titan's seasons. Curiously, the dark northern hood seen by Voyager 1 was transformed into a narrow band by the time of the Voyager 2 encounter.

Fortunately, other instruments were able to relieve some of the imaging team's gloom. Radio signals sliced though the thick atmosphere, revealing different layers all the way to the surface. The occultation data obtained when Voyager passed through Titan's shadow indicated that the satellite is actually a little smaller than its Jovian rival, Ganymede, with a diameter of 5,150 km (3,200 mi). The reason for the error was that Titan had an unexpectedly thick atmosphere.

The satellite itself had a density almost twice that of water, indicating a roughly 50–50 ice–rock composition. In that sense, Titan was a near-twin of Jupiter's largest moons,

Titan—With a diameter of 5,150 km (3,200 mi), Titan is Saturn's largest satellite and second only in size to Jupiter's Ganymede. To the disappointment of Voyager scientists, an orange smog of hydrocarbons meant that no glimpse of the surface was possible. The main features were the bright southern hemisphere and the dark northern polar hood. The image combines separate violet, blue, and green frames taken by Voyager 2 on August 22, 1981, from a range of 4.5 million km (2.8 million mi). (NASA–JPL)

Ganymede and Callisto. Evidence suggested that Titan had a rocky core that extended about two-thirds of the way to the surface. Above this were various high-pressure forms of ice, topped by a crust of rock-hard "ordinary" ice about 100 km (62 mi) thick. However, no evidence of a magnetic field was found.

The puzzle of Titan's atmosphere was solved when Voyager 1 revealed a dense blanket with a surface pressure

---

* A year on Titan lasts about 30 Earth years, so each season is 7½ years long. At the time of the Voyager 1 flyby, spring was just beginning in Titan's northern hemisphere.

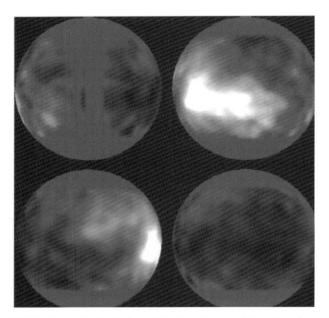

Hubble Views Titan's Surface—The first glimpse of Titan's surface came in 1994, when the Hubble Space Telescope pierced the orange haze. The near-infrared images revealed bright and dark features during a nearly complete 16-day rotation. Particularly prominent was Xanadu, a bright "continent" 4,000 km (2,500 mi) across—roughly the size of Australia. Only Titan's polar regions could not be mapped this way, due to the viewing angle and thick haze near the edge of the disk. The smallest features visible were about 580 km (360 mi) across. (Peter H. Smith [University of Arizona], NASA–STScI)

50% greater than that on Earth. In fact, Titan was only the second world known to have a dense atmosphere of molecular nitrogen—Earth being the first. Well over 90% of the satellite's atmosphere was nitrogen, with most of the remainder in the form of methane.

The dominance of nitrogen on both worlds inevitably led to descriptions of Titan as an early Earth in deep freeze. Titan's present environment could be similar to that on Earth billions of years ago, before life forms began pumping oxygen into the atmosphere. However, a surface temperature of −180°C suggested that the emergence of life—at least on the surface—was most unlikely.

Where did the nitrogen come from? The favored explanation was an early atmosphere rich in ammonia (NH_3). As sunlight split up the ammonia, it was converted to nitrogen and hydrogen.

Another intriguing aspect of the atmosphere was the substantial amount of methane (CH_4)—the primary component of the fuel we know as natural gas. At the temperatures present on Titan, methane could take the form of a gas, liquid or solid—just like water on Earth. This led to speculation that Titan could have ethane–methane oceans and methane rainfall or snow. However, methane is split into other constituents by the action of sunlight, so it should all have been destroyed long ago. The only explanation was that some source—perhaps an underground ocean or aquifer—was continually replenishing the methane supply.

Methane and nitrogen take part in many chemical reactions that create complex organic molecules in the atmosphere and on the surface. Voyager's infrared spectrometer detected many minor constituents in the moon's photochemical smog, most of which were produced by the action of ultraviolet sunlight on methane. Among the hydrocarbons in the global haze were compounds such as ethane (C_2H_6), acetylene (C_2H_2), and propane (C_3H_8). Methane interacting with nitrogen atoms formed "nitriles" such as hydrogen cyanide (HCN).

Voyager 1 obtained images of several tenuous haze layers up to 700 km (435 mi) above the hidden surface. These and the main brownish haze layer at an altitude of 200 km (125 mi) were created by interactions between the nitrogen–methane atmosphere, ultraviolet light from the Sun, cosmic rays, and charged particles trapped in Saturn's magnetic field. The hydrocarbons produced by this methane destruction formed a smog similar to that found over large cities, only much thicker.

Scientists speculated that the smog particles would very slowly sink through the atmosphere, taking about a year to reach the surface. There they could accumulate over millions of years to form a layer of hydrocarbon sludge, perhaps hundreds of meters thick.

As for the obscured surface, scientists would have to

wait for more than a decade before the Hubble Space Telescope was able to pierce the opaque smog at infrared wavelengths and discover intriguing light and dark markings—possible oceans separating large, icy continents. Other observations confirmed that Titan spins once every 16 days, the same time it takes to orbit Saturn, so it always keeps the same face toward the planet.

Icy Satellites

Although the Voyagers succeeded in imaging all of Saturn's major satellites, as well as a few lesser lights, it was not possible to map their entire surfaces in detail, and many mysteries remained. The innermost large satellite, Mimas, displayed the predicted record of heavy bombardment—a certain sign that little had changed since its birth more than 4 billion years ago. The most significant feature was a 130-km (80-mi) impact crater that covered one third of the

moon's diameter and gave Mimas a remarkable resemblance to the "Death Star" in the movie "Star Wars." Now known as Herschel, the crater had a raised rim and a central peak that towered 6 km (4 mi) above its surroundings—comparable to the highest mountains on Earth. Scientists found it difficult to understand how icy Mimas had survived this ancient collision intact.

Since the next satellite out, Enceladus, is only a little larger than Mimas, scientists expected the pair to be near twins. To everyone's surprise, Voyager 2 images showed that, despite its small size and low density, the 500-km (310-mi) satellite seemed to have been resurfaced fairly recently.

Enceladus displayed a surprisingly varied surface, with evidence of geological activity in the not-too-distant past. Some terrains were dominated by sinuous mountain ridges from 1 to 2 km (0.6 to 1.2 mi) high. Other regions were scarred by linear cracks where the crust might be shifting sideways, rather like California's infamous San Andreas fault. Particularly noticeable was the wide variation in the number of impact craters. Some areas were moderately

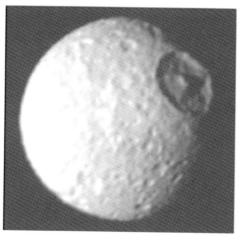

Enceladus and Mimas—Despite its small size (about 500 km or 300 mi in diameter), Enceladus (left) shows evidence of widespread and fairly recent geological activity. This Voyager 2 mosaic shows details about 2 km (3 mi) across. The largest craters are about 35 km (20 mi) in diameter. Also visible are linear fractures and grooves. The Voyager 1 image of pockmarked Mimas (right) was taken by on November 12, 1980, from a range of 425,000 km (264,000 mi). Herschel, the largest impact crater, is 130 km (80 mi) in diameter and displays a prominent central peak. (USGS/NASA–JPL)

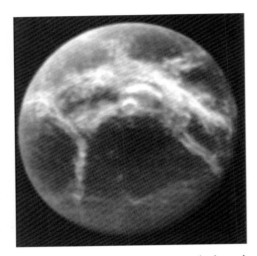

Iapetus and Dione—Iapetus (left) is the outermost of Saturn's large, icy satellites. Like most moons, it always keeps the same face toward the planet. However, it is unique in having a leading (forward-facing) hemisphere that is as black as tar and a bright, icy, trailing hemisphere. This picture was taken by Voyager 2 on August 22, 1981, from a distance of 1.1 million km (687,000 mi). The Voyager 1 image of Dione (right) shows strange wispy streaks on the trailing side, possibly deposits of ice associated with huge fractures. The brighter leading hemisphere shows a large valley and numerous craters. (USGS/NASA–JPL)

cratered with hollows up to 35 km (20 mi) across, some had numerous small craters, while others were grooved plains that were remarkably smooth and uncratered. However, no giant craters such as Herschel were seen—evidence that the entire satellite had been resurfaced at least once.

Its water ice-covered surface also reflected about 90% of the incident sunlight (similar to freshly fallen snow), making it the most reflective satellite in the Solar System. One suggestion was that icy geysers on Enceladus might be ejecting micrometer-sized particles into space that coated the mirror-like surface and replenished the E ring. The only feasible heat engine to power the geological activity appeared to be tidal action as Enceladus is periodically squeezed by the gravity of Saturn and another moon, Dione.

Dione itself was fairly unremarkable, although it, too, showed modest variations in surface appearance. The forward-facing hemisphere was heavily cratered and con-

siderably brighter than the trailing hemisphere. There were several craters on the leading side more than 100 km (62 mi) in diameter, but there were also plains where cratering was less frequent. Bright wisps on the trailing side were thought to represent ice deposits along huge troughs or faults in the moon's icy crust.

With a width of 1,050 km (655 mi), Tethys was only a little smaller than Dione. It, too, was heavily cratered and marked by satellite-girdling cracks in its icy surface. One enormous trench, now named Ithaca Chasma, measured about 65 km (40 mi) wide and stretched three-quarters of the way around Tethys. The other notable feature was a crater 400 km (250 mi) wide, named Odysseus, that displayed a massive central peak and several concentric rings outside the main boundary wall. Its lack of depth indicated that soft ice had filled the hollow after its spectacular excavation.

Rhea, the largest of the icy satellites, generally resembled

Dione, with bright, wispy markings on the trailing hemisphere and heavily cratered terrain pockmarked with hollows more than 40 km (25 mi) across. Some parts of the ancient surface were so saturated with craters that there was no room to make new ones without destroying those already existing.

The most remarkable of Saturn's two-faced moons was Iapetus, the outermost of its large satellites. Scientists were astonished to find that the side that faced forward as the satellite orbited Saturn was blanketed with very dark material, as black as tar or asphalt, whereas the rear-facing hemisphere and the poles were bright and icy.

Many large impact craters were visible all over the surface, suggesting that the surface is ancient. However, the dark layer seemed to be a more recent coating, possibly organic dust derived from a more remote satellite called Phoebe. The leading side of Iapetus would be "painted" as the satellite ploughed through the tenuous dust.

Neither of the Voyagers flew very close to Phoebe. Distant views suggested that it was round and very dark, with a surprisingly large diameter of about 200 km (125 mi).

Not all of Saturn's moons were spherical. Hyperion measured about $400 \times 250 \times 200$ km ($250 \times 155 \times 125$ mi) and was found to resemble a reddish hamburger with one end bitten off. Its irregular shape was testimony to a violent history, and some scientists suggested that it was a rubble pile that had reassembled after a major collision. Unlike its more stately cousins, Hyperion was tumbling chaotically.

Six strange moons were also discovered by the Voyagers. Apart from the F ring shepherds (Prometheus and Pandora), Voyager 1 images revealed Atlas close to the edge of the A ring, and Telesto, located 60 degrees ahead in the same orbit as Tethys. Analysis of Voyager 2 photos led to the discovery of Calypso, orbiting 60 degrees behind Tethys, and Pan, which swept clear the Encke Division in the A ring. Helene, a moon that is co-orbital with Dione, was found in ground-based images in 1980. A further 12 moons were found in 2004 by David Jewitt and colleagues, using the Subaru telescope in Hawaii. These small captured objects brought Saturn's satellite total to 44—with further discoveries by the Cassini spacecraft on the cards.

Cassini–Huygens

The Cassini–Huygens mission was born in June 1982, less than a year after Voyager 2 passed by Saturn, when a joint USA–Europe working group recommended that a Saturn orbiter and a Titan probe should be developed on a cooperative basis. In the early assessment stages it was envisaged that Cassini and its piggybacking probe would be launched toward Saturn in 1994 by a Centaur rocket deployed from the Space Shuttle. This scenario was scrapped after the 1986 Challenger Shuttle disaster, and replaced by a Titan IV–Centaur launch in 1996.

Meanwhile, after a preliminary evaluation of the project, the probe—named in honor of Christiaan Huygens—was selected as part of ESA's Horizon 2000 long-term science plan in 1988. A year later, NASA received the go-ahead from Congress for Cassini—named after another famous seventeenth-century observer of Saturn—but within three years, the multi-billion-dollar program to develop the Mariner Mark 2 spacecraft was canceled. However, the overall mission survived, no doubt partly because of the risk of an international incident involving NASA's European partners if this ambitious "battlestar galactica" endeavor was scrubbed. Scientists and engineers scrambled to redesign the mission and the orbiter.

Despite a vitriolic campaign by "green" groups in America and Europe, who expressed concern over the safety of radioactive power sources on board the spacecraft, Cassini–Huygens finally lifted off from Florida on October 15, 1997.

The Cassini spacecraft was one of the largest and most sophisticated spacecraft ever sent into deep space. With a total mass of about 5,650 kg (6 ton), it was outweighed only by the two Phobos spacecraft sent to Mars by the former Soviet Union. Since it was not possible to send heavyweight Cassini directly to Saturn, the mission team opted for a six-and-half-year trip that included four planetary gravity assists to gain the momentum needed to reach the ringed giant.

Cassini Brakes into Orbit—This artist's view shows Cassini–Huygens during the Saturn Orbit Insertion maneuver on June 30, 2004, just after the main engine had begun firing. The spacecraft is moving toward the viewer and to the right (firing to reduce its spacecraft velocity with respect to Saturn) and has just crossed the ring plane. The maneuver, which lasted approximately 90 minutes, enabled Cassini to be captured by Saturn's gravity. It also offered a unique opportunity to observe Saturn and its rings in great detail. (David Seal/NASA–JPL)

Huygens' Descent—NASA's Cassini spacecraft and ESA's Huygens probe were launched from Earth in October 1997. After a seven-year journey, Cassini–Huygens entered orbit around Saturn on June 30, 2004. The probe was released on December 25 and entered Titan's atmosphere on January 14, 2005. This artist's impression shows the descent sequence of the 2.7-m (9-ft) diameter, wok-shaped Huygens after separation from the Cassini mother ship (top left). In reality, Saturn was not visible through the clouds. (ESA/David Ducros)

All seemed to be going well until February 2000, when engineers identified a design flaw in the Huygens communications system. In-flight tests showed that the receiver on board Cassini would be unable to compensate for the Doppler shift in radio frequency caused by the rapid relative motion of the orbiter and Huygens during the descent over Titan. Unless something could be done, most of the unique data sent back by Huygens during its trip through Titan's atmosphere would be lost.

More than a year passed before engineers produced a revised mission plan to work around the communications problem. To ensure that the probe returned as much data as possible, it was decided to shorten Cassini's first two orbits around Saturn and insert an extra orbit that provided a higher Titan flyby for Cassini.

In order to reduce the Doppler shift in the signal from Huygens, Cassini would have to fly over Titan's cloud tops at an altitude of about 65,000 km (40,625 mi), more than 50 times higher than anticipated at launch. The new geometry meant that Huygens would be released toward Titan on December 25, 2004, for an entry into the moon's atmosphere on January 14, 2005—seven weeks later than originally planned.

Cassini finally swooped over Saturn's rings and braked into orbit on June 30, 2004. On board were 12 scientific instruments that would study every aspect of Saturn and its surroundings, including the planet's changing weather, atmospheric composition, glorious rings, icy moons, magnetic field, and electrically charged particles.

Over its four-year prime mission, Cassini was scheduled to complete 74 orbits around the planet. The key to a successful tour would be 45 Titan flybys, during which the giant moon's gravity would redirect the spacecraft and send it swinging around the Saturnian system. During these encounters, the orbiter's radar and other instruments would penetrate the orange veil over Titan, creating the first detailed topographic maps of the satellite's icy continents and possible methane seas, while unravelling its temperature variations, hazes, and cloud layers.

Meanwhile, after a seven-year piggyback ride from Earth, the Huygens probe would complete a much shorter but equally noteworthy expedition, spending just a few hours descending through Titan's dense atmosphere and reporting back from the mysterious, hidden surface.

In order to carry out its reconnaissance of Titan, the 319-kg (700-lb), wok-shaped Huygens was equipped with a suite of six instruments and 39 sensors that would enable the probe to investigate every aspect of Titan's unique chemical crucible. Not only would it study the haze layers and atmospheric composition, but it would also search for evidence of lightning and measure wind speed, air pressure, and temperature.

No one knew what the imaging system would see when it sent back panoramic pictures from the slowly spinning probe as it dangled beneath a parachute. Would the arrival of Huygens on Titan's frigid terrain be a probe-shattering impact, a squelch into organic ooze, or a splashdown in liquid ethane and methane? If the probe did survive, Huygens carried a Surface Science Package that its creators hoped would make history by sending back the first *in-situ* measurements of an alien sea.

As long as Cassini remained above the horizon, a healthy Huygens would be able to transmit its treasure trove of data, but, after about two hours, humanity's first expedition to land on a satellite orbiting a gas giant would all be over.

Saturn Revisited

On the night of June 30–July 1, 2004, Cassini braked into orbit around Saturn after its seven-year odyssey from Earth. Traveling at a speed of over 20 km/s (45,000 mph),

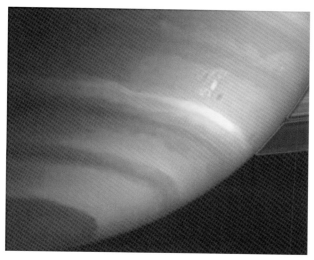

The Dragon Storm—This Cassini false-color view, taken in mid-September 2004, was made using near-infrared filters to show different amounts of methane gas. The strange, complex cloud feature (right), nicknamed the "Dragon Storm," was a powerful source of radio emissions. Short bursts of static were generated by lightning, indicating that it was a giant thunderstorm. Regions with a lot of methane above the clouds are red, indicating clouds deep in the atmosphere. Gray indicates high clouds, and brown indicates clouds at intermediate altitudes. The rings are bright blue because there is no surrounding methane gas. (NASA–JPL)

the spacecraft was reoriented for a 96-minute main engine burn, which slowed it by 626 m/s (1,400 mph) and allowed it to be captured by Saturn's gravitational embrace. Cassini passed through the narrow gap between the F and G rings twice—first while ascending shortly before the burn, then while descending shortly after the burn. During these hazardous passages the high-gain antenna was used as a shield, protecting the spacecraft from potentially lethal dust impacts.

Five science instruments remained active as the spacecraft flew within 15,000 km (9,375 mi) of Saturn's main rings. This was 10 times closer than at any other point in the mission, in a region of space that had never previously been observed. It was also the closest passage to Saturn of the

Southern Hot Spot—This mosaic of 35 exposures was made with the Keck I telescope in Hawaii on February 4, 2004. Taken at infrared wavelengths, it is the sharpest image of Saturn's temperature emissions taken from the ground. Coolest regions are dark red, hottest are white. As expected during the southern summer, the upper troposphere was relatively warm. However, the warm region near the south pole was unexpected, with a prominent hot spot above the pole itself (bottom). Note that the ring particles are coldest after passing through Saturn's shadow (lower left). The black square is due to lack of data. (NASA–JPL)

entire mission—just 19,980 km (12,410 mi) from the cloud tops.

Since this dramatic entry into orbit, Saturn has come under intense scrutiny, with new discoveries in almost every scientific sphere. During the ring crossings, the Radio and Plasma Wave Science instrument on Cassini measured little puffs of plasma produced by dust impacts. The instrument detected up to 680 hits per second, or roughly 100,000 in less than five minutes. Fortunately the particles were small, comparable in size to cigarette smoke.

Cassini's examination of Saturn's atmosphere began while the spacecraft was still approaching the planet. This early data showed that near-equatorial winds raced around Saturn at speeds of about 325–400 m/s (725–895 mph). Curiously, this was considerably slower than the 470 m/s (1,060 mph) jets measured by the Voyagers, but much faster than figures derived from Hubble Space Telescope observations.

Cassini provided an answer to this mystery by discovering that the equatorial winds slowed dramatically with increasing altitude, falling off by as much as 140 m/s (315 mph) over an altitude range of 300 km (185 mi) in the upper stratosphere. The apparent slowing of the wind between the early 1980s and the 1990s was probably a result of the spacecraft measuring storms that raised methane clouds to higher levels where winds are slower.

Outside the equatorial region, winds had remained remarkably stable throughout the Voyager, Hubble, and Cassini eras. How could these energetic jets be sustained on Saturn?

Cassini images revealed a "storm alley" near 35°S, where long-lived convectional storms periodically erupted and a variety of more compact ovals were seen to form and merge with each other. These mergers may help to explain the process by which energy is transferred from Saturn's interior to help to sustain its strong cloud-level winds. Other sequences captured similar dark ovals emerging from the high-level outflow of large convective storms. Such small, dark storms were generally stretched until they merged with the opposing air currents to the north and south.

As suspected, the bright areas of upwelling cloud were found to be associated with electrostatic discharges, thought to originate from lightning. The radio signals from Saturn lightning were incredibly intense, almost one million times greater than terrestrial lightning.

One unusually shaped feature in Storm Alley, known as the "Dragon Storm," was closely observed during July and September 2004, and found to be associated with a powerful source of radio emissions. Cassini detected the radio bursts only when the storm was rising over the horizon on the night side of the planet as seen from the spacecraft. The bursts stopped when the storm moved into sunlight.

Other results, using Cassini and Hubble data, showed that Saturn's magnetosphere is unique. In particular, the planet's auroras (Northern and Southern Lights) were seen to respond to the solar wind in a different way from either Earth or Jupiter. Instead of being dominated by the magnetic polarity of the charged particles arriving from the Sun, Saturn's auroras were heavily influenced by fluctuations in the pressure of the solar wind.

Non-imaging instruments also returned some interesting results. Cassini crossed Saturn's bow shock on June 27 when it was still 2.97 million km (1.85 million mi) from the planet, half as far again as the distance recorded by the Pioneer and Voyager spacecraft. The difference was explained by the different flight trajectories, since Cassini was approaching the magnetosphere from the side, rather than "head on." The spacecraft actually crossed the bow shock seven times during its approach to Saturn, as the magnetosphere ballooned in and out under the pressure of the solar wind.

Another pre-encounter observation left scientists puzzling over Saturn's rotation period. Natural radio signals from the planet gave a result of 10 hours, 45 minutes, 45 seconds (plus or minus 36 seconds)—about six minutes, or 1%, longer than the radio rotational period measured by the Voyager spacecraft in 1980 and 1981. This seemed to indicate that significant changes were taking place in the planet's deep interior.

Another instrument on Cassini took the first image of the bubble of energetic particles trapped in the magnetic field. It later discovered a new radiation belt that stretched almost from the cloud tops to the inner edge of the D ring. This showed that the radiation belts extend far closer to the planet than previously thought.

Wriggling, Undulating Rings

As Cassini closed in on scintillating Saturn, more and more details became visible in the rings so that, by early 2004, clumps in the narrow F ring and the nearby shepherd moons were clearly visible. Curiously, the expected ring spokes failed to appear, causing scientists to speculate that their formation was more likely when the Sun was shining on the rings at a low angle.

Excitement mounted as the spacecraft swept beneath the sparse E ring and headed for the "empty" gap between the F and G rings. Cassini then crossed the ring plane and looped over the vast disk—its closest approach to the rings in the entire four-year mission—before diving southward for a second passage through the F–G gap.

While crossing the plane of Saturn's rings, the Radio and Plasma Wave Science instrument detected up to 680 hits per second from dust particles comparable in size to cigarette smoke. Roughly 100,000 dust impacts were recorded in less than five minutes.

New information on the structure, temperature, and composition of the rings was obtained during the unique fly past. Temperatures on the unlit side of Saturn's rings were found to vary from a relatively warm −163°C in more transparent sections, such as the Cassini Division or the inner C ring, down to −203°C in denser, more opaque regions that reflected more sunlight, such as the outer A ring and the middle B ring. Even individual ringlets in the C ring and the Cassini Division were observed to be cooler than the surrounding, more transparent regions.

Scallops and Streamers—This Cassini image of the sunlit side of Saturn's rings shows the 325-km (200-mi) wide Encke Division in the A ring, which is swept clear by the moon Pan. The inner edge of the gap shows a pattern of undulations, or "scallops," and streamers created by the passage of the small moon. The brightest ring in the gap coincides with Pan's orbit. Also visible are three faint ringlets on either side of the main Encke ring. The image was taken on June 30, 2004. (NASA–JPL/Space Science Institute)

Thieving Prometheus—This Cassini view captures the shepherd moon Prometheus (center) alongside five separate strands of the F ring. The 102-km (63-mi) diameter moon appears to be connected to the ringlets by a faint strand of material. Scientists speculated that the moon might be pulling material away from the ring. Prometheus is also creating discontinuities or "kinks" in the ringlets and gaps in the sparse inner strands. The image was taken on October 29, 2004, at a distance of about 782,000 km (486,000 mi) from Prometheus. (NASA–JPL/Space Science Institute)

It was no surprise to find that Saturn's rings are almost exclusively composed of water ice, but ultraviolet and visual-infrared observations also revealed "dirt" mixed with the ice in the Cassini Division and in other small gaps in the rings. The F ring was also found to contain more dirt, although, in general, the outer part of the icy rings contained fewer non-icy contaminants.

Particles between the rings seemed to closely resemble dark material seen on the sizeable outer moon, Phoebe, supporting the theory that the rings are remnants of moons that broke apart when they approached too close to Saturn. In addition, the continuous bombardment by meteorites and small objects has probably spread dirty material over much of the rings. Cassini also detected a sudden and surprising increase in the amount of atomic oxygen (derived from water ice) at the edge of the rings. This led scientists to speculate that the oxygen may have been produced when a small object collided with the rings only a few months earlier.

For many ring scientists, the devil was in the detail provided by the 61 close-up images taken soon after Cassini braked into orbit. Since the spacecraft was hurtling past at 15 km/s (about 34,000 mph), only parts of the rings were targeted, with most images taken on the dark, shaded side. The dramatic results left researchers struggling to describe the great variety of features.

Some images show patterned waves in the rings, resembling light and dark stripes of varying width. Density waves that were seen spiraling away from Saturn were associated with gravitational resonances created by small moons embedded within the rings. Also linked with nearby satellites that followed inclined orbits were corrugated bending waves that resembled ocean waves spiraling in toward the planet. Other structures were described as "mottled," "ropy," or "straw."

One amazing image showed the Encke gap in the A ring, home to the small moon Pan. As it patroled the 325-km (200-mi) wide division, Pan created regular "scallops" and "streamers" along the gap's inner edge, where particles bunched together like a traffic jam on a highway. Also visible were three faint ringlets alongside the brighter main ringlet, which ran like a thread along Pan's orbit. Another new finding was that Pan's orbit is slightly inclined with respect to the A ring.

Weak, linear density waves in Saturn's rings caused by Atlas and Pan were used to obtain more reliable calculations of their masses and their orbits. The masses imply that the moons have very low densities and may be constructed like rubble piles.

Another satellite, Prometheus, was shown to play a major role in shaping the narrow F ring. Drapelike scallops created by the moon's passage, were found in the delicate, diaphanous material interior to the ring. Also seen were "drapes" that extended from the main core of the F ring, through the inner strand and into the interior dust sheet. One graphic image showed Prometheus pulling a faint ribbon of material from the inner strand. Zigzagging kinks and knots in the ring were almost certainly caused by the little shepherd.

Various faint, narrow ringlets were also discovered

between the A and F rings, as well as within some of the gaps. One of these, a 300-km (185-mi) wide ring known as S/2004 1R, was found along the orbit of the small satellite, Atlas. Several of the ringlets were kinked, and it seems likely that at least some of the ringlets occupying gaps are shaped by tiny, unseen moons acting as shepherds.

Several faint "spikes" were discovered on the outside of the 42-km (26-mi) wide Keeler gap, which lies about 250 km (155 mi) inside the outer edge of the A ring. The suspicions of scientists were confirmed on May 1, 2005, when a new moon, now called Daphnis, was discovered in the center of the gap. Scalloped ring features on either side of the satellite closely resembled those in the Encke gap.

More discoveries are sure to be made in the years ahead as Cassini gets a better view of the rings. Until May 2005, the spacecraft stayed in the plane of the rings, so the rings were mainly seen edge-on. Its path was then adjusted to enable it to pass above and below the ring plane. By the summer of 2008, the inclination of the orbit will reach nearly 80 degrees, allowing Cassini to see almost the entire ring system laid out in all its glory.

Moons in Close-up

During its four-year orbital survey, Cassini was scheduled to execute 52 close encounters with seven of Saturn's 31 known moons. It began in spectacular style with a close flyby of Saturn's intriguing outer moon, Phoebe, on June 11, 2004—several weeks before arrival at the planet. Instead of a near-spherical shape, the satellite turned out to be an irregular chunk of ice, about 220 km (137 mi) across and heavily potholed with craters both large and small. Images revealed bright streaks in the ramparts of the largest craters, bright rays emanating from smaller craters, and long, uninterrupted grooves. The plethora of 50-km (31-mi) craters suggested that debris from past impacts may have created many of the tiny moons that follow elongated orbits much like Phoebe's.

The moon was composed of water ice, carbon dioxide

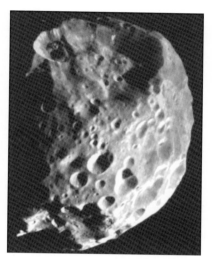

Phoebe—Cassini obtained this close-up view of Phoebe on June 11, 2004, from a distance of 32,500 km (20,200 mi). Phoebe appears to be an ice-rich body coated with a thin layer of dark material. Small bright impact craters reveal fresh ice and are probably fairly young. On some crater walls, dark material appears to have slid downward, exposing brighter ice. The craters are named after the Argonauts, who sought the golden fleece. The largest of these, about 100 km (62 mi) across, is named Jason (top). Some areas that appear particularly bright—especially lower right—are over-exposed. (NASA–JPL/Space Science Institute)

ice and rock, overlain with a layer of darker material—probably carbon-rich compounds—perhaps 300–500 m (1,000–1,650 ft) thick. This primordial mixture suggested that Phoebe may have formed as a Kuiper Belt object in the outer Solar System before it was nudged inward and captured by Saturn.

A close flyby of piebald Iapetus provided a clear view of the moon's coal-black hemisphere, known as Cassini Regio, and its transition to bright, clean ice above 40 degrees latitude. Close inspection revealed that the surface was coated by wispy streaks of darker material, typically a few kilometers wide and sometimes tens of kilometers long, which were roughly aligned north–south.

An ancient, impact basin, 400 km (250 mi) wide, was

The Equatorial Ridge of Iapetus—This image of Iapetus shows a number of surface features whose origins are unknown. One unique feature is a long, narrow ridge that lies almost precisely on the equator, bisecting the entire dark hemisphere. The ridge reaches 20 km (12.5 mi) high and extends over 1,300 km (800 mi). In places, the ridge soars to some three times the height of Mt Everest. The image was taken by Cassini on December 31, 2004. (NASA–JPL)

Enceladus fountains—This false-color Cassini image shows Saturn's moon, Enceladus, backlit by the Sun. Fountains of icy particles and water vapor are seen spraying from the south polar region to an altitude of at least 500 km (300 mi). The image, taken on November 27, 2005, is looking more or less broadside at the satellite's "tiger stripe" fractures where warmer ice has been detected. (NASA/JPL/Space Science Institute)

visible in Cassini Regio. The basin was overlain by smaller, more recent, impact craters. Around its rim were steep scarps, many of which appeared bright, probably due to exposed outcrops of relatively clean ice. The scarps and walls of nearby craters were brightest on north-facing slopes, while the opposite scarps were stained with the darker material. The origin of this mysterious material remained unsolved.

The most surprising discovery was an elongated mountain ridge that stretched for at least 1,300 km (800 mi) along the satellite's equator. No more than 20 km (12 mi) wide, the ridge rose at least 13 km (8 mi) above the surrounding terrain—and would therefore tower above Mt Everest. Scientists were left to puzzle over whether it is a mountain belt that has folded upward, or a crack in the surface through which material erupted onto the surface and then piled up, rather than spreading outward.

Bright, icy Enceladus proved to be just as intriguing as the Voyager data had suggested. Cassini's first two close

flybys revealed that it has a tenuous atmosphere of ionized water. Since the satellite's gravity is not strong enough to hold onto an atmosphere for very long, this suggested the presence of a continuous source, possibly ice volcanism, geysers, or gases escaping from the surface or the interior. Cassini also discovered a stream of tiny particles that could either belong to a dust cloud around Enceladus or have originated from Saturn's outermost ring.

The surface of Enceladus displayed evidence of prolonged crustal activity. Apart from the usual dense, polka-dot pattern of impact craters, detailed images also showed a complex web of fractures and faults, some only a few hundred meters across, others up to 5 km (3 mi) in width.

The rims and interiors of many ancient craters seemed to be sliced through by narrow, parallel grooves that created lanes around 1 km (0.5 mi) in width. The oldest fractures had a softer appearance and had been modified by numerous, superimposed craters. The youngest terrain

Hyperion—This false-color view of the moon Hyperion reveals a surface dotted with craters and modified by some unknown process to create a strange, "spongy" appearance. Cassini flew by Hyperion at a distance of only 500 km (300 mi) on September 26, 2005. Hyperion is 266 km (165 mi) across and spins in a chaotic rotation. Its low density suggests that it resembles a pile of rubble with much of its interior as empty space. (NASA/JPL/Space Science Institute)

was marked by a paucity of craters among a twisted network of ridges and troughs.

A distinctive pattern of slightly curved and roughly parallel faults close to the moon's south pole was dubbed "tiger stripes." The most detailed images of the southern terrain showed ice blocks about 10–100 m (33–328 ft) across. Small areas near the fresh "tiger stripe" fractures were more than 30°C warmer than expected, possibly because of heat escaping from the interior.

The link between the tiger stripes, the warmer ice and the supply of material to the E ring was confirmed when Cassini looked back at the night side of Enceladus on November 27, 2005. The back-lit view showed sunlight shining through bright fountains of fine particles that were streaming from geysers close to Enceladus' south pole. This

made the small moon only the third world where such activity has been confirmed, leaving scientists scratching their heads about the cause.

Another mid-sized moon to come under scrutiny was Dione. To the surprise of scientists, the wispy features seen by the Voyager spacecraft on the hemisphere facing away from Saturn were found to be bright ice cliffs and braided fractures rather than thick ice deposits.

Tumbling Hyperion was shown to have a heavily cratered, sponge-like surface. With a density only about 60% as dense as water ice, it appeared to be a mixture of icy rubble and empty space.

One of the smallest objects to be observed was Epimetheus, a satellite that exchanges orbits every four years with another small moon, Janus. Epimetheus was found to be 116 km (72 mi) across, some 65 km (41 mi) smaller than Janus. Irregularly shaped and dotted with many large, soft-edged craters, Epimetheus exhibits a battered surface that is several billion years old. Janus is also angular in shape and heavily pockmarked, with several craters 30 km (17 mi) in diameter. Spectra of Epimetheus indicated that its surface is mostly water ice, but the densities of the co-orbital pair are so low that they may be mere "rubble piles."

Early in its tour, Cassini uncovered two new moons. Approximately 3 km (2 mi) and 4 km (2.5 mi) across, they were among the smallest bodies so far seen around Saturn. Named Methone and Pallene, they were found 194,000 km (120,500 mi) and 212,000 km (131,700 mi) from the planet's center, between the orbits of Mimas and Enceladus. Methone may be an object spotted in a single image taken by the Voyager spacecraft 23 years ago, when it was labeled as S/1981 S14.

A second co-orbital "Trojan" moon of Dione was also discovered. Named Polydeuces, it, too, was tiny—only about 5 km (3 mi) across. Unlike Helene, Dione's other Trojan moon, Polydeuces can get as close as 39 degrees to Dione and then drift up to 92 degrees away, taking over two years to complete its journey around the gravitationally stable Lagrange point. The extent of this wandering is the largest detected so far of any Trojan moon.

The Titan Tour

In many ways, giant Titan was the key to the success of Cassini's tour of the Saturn system. Not only was it one of the primary targets of the scientific investigations, but 45 gravity assists during the four-year primary mission would enable the spacecraft to change orbit and bounce from object to object.

Scientists were hoping that Cassini's suite of state-of-the-art instruments would be able to pierce the dense cloud blanket and reveal the true nature of its hidden surface. By combining data obtained by the onboard radar, camera, and spectrometers with the "ground truth" information sent back by Europe's Huygens probe, it was hoped that frigid Titan could be brought in from the cold.

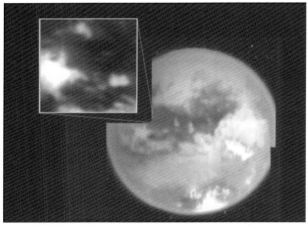

Huygens Landing Site—This mosaic was obtained during Cassini's first low-altitude flyby of Titan on October 26, 2004. The visual and infrared images were taken at altitudes ranging from 100,000 to 140,000 km (88,000 to 63,000 mi), less than two hours before the spacecraft's closest approach. Surface features are clearly visible, as well as a large methane cloud close to the south pole (bottom). The green ring marks hazes in the upper atmosphere. The inset picture shows the landing site of the Huygens probe, which touched down on the boundary of the bright and dark regions on January 14, 2005. (NASA/JPL/University of Arizona)

The first flyby took place on July 2, 2004, when Cassini soared over the south polar region at a distance of 339,000 km (210,600 mi). Scientists were relieved to see that the use of near-infrared filters and polarizers made it possible to observe surface features through the hazy atmosphere.

A patchwork of bright and dark features, some only 10 km (6 mi) across, hinted at a varied terrain modified by active weather and erosion. There were also vague suggestions of sinuous, channel-like markings—possibly river beds or deep canyons. However, the most prominent feature was an almost circular scattering of white clouds—presumably made of methane—that changed shape over a period of five hours. It was thought likely that the methane cycle of rain, channel erosion, runoff, and evaporation was most active in the relative warmth of the southern summer.

Some of the mysteries of the hidden surface were unravelled during three subsequent close encounters in October and December 2004, and April 2005. For the first time it was possible to confirm that, despite its icy temperature, the surface of Titan has been shaped by processes very similar to those found on Earth.

Several discontinuous circles were seen near the western edge of the bright Xanadu region. It was thought that these might be impact craters between 30 km (19 mi) and 50 km (30 mi) in diameter. Another 80-km (50-mi) wide crater with a dark floor and an apron of bright ejecta was seen to the east of Xanadu, while the radar captured a 440-km (273-mi) diameter crater or ringed basin. Despite its size, there was no sign of a central peak, possible evidence of erosion. Even more significant was the relative absence of such depressions—evidence that Titan's surface has been modified quite recently by active geological processes.

Other pictures showed bright features that appear to be streamlined as if were they formed by winds in Titan's atmosphere moving from west to east. The boundary between the bright and dark materials was sometimes quite complex, with some of the bright patches resembling thin surface plates that had been split and spread apart over underlying dark material. Occasional "islands" were seen in the dark "sea."

No direct evidence for the presence of fluids was found,

although long, narrow lines running across bright terrain suggested larger examples of the "rivulets" and tributary channels seen by the Huygens probe. These linear features were about 2 km (1 mi) wide, and tens of kilometers long. There was some debate over whether these were more likely to be fed by underground springs than by rainfall. There was also a sinuous feature that could be a river, roughly 1,500 km (930 mi) in length. Other dark linear features within the bright area to the west of Xanadu appeared to be the result of faulting. Their hue might be due to modification by erosion and/or deposition.

Cassini also found places where "cryovolcanism" might be erupting a slurry of liquid ammonia and water onto the frozen landscape. One circular feature, roughly 30 km (19 mi) in diameter, is thought to be a possible "ice volcano" that is releasing methane into Titan's atmosphere. Another intriguing feature is a "hot spot," 483 km (300 mi) wide, southeast of Xanadu—an area possibly warmed by a recent asteroid impact or by a mixture of water ice and ammonia oozing from an ice volcano. Also detected was a dark "lake" with a smooth shoreline that was comparable in size to Lake Ontario.

One of the most surprising findings was that weather patterns changed significantly over a couple of months. During Cassini's first close flyby in October 2004, many tall convectional clouds were seen near the south pole. By December, all that remained were a few small remnants, whereas many clouds were seen at mid-latitudes. By studying clouds imaged at different times, scientists were able to study wind patterns. Close monitoring of 10 discrete clouds revealed westerly winds that reached 34 m/s (about 75 mph or hurricane force) in Titan's middle atmosphere. This was the first direct evidence that—as on Venus—winds blow faster than the surface rotates, a phenomenon known as "super-rotation."

Much larger cloud streaks, typically 1,000 km (620 mi) long and aligned east–west were seen to drift along at only a few meters per second. Apparently these clouds originated closer to Titan's surface, perhaps from places where methane was being released into the atmosphere from below Titan's surface, or where the gas was rising over higher ground.

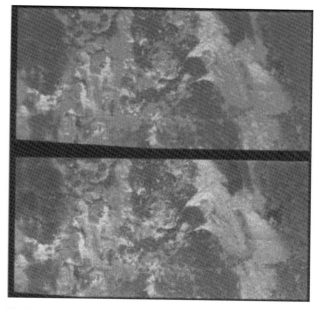

The Rough and the Smooth—This radar image of Titan's northern hemisphere, taken on October 26, 2004, shows a sharp contrast between smooth and rough terrain. The false-color version is used for easier recognition of different surface features. Brighter areas may correspond to rougher terrains, slopes facing the radar, or different materials. The pink colors enhance smaller details, while the green color represents smoother areas. Winding, linear features that cut across dark areas may be ridges or channels. A large dark circular feature is visible in the west (top left), but very few fresh impact craters are visible. The area shown measures about 150 km (93 mi) by 300 km (186 mi). (NASA/JPL)

Images of Titan's night side, in which the stratosphere was backlit by the Sun, surprised scientists by showing evidence of an entire series of haze layers. These may be evidence of gravity waves, the atmospheric equivalent of ripples on a pond, propagating upward. One detached haze layer was found to be 150–200 km (90–120 mi) higher than when it was originally detected by Voyager 1, possibly due to seasonal effects. The hazes themselves were found to be made of hydrocarbons and carbon–nitrogen compounds (nitriles). Since these chemicals are probably raining down toward the surface, they must be continually replaced by the action of sunlight or energetic particles from Saturn's magnetosphere.

Titan Comes in From the Cold

On December 25, 2004, after a 4-billion-km (2.5-billion-mi) piggyback ride that lasted almost seven years, ESA's Huygens probe slowly separated from the Cassini mother ship and headed for a near-equatorial landing site on the day side of Saturn's smog-shrouded satellite, Titan. The morning of January 14 (European time) saw the saucer-shaped probe slam into the moon's hazy upper atmosphere at a speed of 6 km/s (13,500 mph).

Protected by an ablative heat shield, Huygens' descent rapidly slowed, enabling it to deploy a 2.6-m (8.5-ft) pilot chute at an altitude of about 160 km (100 mi). After 2.5 seconds, this chute pulled away the probe's aft cover so that the main 8.3-m (27-ft) parachute could billow out to stabilize its swinging motion. Once the front shield was released, the six scientific instruments swung into action, sampling the atmosphere, taking pictures, and savoring the strange, alien environment. Microphones on board Huygens even recorded the sound of wind rushing by the probe during its descent.

Huygens was designed to float in case it landed in a sea or lake, but—to the dismay of some mission scientists—its 2 hour 28 minute descent ended with more of a squelch than a splash. It was the first time that a spacecraft had touched down on such a remote world. Huygens not only survived the jolting impact but it continued transmitting from the surface for several hours, long after the Cassini orbiter dropped below the horizon and was no longer able to record the data.

More than 474 megabits of data were received, including some 350 pictures taken during the descent and on the ground. The atmosphere was probed and sampled from an altitude of 160 km (100 mi) all the way to the surface, revealing a uniform mix of methane with nitrogen in the stratosphere. Methane concentration increased steadily in the troposphere (lower atmosphere), with clouds of

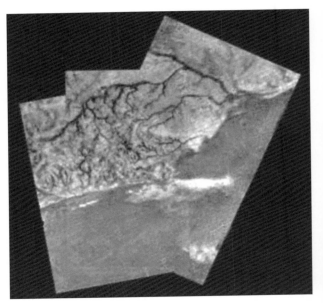

Methane Rivers?—These three frames were taken from an altitude of 16 km (10 mi) during the descent of ESA's Huygens probe to the surface of Titan on January 14, 2005. A light, high ridge area is marked by a series of black, branching channels. Similar to river patterns on Earth, this suggests that the channels have been cut by flowing liquid—presumably methane. The dark hue of the channels and the nearby lowland supports the theory that hydrocarbon dust has been washed into the dry river beds and the low-lying terrain—possibly an evaporated sea or lake. (ESA/NASA/JPL/University of Arizona)

methane at about 20 km (12.5 mi) altitude and methane or ethane fog near the surface. There were even indications of rain falling from some of the clouds. New evidence based on finding atmospheric argon 40 indicated that Titan has experienced volcanic activity generating not lava, as on Earth, but water ice and ammonia.

The only disappointment was the failure to receive Doppler data and some images through one of the receiver channels, due to human error. However, the probe's signal was closely monitored by a global network of radio telescopes on Earth, making it possible to analyze varia-

Tangerine Titan—The first color image from the surface of Titan was returned by the Huygens probe on January 14, 2005. The surface was darker than expected, consisting of numerous pebble-sized stones made from a mixture of water ice and hydrocarbons. The largest boulder, lying about 85 cm (34 in) from the camera, was about 15 cm (6 in) across. Hollows at the base of these ice pebbles resemble those caused by erosion on river beds. The hazy appearance of the image suggested the possible presence of methane or ethane ground fog. (ESA/NASA/ University of Arizona)

tions in Titan's winds by reconstructing the descent trajectory with a remarkable accuracy of 1 km (0.6 mi).

Results showed that the winds generally blow in the direction of Titan's rotation (from west to east) at nearly all altitudes. They were weak near the surface, increasing slowly with altitude up to about 60 km (37.5 mi). Higher in the atmosphere, Huygens was given a rough ride, probably due to significant vertical wind shear. The maximum speed of roughly 430 km/h (265 mph) was measured about 10 minutes after the start of the descent, at an altitude of about 120 km (75 mi).

Panoramic images returned by the Descent Imager-Spectral Radiometer revealed strong evidence for a methane cycle similar to the water cycle that prevails on Earth. Overlapping snapshots taken as the probe rotated beneath its parachute showed a complex network of black, narrow, steep-sided, drainage channels running from brighter highlands to lower, flatter, dark regions.

The branching channels—similar to drainage patterns on Earth—merged into river systems that seemed to run into lake beds featuring offshore "islands" and "shoals." Although any surface liquid seemed to have evaporated or seeped underground, apparently leaving the alien rivers and lakes dry, it was thought that rain may have occurred in the recent past. White patches seemed to show tendrils of ground fog, made not of water but perhaps of ethane or methane.

The much lighter uplands visible in the images were actually hills that rose about 200 m (660 ft) above the dark dry lakes. They were made of water ice (which is rock-hard at Titan's temperature of about −180°C), covered by deposits of organic drizzle that settles out of the atmosphere. Scientists speculated that methane rains washed the hydrocarbon dust off the hillsides into rivers, eventually depositing it on the lowlands.

A penetrometer on the probe recorded the moment when it hit solid ground at a speed of 4.5 m/s (10 mph), experiencing a brief deceleration that scientists attributed to striking a boulder on landing. The surface was described as resembling a wet sand or clay overlain by a thin, solid crust, and mainly composed of a mixture of water ice and hydrocarbons. Instruments sniffed bursts of methane gas boiling out of the surface as the result of heat generated by Huygens.

Bathed in subdued orange light, the landing site resembled a dry river bed, overlain by small, rounded pebbles between 3 mm (0.15 in) and 15 cm (6 in) across. The absence of sharp edges and hollows at the bases of some rocks suggested that the area was sometimes inundated by flash floods of liquid methane. The hazy atmosphere suggested the possible presence of methane or ethane ground fog.

Frederick William Herschel—Frederick William Herschel (1738–1822) is perhaps most famous for his discovery of Uranus, the first planet found since antiquity, on March 13, 1781. Herschel was born in Hanover, Germany, and became well known as both a musician and an amateur astronomer. He emigrated to England in 1757, and, with his sister Caroline, began making the most advanced astronomical instruments of the time. The discovery of Uranus was made using a home-made 15.7-cm (6.2-in) reflector. His later creations included the largest telescope of the day—a 12-m (40-ft) long instrument with a 1.9-m (48-in) mirror.

Appointed the personal astronomer to King George III (after whom he named the new planet), he later discovered two satellites of Uranus (Titania and Oberon) in 1787, followed by two moons of Saturn (Mimas and Enceladus) in 1789.

In 1800 he discovered what he called "calorific rays" (now known as infrared radiation) during studies of the "rainbow" created when light is divided into its colors by a prism. It was the first time that someone had shown the existence of forms of light that our eyes cannot see. At the very end of his life he was elected to be the first president of the newly founded Royal Astronomical Society.

9 URANUS: THE TOPPLED GIANT

The Georgian Star

For more than 5,000 years it was generally accepted that there were just five planets (in addition to the Earth) circling the Sun. There had been some suggestions of another planet waiting to be found between the orbits of Mars and Jupiter, but no one seems to have thought of any worlds lurking beyond Saturn. Imagine the impact on the scientific world of the chance discovery of a seventh planet by an amateur astronomer.

The man responsible for this upheaval, William Herschel, had been born in Hanover, one of the small states that made up Germany at that time, but settled in Bath, England, in 1757. He earned a living first as a musician and music teacher, then from 1767 as an organist, composer, and choirmaster. However, he also began taking a keen interest in astronomy and devoted much of his spare time to making highly efficient telescopes and observing the stars.

The event that shook the astronomy establishment and changed Herschel's life took place on March 13, 1781. As part of an ambitious plan to review all stars down to magnitude 8, Herschel was surveying the constellation of Gemini using a 7-ft (2-m) long reflector fitted with a 15.7-cm (6.2-in) primary mirror. Near the star Zeta Geminorum, he noticed an unusually large, bright object, which he recorded in his journal as "either a nebulous star or perhaps a comet."

Further observations made over the next few days seemed to confirm that the newcomer was a comet, since it moved against the background of "fixed" stars. However, the object's sharp outline and non-cometary path across the sky soon led the leading astronomers of the day to conclude that Herschel's "comet" was, in fact, a new planet.

Calculations of its orbit were simplified by the realization that the planet had been—unknowingly—recorded by previous observers. This is not too surprising, since the planet is just visible to the naked eye, if you know exactly where to look. The first Astronomer Royal, John Flamsteed, had noted it no fewer than six times between 1690 and 1715, without realizing its true nature or significance.

By taking these observations into account, it became clear that the latest addition to the planetary roster was so remote that it took more than 80 years to circle the Sun once. (The duration of the Uranian year is now known to be 84 Earth years.) Its mean distance from the Sun turned out to be a staggering 2.87 billion km (1.78 billion mi), twice as far as Saturn, which was previously the most distant planet. At a stroke, Herschel had doubled the diameter of the Solar System.

Needless to say, the German immigrant became a global celebrity overnight. Every possible honor came his way, including a knighthood. King George III gave him a stipend of £200 a year, and appointed Herschel as his private astronomer.

The grateful Herschel wanted to name the planet after his royal sponsor, but, not surprisingly, this was far from popular outside the British empire. French astronomer Joseph Lalande, one of the first to calculate its orbit, generously suggested that the planet be named in honor of its discoverer, while Johann Bode, editor of the *Astronomical Yearbook of Berlin*, put forward the name Uranus. For the next 60 years, the various competing names were used in different countries, but, eventually, common sense prevailed and Bode's proposal was finally accepted. The seventh planet was called Uranus, after the first mythological ruler of Olympus, the father of Saturn and grandfather of Jupiter.

Meanwhile, established as the most famous astronomer of his era, Herschel continued to be fascinated by the world he had discovered. Using his new reflector, which had a length of 6 m (20 ft) and a 46-cm (18.2-in) mirror, he was able to discover two satellites of Uranus—Titania and Oberon—in 1787.* Through meticulous study of their motion, he came to the surprising conclusion that the satellites' orbits were steeply inclined to the ecliptic.

By careful estimation of the planet's diameter and of Oberon's orbital period, Herschel was able to make the first reasonably accurate determination of Uranus' mass, volume, and density. He also thought that he could see some evidence of a ring system, but later dismissed it as an optical illusion. The impression that Uranus was flattened at the poles, however, was not an illusion. In 1794, when the equatorial region of Uranus was directed toward Earth, he noted, "The planet seems to be a little lengthened out, in the direction of the satellites' orbits."

Meanwhile, the introduction in 1789 of a gigantic 12-m (40-ft) telescope with a 1.2-m (48-in) mirror paid immediate dividends with the discovery of two satellites of Saturn: Enceladus and Mimas. The giant proved very cumbersome, however, and most of Herschel's observations continued to be made with its smaller predecessor.

Herschel continued to make new scientific breakthroughs, including the discovery of invisible infrared light beyond the red region of the rainbow. He died on August 25, 1822, just short of his eighty-fourth birthday. His remarkable life had

* The names were chosen 65 years later by John Herschel, William's son, who was also a noted astronomer. Avoiding Greek and Roman mythology, the normal source for Solar System names, he chose fairies and gnomes from Shakespeare's plays, *A Midsummer Night's Dream* and *The Tempest*, and Alexander Pope's *The Rape of the Lock*.

lasted almost exactly the period of time it takes Uranus to complete one revolution around the Sun.

The Seventh Planet

Uranus was well named, as it is, indeed, a giant. Early calculations soon indicated that the planet is about four times the Earth's diameter, though its great distance meant that the accurate measurement of Uranus' size remained a problem until quite recently. The current figure for its equatorial diameter is 51,118 km (31,763 mi), meaning that 64 Earths would fit inside Herschel's world. Only Jupiter and Saturn are larger than Uranus in the Sun's domain.

Despite its impressive size, the newcomer was found to be a middleweight rather than a heavyweight. By studying the motions of its satellites, its was relatively easy to calculate that Uranus is more than 14 times as massive as Earth, resulting in a high escape velocity of 21 km/s (13 mi/s). However, the overall density is only a little higher than that of water, suggesting that it consists mainly of light materials, probably hydrogen, water, helium, ammonia, and methane.

Uranus' low density and bloated size mean that gravity at its visible surface is actually lower than on Earth. Assuming that someone could stand on its clouds, they would be so far from the center of the planet that their weight would be about one-fifth lower than on Earth.

In a telescope Uranus appears as a small blue–green disk. Until recently, it was very difficult to see any surface detail with even large instruments. However, ground-based observers were able to report irregular changes in brightness, along with relatively dark poles and a light equatorial belt sandwiched between two dark bands.

Since the surface was so difficult to resolve, telescopic observers were frustrated when they tried to measure the planet's period of rotation. Indications that the planet's shape was flattened, like Jupiter and Saturn, led to the suggestion that the rate of spin was fairly rapid—probably between 7¼ and 12½ hours.

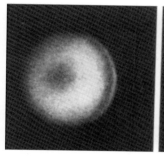 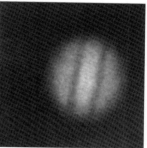

Tilted Uranus—Until the late twentieth century, ground-based observations of Uranus showed very little detail. In a telescope, the planet appears as a bluish-green disk, about 4 arc seconds in diameter (450 times smaller than a full Moon). Although large telescopes often showed no features at all, the best visual observers commonly recorded two faint belts on either side of a bright equatorial zone and darker poles. These sketches by Eugène Antoniadi, showing the north pole (left) and equator (right) were made during observations with the 0.8-m (33-in) refractor at Meudon Observatory. (From *L'Astronomie*, vol. 50, 1936)

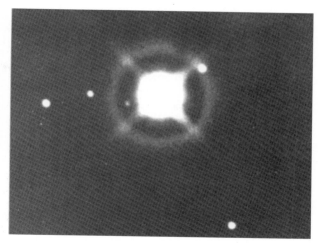

The Discovery of Miranda—Until the flyby of Voyager 2 in 1986, only five moons of Uranus were known. Above is the photograph in which the fifth of these, Miranda, was discovered by Gerard Kuiper on March 1, 1948. The picture, taken with the 2.1-m (82-in) reflector at McDonald Observatory, shows an overexposed planet (center) and the satellites (left to right) Titania, Umbriel, Miranda (barely visible to the left of Uranus), Ariel, and Oberon. The apparent ring was caused by the telescope's optics. (McDonald Observatory)

Not until the early twentieth century was it possible to search for a Doppler shift in the spectrum of the planet's atmosphere. Calculations by Percival Lowell and Vesto Slipher suggested that its rotation was 10 hours 50 minutes retrograde (east to west). Later spectral studies indicated that Uranus spins on its axis once every 15 hours 30 minutes, though it was accepted that this figure could be in error by several hours. Other research using brightness changes suggested a period of 23.9 hours.

If the rotation period was a mystery, observers found that there was nothing unusual about the orbit, which is fairly circular and very close to the plane of the ecliptic. The great distance of Uranus from the Sun means that it needs to progress along its orbit at a mere 6.8 km/s (4 mi/s) so that it takes 84 years to travel once around the Sun. Consequently, Uranus has only completed two orbits since Herschel discovered it—or, put another way, a Uranian born in 1781 is still less than 3 "years" old!

The discovery of satellites was the key to another discovery. By recording their paths around the planet, it was soon established that the axis of Uranus is inclined at about 98 degrees to the plane of its orbit. This unusual arrangement means that it lies on its side and appears to "roll" around the Sun. Sometimes the axis points more or less directly at the Sun, so that one or other of the poles receives more heat than the equator. This means that the poles take it in turns to be the warmest areas on the planet.

However, each pole also remains in total darkness for 21 years during every orbit. In the mid-1980s, around the time of the Voyager 2 encounter, the south pole was pointing toward the Sun while the north pole was in the middle of its long winter. Today, Uranus has moved around its orbit, so that we now see the equatorial zone.

The reason for this peculiar inclination is unknown, but one favored theory is that an Earth-sized object smashed into a young Uranus, tipping it over on its side. The collision with such a large planetoid may also have led to the creation of the Uranus satellites.

Deprived of warmth from the distant Sun, the planet's atmosphere would resemble a deep freeze where a thermometer registered –210°C. However, as with so many other aspects of the planet, the composition and internal structure of Uranus remained rather speculative.

Spectroscopic studies showed that the atmosphere appeared to consist of the same primordial gases as the other giant planets, though in differing proportions. The detection of hydrogen indicated that this was probably the dominant constituent of the atmosphere, though perhaps less so than on Jupiter and Saturn.

The only other gas definitely identified was methane, whose strong spectral absorption lines suggested that Uranus possessed up to 10 times as much of this gas as Jupiter. The methane accounted for the blue–green color of Uranus, since it absorbs most of the incoming red light and scatters a high proportion of the blue light. In addition, helium was thought to be present, together with some ammonia which probably formed clouds at greater depth than on Jupiter and Saturn, thus avoiding detection.

Observations also showed that the upper atmosphere is remarkably clear compared with the dense haze and clouds of ammonia crystals above Jupiter and Saturn. Sunlight is, therefore, able to penetrate to a considerable depth before being reflected.

The internal structure of Uranus was far from clear. Early models suggested a large rocky core overlain by water ice and an atmosphere of hydrogen, methane, and ammonia. By the early 1980s, this was refined to a three-layered planet with a fairly small rocky core, overlain by a liquid mantle of water, methane, and ammonia, and topped by a thick hydrogen–helium atmosphere. Whether Uranus possessed a significant internal heat source remained a mystery.

Rings and Satellites

Until 1977, astronomers assumed that the only planet to boast a set of rings was Saturn. Although various observers, including William Herschel, had reported rings around

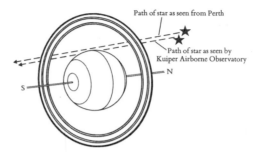

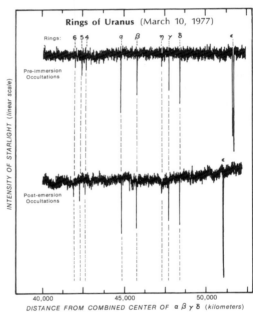

Rings of Uranus (March 10, 1977)

The Discovery of Rings—On March 10, 1977, astronomers around the world prepared to observe a rare occultation of a star by Uranus. One team using the infrared telescope on board the Kuiper Airborne Observatory (KAO) noticed a number of dips in the starlight when the planet passed in front of the ninth magnitude star SAO 158687. By combining observations from other locations, it was eventually agreed that nine rings had been recorded during the occultation. (James L. Elliot, *Annual Reviews of Astronomy and Astrophysics*, 1979)

Uranus and Neptune at various times, these claims had always been withdrawn or lacked confirmation.

The event that caused the text books to be rewritten took place on March 10, 1977. Taking advantage of a prediction that Uranus would pass in front of a star known as SAO 158687, astronomers in Australia and around the shores of the Indian Ocean prepared for the rare event. Their intention was to time very precisely the moments when the star was extinguished by the intervening bulk of the planet and when it reappeared. These measurements, taken from many different locations, would enable astronomers to refine their data for the planet's size and orbit.

Flying over the Indian Ocean in the Kuiper Airborne Observatory (named after the discoverer of Miranda), an American team led by James Elliot turned the 91-cm (36-in) telescope toward the modest red star. Since the exact time of the occultation was not known, they began to record the brightness of the infrared light coming from the star well in advance of the predicted occultation. To their astonishment, they noticed a series of blips in the signal before and after the star was eclipsed. Hundreds of kilometers away, a team at Perth Observatory observed similar blips in the starlight, even though Uranus passed to the north of the star. Other groups as far away as India, Japan, and South Africa also reported unexpected events.

The initial knee-jerk reaction of the Kuiper team was to suspect instrument malfunctions or the presence of one or more unknown satellites. However, when Elliot returned to the United States, he and his wife Elaine spread the full data record across their living room floor. Only then did he realize that five notable dips in the starlight appeared as a mirror image on either side of the planet. There was no doubt that his team had discovered a set of narrow rings around Uranus.

When the data from other sites were taken into account, it eventually became clear that no fewer than nine rings had been detected. The three rings closest to Uranus and first recognized by the Perth group, led by Robert Millis, were known as 6, 5 and 4. The five original rings announced by Elliot's team were given Greek letters (alpha, beta, gamma, delta and epsilon) in alphabetical order that corresponded

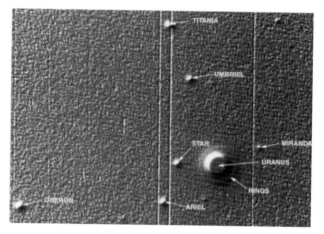

The First Ring Image—Infrared images of Uranus' rings were obtained soon after their discovery. However, the first photograph in visible light was taken in 1984 by Richard Terrile (JPL) and Bradford Smith (University of Arizona) at the Las Campanas Observatory in Chile. The digital image was computer processed to create a false three-dimensional texture in the image. The ring system shows up next to the brighter planet, but the individual rings were not visible. (NASA–JPL)

to their distance from the planet. A less obvious sixth ring that was identified at a later date was labeled eta. No explanation for the omission of the sixth Greek letter, zeta, was given. This complex and haphazard system of nomenclature has survived to the present day.

From the very short blips in the light record, it was obvious that the rings are very narrow, generally no more than 10 km (6 mi) across. The exception was the outer, epsilon ring, whose width increased from 20 to 100 km (12.5 to 62 mi) with distance from Uranus. Even more surprising was the fact that six of the rings are slightly elliptical, rather than circular. In particular, the distance of the epsilon ring from Uranus was found to vary by 800 km (500 mi).

The rings were shown to begin about 16,000 km (10,000 mi) above the planet, with their outer edge about 25,000 km (15,500 mi) from Uranus' cloud tops. Unlike most of the rings around Saturn, they seemed to have very sharp edges and were very dark, reflecting less light than coal dust.

None of the rings completely blocked the light from the star, indicating that they consist of small boulders, stones or even dust. Scientists suggested that they were created when tidal forces tore apart one or more former moons after approaching too close to Uranus. Exactly what confined the ring particles in such well-defined paths remained a mystery, although one theory suggested that the gravity of small satellites within the ring system kept them under control.

If the rings were a surprise, Uranus' small retinue of satellites seemed relatively uninspiring. Their small sizes and great distance from Earth meant that, prior to the Voyager 2 flyby of 1986, little was known about them. Perhaps the most notable characteristic was that, despite the topsy-turvy rotation of Uranus, all five moons followed circular orbits above the equator, traveling in the same direction as the planet below. This relationship led most astronomers to assume that they were formed during a massive collision that caused the planet to topple over on its side.

Astronomers could only estimate their diameters according to their apparent brightness and measured albedos (light reflectivity). They were all modest in size, between 300 and 1,100 km (190 and 690 mi) across. Their distances from the center of the planet varied from 130,000 km (81,000 mi) for Miranda to 586,000 km (364,000 mi) for Oberon. Miranda scooted around Uranus in less than 1½ days, while Oberon took a more leisurely 13 days.

Oberon and Titania—the moons discovered by William Herschel—were the brightest and considered to be very similar in size. Ariel and Umbriel, second and third out from Uranus, were discovered by an English amateur astronomer, William Lassell, in 1851. Little was known about Miranda, the smallest satellite, which had been found during photographic observations by Gerard Kuiper in 1948.

Spectroscopic studies showed that Oberon, Titania, and Ariel were probably covered with water ice or frost. Umbriel seemed less icy, suggesting bare rock or soil on its surface. None of them was large enough to have any atmosphere, and they all seemed to resemble many of Saturn's intermediate-sized moons.

Voyager's Visit

Although Uranus had been known and studied for 200 years, our knowledge of this remote world was still very scanty until a hardy robotic ambassador from Earth succeeded in going where no spacecraft had gone before. However, the third planetary encounter during Voyager 2's "Grand Tour" of the outer Solar System was far from a foregone conclusion.

When Voyager 2 began its odyssey in August 1977, its mission was straightforward—to combine with its sister craft in exploring the realms of Jupiter and Saturn. Thanks to the painstaking work of JPL graduate student Gary Flandro in the 1960s, planners were aware that a rare alignment of the outer planets opened up the possibility of an extended voyage to Uranus and Neptune, even though the Voyagers were optimized for operations no further out than 10 AU. The key to Voyager 2's future lay with its twin. If Voyager 1 could successfully complete its reconnaissance of Jupiter and Saturn, the second spacecraft could be let loose in the dark depths of the Solar System.

The moment of truth came in November 1980, when Voyager 1 swept past Titan and was diverted upward, away from the ecliptic plane, by Saturn's mighty gravity. This triumph of celestial navigation and technical prowess opened the door of opportunity for Voyager 2's Grand Tour. Two months later, NASA announced its decision to add Uranus to the spacecraft's itinerary—despite an assessment which concluded that the chances of survival for another five years were only 60–70%.

During its closest approach to Saturn on August 25, 1981, Voyager 2 received a gravity assist from the giant ringed planet. The wind seemed set fair for humanity's first visit to Uranus, but almost immediately a problem arose when the scan platform, which pointed the cameras and a number of other instruments toward their targets, suddenly seized. From now on, the platform would have to be used sparingly and swiveled at low speed.

Voyager 2 was already suffering from a number of

Voyager 2 at Uranus—This artist's impression was painted before the Voyager 2 encounter with Uranus on January 24, 1986. It shows the nine known rings, which were almost fully open to Earth-based observers. Voyager is approaching the planet's southern hemisphere. Since the polar axis of Uranus lies "side on" to the plane of its orbit, the planet resembles a spinning top that has toppled over. As a result, the dark northern hemisphere was always hidden from Voyager's cameras. (NASA–JPL)

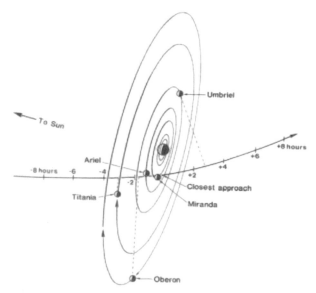

The Voyager 2 Flyby—8½ years after leaving Earth, Voyager 2 completed its third flyby of a giant planet, sweeping just 81,600 km (50,700 mi) above the cloud tops of Uranus. Like an arrow flying toward a target, the spacecraft saw Uranus, its rings and orbiting satellites arranged like concentric rings before it. This weird alignment meant that the best photo opportunities for the planet, rings, and moons were squeezed into a period of less than six hours around the time of closest approach. (NASA–JPL)

malfunctions, most notably the loss of its primary radio receiver and damage to the backup receiver, which could now only pick up signals in a single, narrow waveband. In addition, its nuclear power source was inexorably draining away with the passage of time. A power output of 470 watts at launch would be limited to about 400 watts when Voyager arrived at Uranus—not enough to operate all spacecraft subsystems simultaneously.

Flyby conditions also left a lot to be desired. Since Uranus orbits about twice as far from the Sun as Saturn, the spacecraft's cameras would have to cope with light levels four times lower during the forthcoming encounter. This increased the likelihood of image smear during the high-speed flyby, particularly if the spacecraft suddenly jolted while the shutters were open. Matters were not helped by the intense darkness of some of the objects around

Uranus. "Taking pictures of the rings and some moons is like trying to photograph a piece of charcoal against a black backdrop," said Voyager scientist Richard Terrile.

Engineers managed to prevent any unwanted motion by programming the attitude control system to fire the thrusters and gently correct its position. Instead of relying on the balky scan platform to keep a target in view, the entire spacecraft was programmed to turn at the correct rate, like a camera panning on a fast-moving vehicle—even though this caused the main antenna to point away from Earth.

Greater distance also meant reduced data rates from Voyager's instruments. In order to minimize the loss of information, the spacecraft was programmed to compress the image data, transmitting only the change in brightness between neighboring pixels rather than the absolute brightness values.

Finally, a decision was made to upgrade the ground facilities by electronically combining the signals received by several antennas. For example, the 64-m (210-ft) and two 34-m (105-ft) dishes of NASA's Deep Space Network in Australia were combined with the 64-m Parkes radio telescope, more than doubling the potential data rate from Uranus.

Just as everything seemed to have been prepared for the historic encounter, photographs from Voyager began arriving with large blotches. With only four days to go before closest approach, the ground team succeeded in sending a patch to overcome the computer's memory failure.

The hard work paid off. On January 24, 1986, after a hazardous passage lasting almost nine years, Voyager 2 began its brief reconnaissance of the seventh planet from the Sun. More than 200 years after its discovery, Herschel's "Georgian Star" was about to be transformed from a pale blue enigma into a three-dimensional world with its own unique quirks and characteristics.

Uranus Unveiled

Excitement and anticipation were filling the corridors of the Jet Propulsion Laboratory in Pasadena long before Voyager 2 arrived in the vicinity of Uranus. Starting on November 4, 1985, 81 days before closest approach, the spacecraft began a long-range survey of the fast-approaching planet, its rings, and satellites.

Curiously, as the days passed and Uranus grew larger in the images, the planet refused to reveal any atmospheric features. Frustrated scientists began to tire of staring at a world that was "as featureless as an airbrushed Ping-Pong ball." With nothing noteworthy to report, the imaging team resorted to arguments over whether Uranus was bluish-green or greenish-blue.

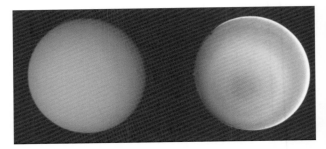

Pale-Blue World—These two views of Uranus were returned on January 17, 1986, when Voyager 2 was 9.1 million km (5.7 million mi) from the planet. The planet's blue–green color results from absorption of red light by methane gas in the atmosphere. By using false color and extreme contrast enhancement, details become visible (right), including a dark polar hood surrounded by concentric bands. (Compare the drawings by Antoniadi on p. 219.) One explanation is that a brownish haze or smog concentrated over the pole is arranged into bands by circulation in the upper atmosphere. (NASA–JPL)

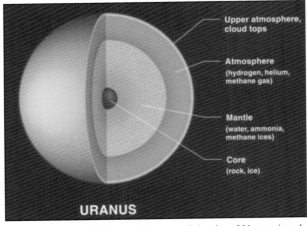

Internal Structure of Uranus—The overall density of Uranus is only a little higher than that of water, making it less substantial than all the planets except Saturn. For many years the interior of Uranus remained a mystery, although ground-based instruments revealed the presence of methane and hydrogen gases in the atmosphere. By the 1980s, models favored a world with a hot rocky core, surrounded by an icy mantle of frozen water, methane, and ammonia. (LPI)

Finally, in late December, shy Uranus began to show that it had a character after all as computer-enhanced pictures revealed elusive bands of cloud around a dark polar hood. But what was happening beneath the bland exterior? Not until 10 days before the flyby did discrete clouds pop out of the haze, providing an opportunity to assess the atmospheric circulation.

The rare methane clouds resembled much larger versions of the "anvil" clouds that grow at the top of thunderstorms on Earth. One 4,000-km (2,500-mi) cloud near 33°S, stretched into a narrow ribbon by wind action, testified to the presence of convection as warmer air bubbled up from below.

The first important conclusion was that the winds between 20 and 45°S roar around the planet from east to west—the same direction as the planet's rotation at speeds of 40–160 m/s (90–360 mph).

Despite Uranus' tilt and the summer warmth at the south pole, the winds were blowing parallel to the equator, just as they do on Jupiter and Saturn. Clearly, the planet's rapid rotation was the dominant factor in controlling the visible motion of the atmosphere.

Studies of cloud shapes and movement suggested that wind velocity increased with altitude between 27 and 70°S. This surprising conclusion contrasted with theoretical expectations, based on the fact that the polar regions receive more heat from the Sun than the equator during the long Uranian year. Scientists assumed that the expected flow from a "warm" pole to a cooler equator must be occurring in a manner unseen by Voyager's instruments.

The spacecraft's infrared spectrometer confirmed that Uranus is an extremely cold place, with an average temperature of –210°C. In all directions the temperatures were about the same at similar levels in the atmosphere, with the exception of a slightly colder band between 15 and 40°S. Even the sunny south pole had a temperature comparable to the north pole, which had not seen the Sun for two decades.

As expected, the main constituents of the atmosphere were found to be hydrogen and helium, although the abundance of helium—about 15%—was much lower than some Earth-based studies had suggested. On the other hand, it was comforting to know that this figure was comparable to the amount of helium thought to be present in the solar nebula from which the planets formed. Other minor constituents were methane, deuterium, and ammonia. (Acetylene, a more complex hydrocarbon, had been found by ground-based observations.)

Until the Voyager encounter, no one was sure about the planet's internal rotation rate. The mystery was finally solved by studying the magnetic field and radio signals from the interior. A day on Uranus was found to last for 17 hours 14 minutes, 1½ hours longer than expected. Curiously, there was no significant internal heat source—even though one was later found at similarly sized Neptune.

Spacecraft measurements of the planet's elusive magnetic field and its associated magnetic tail showed that it extended at least 10 million km (6.2 million mi) behind the planet. The strength of the field at Uranus' surface was found to be roughly comparable to that on Earth.

Trapped hydrogen ions (rather than heavier particles) populated the radiation belts, which were similar in intensity to those of Saturn. As the largest moons of Uranus plough through these belts, irradiation by charged atoms rapidly darkens any methane on their icy surfaces, possibly helping to explain why they are some of the blackest objects known.

Although a strong magnetic field was no surprise, its unique alignment and location were unexpected. Most planets, including Earth, resemble bar magnets, with the "bar" positioned close to their centers and offset by a modest amount from the north–south axis. In contrast, Uranus possessed a magnetic field that was tilted at a 60-degree angle to the spin axis. Struggling to explain this oddity, some scientists speculated that Uranus was undergoing a reversal of its magnetic field.

The center of the dipole magnet was offset from the planet's center by almost 7,700 km (4,800 mi)—about one-third of the distance to Uranus' cloud tops. This meant that (like Earth's magnetic field) the north magnetic pole lay in the southern hemisphere, but at a latitude of 15.2°S. The south magnetic pole was at 44.2°N, almost halfway between the equator and the planet's north pole.

The odd tilt of the magnetic field, coupled with the fact that the planet lies on its side, causes the magnetic field lines to twist like a corkscrew as Uranus spins. Since the field cannot be created in the planet's central rocky core, scientists concluded that it is generated by currents in the mantle, an electrically conductive, superpressurized "ocean" of water and ammonia.

Since scientists were not expecting the magnetic poles and auroral zones to be so close to the equator, most scans by Voyager's instruments were aimed toward the wrong places. However, the ultraviolet experiment did find some evidence of an aurora—on the night side only. It also found a mysterious "electroglow," somehow caused by energized electrons above the day side.

Surprising Satellites

Perhaps the Voyager team had been spoiled by the remarkable revelations from the twin spacecraft's visits to Jupiter and Saturn, but expectations were not too high as the third giant planet drifted into view. In particular, the modest system of five medium-sized satellites seemed less promising than the four Galilean giants, smoggy Titan, and the various icy worlds that circled the inner gas planets.

The consensus was that the moons of Uranus must be heavily cratered, long-dead relics left over from the formation of the planet. As such, no evidence of geologic activity was anticipated. However, once again, Voyager 2 challenged these assumptions and left the scientists scrambling to explain the strange landscapes that were displayed to human eyes for the first time.

Since the spacecraft was diving toward the Uranian system like a dart spearing into a dartboard, there would be little opportunity to shift its gaze from one satellite to another. Although long-range images were available for weeks beforehand, the most detailed pictures of all five

Uranus' Largest Moons—This mosaic of Voyager 2 images shows Uranus surrounded by the five largest satellites and Puck, the first of the 10 small moons discovered during the spacecraft's flyby. Clockwise from lower left: Ariel, Umbriel, Oberon, Titania, Miranda, and Puck. The satellites are grayish in color and tend to increase in density with distance from Uranus. Miranda is mainly ice, but its larger neighbors must be at least 50% rock. The planet and satellites are not to scale. (NASA–JPL)

satellites had to be squeezed into a few hours on January 24, 1986, the day before closest approach.

The trajectory was also influenced by the need to pass close to Uranus for a gravitational assist toward Neptune. This meant that the scientists' preference for images of the larger moons had to take second place, while Miranda, the junior member of the family, would receive the most detailed scrutiny.

The first satellite to be unveiled was Oberon, the outermost of the familiar five. The best pictures taken by Voyager's cameras were gathered from a distance of 660,000 km (410,100 mi), about nine hours before closest approach to Uranus. Since this was the most distant of the satellite flybys conditions were not very favorable, but, despite a spatial resolution that was restricted to about 12 km (7.5 mi), scientists were delighted with the results.

With a diameter of 1,523 km (946 mi), Oberon was found to be only slightly smaller than Titania. As expected,

Amazing Miranda—Miranda displays a unique landscape never seen on any other Solar System body. Only 485 km (300 mi) across, its surface is dominated by three "coronae"—a bright, angular "chevron" and two "ovoids" that resemble race tracks or layered cake. This Voyager 2 mosaic shows the moon's southern hemisphere. The satellite's south pole lies in the center, next to the chevron. Brightly illuminated on the limb at top right is Verona Rupes, a 20-km (12.5-mi) high cliff that probably marks a major fracture in the surface. (NASA–JPL)

much of the surface was ancient, cratered terrain. Several large craters had dark floors and were surrounded by rays of lighter material ejected during their formation. The largest of these (named Hamlet) had two dark spots on either side of a bright central peak. It seemed that some isolated, icy volcanism had occurred in the distant past. Another peak, probably rising from a large impact crater, was visible on the limb, rising 11 km (7 mi) above its surroundings. Further evidence of a more active period included linear features that appeared to be traces of enormous faults—crustal cracks that marked a global upheaval on the gray, icy world.

Titania proved to be remarkably similar to Oberon in size, density, color and reflectivity. However, its cratering record proved to be quite different. Voyager's cameras showed few large craters, suggesting that the basins gouged during the early bombardment by planetesimals had mostly been erased. Scientists speculated that Titania had become warm enough to melt and resurface itself, wiping out its original large craters. Only later did smaller, relatively slow-moving debris in orbit around Uranus pepper the fresh ice to produce a population of rounded hollows. Also noticeable was an extensive network of branching faults, including several rift valleys 20–50 km (12–31 mi) in width and 2–5 km (1–3 mi) deep. These tectonic features appeared to have been created during a late stage of crustal stretching.

In early Voyager images, Umbriel displayed a surprisingly uniform surface. As the satellite drew closer, it maintained this bland exterior, with the notable exception of a small, bright marking near its equator. The highest resolution images, capable of picking out objects about 10 km (6 mi) across, showed a uniformly dark object, apart from the bright deposits on the floor of one crater (now called Wunda), a bright crater peak not far away, and some cliffs on the moon's opposite side.

Why is Umbriel so much darker than its neighbors, particularly Ariel? With a diameter of 1,169 km (726 mi), it is possible that Umbriel was too small to melt completely in its youth. One suggestion is that global ice volcanism resurfaced its outer layers, coating them in carbon-rich materials that were too thick for meteorite impacts to penetrate.

Compared with its staid companions, 1,158-km (720-mi) Ariel seemed to have enjoyed a number of wild episodes during its lifetime. Faults and rift valleys over 10 km (6 mi) deep and hundreds of kilometers long were much larger and more widespread than on Titania, suggesting major upheavals. A relatively bright surface and extensive areas of smooth terrain suggested that a mix of water ice, ammonia, and methane had risen to fill in many of the major valleys and impact craters.

Against all expectations, little Miranda turned out to be the star of the show. Early Voyager images revealed an intriguing V-shaped feature unlike anything ever seen before, but no one was prepared for the images taken less than two hours before closest approach. Miranda's southern hemisphere displayed a bewildering variety of unique landforms, including the "chevron" seen from afar and two huge oval features nicknamed "circi maximi" after Roman chariot-racing tracks. All three were marked with parallel faults and ridges. On the day–night terminator could be seen a 20-km (12-mi) cliff, now known as Verona Rupes. In the weak gravity of Miranda, an object would take almost 10 minutes to fall to the valley floor below.

Scientists are still struggling to explain this tortured terrain. One early suggestion was that the moon was shattered by a huge impact, then slowly reassembled itself. Today, scientists believe that the three "coronae" represent parallel ridges produced by stretching of the crust and upwelling of icy material. The internal separation of ice and rock ceased prematurely when the satellite cooled before its differentiation was completed.

Shepherd Moons

Before the arrival of Voyager 2, Uranus was known to have five modestly sized moons, but, based on their experiences during the Jupiter and Saturn encounters, scientists expected that count to rise significantly. They were not disappointed. The first newcomer, temporarily designated 1985 U1 and now known as Puck, was found on December 30, lurking almost precisely midway between Miranda and the epsilon ring. From January 3 onward, the smaller, more elusive members of Uranus' family swam into view, so that, by the end of the Voyager flyby, no fewer than 10 small satellites had been discovered, all of which were located inside the orbit of Miranda.

The early discovery of 1985 U1 gave the mission team sufficient time to modify the imaging sequence so that Voyager could capture one image of the new satellite before turning its attention to Miranda. A problem with a ground antenna in Australia meant that the first transmission of 1985 U1's picture was not received, but a second attempt succeeded before the data were overwritten on the space-

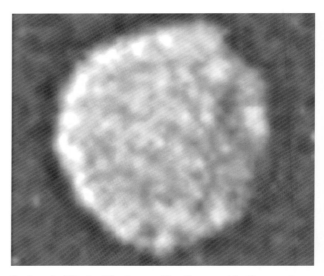

Pockmarked Puck—The first satellite discovered by Voyager 2 was 1985 U1. It was promptly dubbed "Puck" by imaging team member Robert Brown—after a character from Shakespeare's play *A Midsummer Night's Dream*. Scientists were surprised to see that Puck is roughly spherical and about 160 km (100 mi) across, making it the sixth largest of Uranus' moons. Its surface was very dark—almost as black as the planet's rings—and showed a 45-km (28-mi) impact crater near the limb. (NASA–JPL)

craft's busy tape recorder. The moon was shown to be 162 km (101 mi) across, a dark, battered world that took a little over 18 hours to orbit Uranus. The satellite reflected just 7% of the incoming sunlight—much less than the larger moons but more than the dusky rings.

Voyager's cameras could only show the other minor moons as smears against a background of stars and rings, so few details of their physical nature were obtained. Estimates of their sizes were derived from their brightness and distance. Assuming that they reflected as much sunlight as 1985 U1, their diameters were calculated to be between 20 and 135 km (12 and 84 mi). Voyager obtained no information about their masses, but their dark surfaces suggested a mixture of organic (carbon-based) compounds and water ice.

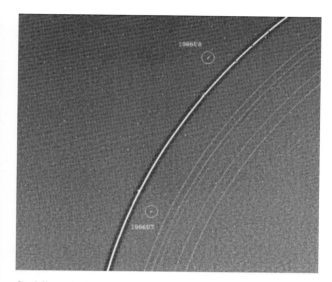

Cordelia and Ophelia—Cordelia (1986 U7) and Ophelia (1986 U8)—the tiny specks inside the circles—were two of the small moons discovered during the Voyager 2 flyby. Located only about 50,000 km (31,250 mi) from the center of Uranus, the satellites are obviously strong enough to resist the efforts of Uranus to pull them apart. Orbiting either side of the relatively bright epsilon ring, they act as shepherds, keeping the particles confined in a narrow ring. (NASA–JPL)

This lack of hard facts made it hard to tie down their origins. The satellites' standard, circular orbits close to Uranus' equator indicated that they were not rogue objects captured by the giant planet's gravity. Were they long-term companions of Uranus or fragments of a much larger body that was subsequently shattered?

Most of the moons were more or less uniformly distributed between the rings and Puck. The exceptions were 1986 U7 and 1986 U8, two moons that traveled around Uranus on either side of the epsilon ring. Well before Voyager found these newcomers, scientists had predicted that the narrow, sharply defined rings could only

be explained by the presence of shepherding satellites. (Similar discoveries had been made during the Voyager flybys of Saturn.) Repeated tugs on the ring particles by these near neighbors were preventing the tiny pieces of debris from straying.

Other shepherds were expected to be carrying out similar duties alongside other rings, but none was found. If these elusive ring occupants do exist, they must be smaller than 20 km (12 mi) across.

The task of naming the new satellites was, as always, assigned to the International Astronomical Union (IAU). Everyone expected that the IAU would eventually agree to names that were in keeping with the existing system of nomenclature, but the Challenger disaster of January 28, 1986, just four days after Voyager's closest approach to Uranus, led to the pressure from many quarters to name seven of the moons after the crew members who perished only 71 seconds after lift-off.

In the end, the crew's names were given to craters on the lunar far side, while the IAU followed the tradition of naming Uranian satellites after characters from the works of Shakespeare and Pope.

1985 U1 was named Puck, after a fairy in *A Midsummer Night's Dream*. The others were: Portia (1986 U1), the rich heiress who becomes the wife of Bassanio in *The Merchant of Venice*; Juliet (1986 U2), the heroine of *Romeo and Juliet* and a lady loved by Claudio in *Measure for Measure*; Cressida (1986 U3), one of the main characters in *Troilus and Cressida*; Rosalind (1986 U4), the daughter of the exiled duke in *As You Like It*; Desdemona (1986 U6), the wife of *Othello*; Cordelia (1986 U7), the youngest of the royal daughters in *King Lear*; Ophelia (1986 U8), the Lord Chamberlain's daughter in *Hamlet*; and Bianca (1986 U9), the younger sister of Katherine in *The Taming of the Shrew*.

The only newcomer not named after a Shakespearean character was Belinda (1986 U5). Like its much larger neighbor Umbriel, it was named after a character from Alexander Pope's *The Rape of the Lock*.

Eleven Rings

One of the most intriguing features of the Uranian system was the elusive ring system. Unable to rival the visual spectacle provided by Saturn's disk, they had, nevertheless, fascinated scientists ever since their discovery in 1977. Although nine years of ground-based observations had tied down the shapes and relative locations of the nine rings, Voyager 2 promised to unveil some of their secrets.

As expected, the non-reflective rings were extremely difficult to image against the dark sky—it was like photographing charcoal on a black velvet background. With the exception of the epsilon ring at the outer edge of the system, about 25,600 km (16,000 mi) above the cloud tops, all of the rings were extremely thin—generally less than 10 km (6 mi) across.

Smoke Rings—Looking back toward the Sun, Voyager 2 found a continuous distribution of smoke-sized particles between the well-known rings. Also showing brightly on the inside of the epsilon ring is the dusty lambda ring (1986 U1R), which was discovered by Voyager. The image was taken in the shadow of Uranus, at a distance of 236,000 km (142,000 mi) with a resolution of about 33 km (20 mi). The 96-second exposure also revealed fine dust not visible from other viewing angles. Streaks are trailed stars. (NASA–JPL)

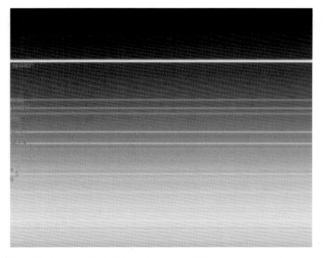

Ring Variations—This false-color view of the rings was made from images taken by Voyager 2 from a distance of 4.17 million km (2.59 million mi). All nine pre-Voyager rings are visible; the fainter, pastel lines between them are the result of computer enhancement. The brightest, or epsilon, ring (top) appears neutral in color, while the eight fainter rings show slight color differences. The lambda ring is too faint to be visible. (NASA–JPL)

At least six of the rings were found to be aligned at a very slight angle to the planet's equator. They were also very slightly out of round and sometimes variable in width. Several seemed to wobble and "pulsate" in and out, as if some unseen giant was alternately squeezing, then releasing them. It was thought that this motion could be caused by Uranus' slightly flattened shape, along with the gravitational influence of its many moons.

The epsilon ring, the broadest and brightest of the family, was one of the few rings found to lie above the equator, but its orbit was also noticeably eccentric. Not only did its distance from Uranus vary by about 800 km (500 mi), but its width changed proportionately—at its closest to the planet, it was only 20 km (12.5 mi) across, whereas at its furthest the ring expanded to 100 km (62 mi).

Voyager's cameras and other instruments did reveal some structure as the epsilon ring passed in front of distant stars. Indeed, the ring appeared twinned in some camera images. Studies of the way light passed through the ring suggested that it was made of boulders up to 1 m (3 ft) across.

However, the most important discovery about the epsilon ring was the presence of two companion satellites. Orbiting only 4,000 km (2,500 mi) apart, 1986 U7 (Cordelia) and 1986 U8 (Ophelia) lay on either flank of the ring. Despite their modest sizes, they were able to act as shepherds, forcing wayward particles and boulders back into the ring and maintaining its sharp edges. Scientists suspected that these two denizens of the rings might also modify the behaviour of some of the other rings, though other small shepherds, unseen by Voyager, could also linger between the lanes of particles.

Voyager also discovered two more faint rings hidden inside the familiar system. Image-processing revealed 1986 U1R—now known as the lambda ring—as a narrow, dusty ribbon between the epsilon and delta rings. The ring had actually been discovered earlier during ground-based observations of stellar occultations, but it had been ignored because it had not appeared on both sides of the planet. 1986 U2R (still not named) showed up as a band of fine material, 2,500 km (1,560 mi) wide located about 2,000 km (1,250 mi) closer to the planet than the 6 ring.

The most telling image of the ring system came shortly after the closest approach, when Voyager's cameras looked back toward the Sun. Gone was the sharply delineated pattern of narrow rings and in its place was a scene more reminiscent of Saturn, with numerous dusty ringlets, some broad and fuzzy in appearance, merging to form a single disk of material.

However, appearances can be deceptive. The amount of dust in the rings is actually very small. If it could all be swept into a single pile, it would still only be sufficient to make a ball no more than 100 m (330 ft) in diameter.

The tenuous hydrogen gas that pervades the rings would be expected to remove the smoke-sized dust in no more than one million years—a relatively short time by Solar System standards—so some mechanism must be replacing the lost material. The main renewal mechanism seems to be collisions between ring boulders and larger particles.

Voyager also surprised the mission scientists when it found a sparse cloud of dust during its crossing of the ring plane between the epsilon ring and Miranda. Thirty hits per second were recorded from particles striking the spacecraft. Although this was very modest compared with 600 hits per second during the ring plane crossing at Saturn, the presence of this amount of material 90,400 km (56,500 mi) above the planet was still unexpected.

Uranus and Hubble

Voyager 2's departing shot of a pale blue, crescent Uranus was humankind's last close-up view of the seventh planet. Since no more missions are scheduled to visit Uranus, astronomers must now rely on Earth-orbiting observatories and the new generation of large ground-based telescopes—often fitted with adaptive optics to overcome atmospheric turbulence—to probe the secrets of the tilted world.

One of the most fruitful avenues of research remains the study of stellar occultations—which have now confirmed the existence of the lambda ring—and spectroscopic observations that can search for additional trace gases in the atmosphere.

Instruments such as the Hubble Space Telescope have also been invaluable for witnessing the changes taking place as the planet crawls around the Sun. Its cameras have revolutionized telescopic studies of Uranus, giving views of the planet comparable in resolution to the Pioneer space-craft photographs of Jupiter. Imaging in near-infrared light is also important, since it enables telescopes to probe deeper into the planet's atmosphere than was possible with Voyager's cameras.

As Uranus turns sideways to the Sun, thus presenting its equator and northern hemisphere to solar warmth for the

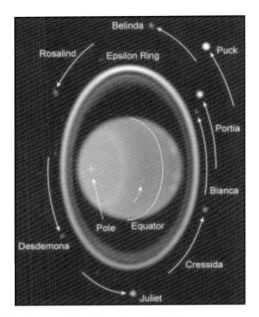

Changing Seasons—By November 1997, Uranus was presenting part of its northern hemisphere to Earth-based observers. This enabled the Hubble Space Telescope to detect clouds in this hemisphere for the first time since the early twentieth century. This enhanced, false-color image shows banded structure and numerous bright clouds, each almost as large as Europe. Another faint white cloud is barely visible near the bottom of the blue band. Such images were, for the first time, used to measure wind speeds in the northern hemisphere. (Erich Karkoschka/NASA)

time-lapse movie of huge spring storms as Uranus' northern hemisphere awoke from its decades-long winter. In place of the bland blue ball observed by Voyager, Uranus was transformed into a dynamic world with the brightest clouds in the outer Solar System. The towering storms were probably made of methane crystals which condensed at −200°C as warm bubbles of gas welled up from deep in the atmosphere.

The movie also revealed for the first time how the ring system wobbles like an unbalanced wagon wheel. This wobble may be caused by the shape of Uranus, which is like a slightly flattened globe, along with the gravitational tug from its many moons.

Meanwhile, ground-based observers were able to almost double the number of known Uranian satellites. The breakthrough came in October 1997 when Brett Gladman's team spotted two small moons with the 5-m (200-in) Palomar telescope. The newcomers—eventually named Caliban and Sycorax—were found to follow retrograde, elliptical orbits far beyond their better-behaved cousins. They were also noticeably red, resembling the Kuiper Belt objects that lurk beyond Pluto.

Three more junior members of the Uranian family surfaced in 1999, followed by three in 2001 and a further trio in 2003. With two exceptions, all these icy objects were found millions of kilometers from Uranus, traveling the "wrong way" around the planet. They probably represent wandering debris that passed too close to the giant planet and were captured in its gravitational web.

The exceptions were 2003 U1 (Mab) and 2003 U2 (Cupid), which were discovered in images taken with the Hubble Space Telescope's Advanced Camera for Surveys. Both moons lie inside the orbits of the five major satellites and are only about 10 km (6 mi) across, making them smaller than any other Uranian satellites yet found. Mab orbits in the "empty" region between Puck and Miranda, while Cupid lies only 500 km (300 mi) inside the orbit of Belinda, within the swarm of small satellites that crowd close to the planet.

The HST observations also helped astronomers to rediscover 1986 U10 (Perdita), another tiny moon that

first time in a generation, the HST has observed how Uranus adapts to seasonal changes. During the early 1990s the Wide Field and Planetary Camera was finding it relatively easy to spot storm clouds, suggesting that the atmosphere was becoming more transparent or that increased convection was causing methane clouds to bubble up more readily.

By 1997, Erich Karkoschka, Heidi Hammel, and others were able to detect bright clouds in the planet's northern hemisphere for the first time since the early twentieth century. Two years later, the scientists released a dramatic

232

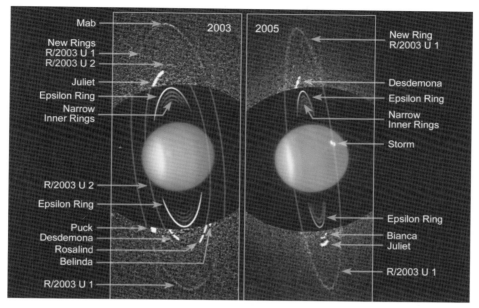

More Small Moons—These composite images, taken with the Hubble Space Telescope in 2003 and 2005, reveal two newly discovered rings encircling Uranus. The outermost ring (R/2003 U1) is probably replenished by dust from impacts with a satellite called Mab, which is embedded in it. The new outer ring (R/2003 U2) is twice the radius of the main ring system around Uranus. Only a faint segment is visible at the 12 o'clock position. [NASA, ESA, and M. Showalter (SETI Institute)]

had originally been found by Erich Karkoschka (University of Arizona) in 1999 during a search of archived Voyager pictures. Perdita orbits about 1,100 km (680 mi) beyond Belinda.

After all of these recent discoveries, Uranus now boasts a grand total of 27 moons, a family outnumbered only by Jupiter and Saturn. The planet also holds the record for close satellites, with 18 in its inner system. They are so closely packed that the larger moons must inevitably perturb the motions of their smaller neighbors.

The region is so crowded that the orbits of these moons are chaotic and dynamically unstable over timescales of only a few million years, so scientists are trying to understand how the moons can coexist with each other.

Clearly, not all of Uranus' satellites formed at the same time as the planet, more than 4 billion years ago. Cupid and Perdita, for example, are probably shards that broke away when a comet smashed into Belinda.

Astronomers are working to refine the orbits of the newly discovered moons with further observations. This will show how they interact with one another, perhaps providing evidence of how such a crowded system of satellites can remain stable. It could also reveal whether these moons have any special role in confining or "shepherding" Uranus' system of rings.

Hubble observations have already discovered a faint ring that shares its orbit with Mab. Provisionally known as R/2003 U1, it is probably replenished by dust created by impacts with the moon. Another recently discovered ring lies between the orbits of Portia and Rosalind, in a region with no known source bodies.

What about future opportunities? With less than half of each major satellite so far imaged, scientists are eager to discover what lies in their unseen northern hemispheres.

Perhaps the best window of opportunity occurs during the spring and autumn equinoxes, when the entire surface of each moon is illuminated as they orbit the planet. Unfortunately, after the equinox of 2006 the next favorable photo opportunity for an orbiting spacecraft will be half a Uranian year later—in 2048.

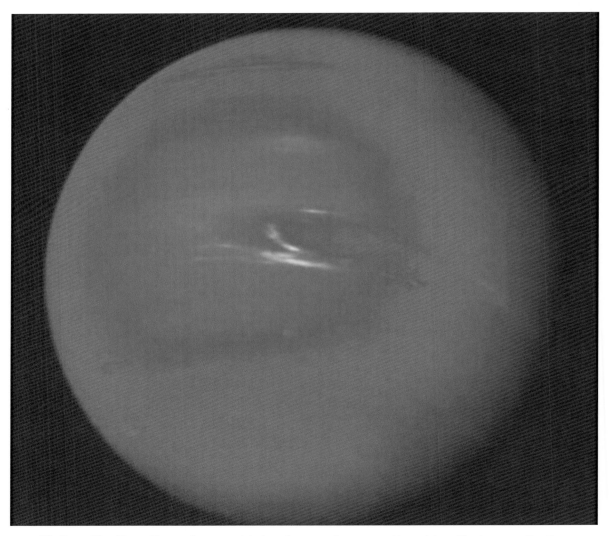

The Spotty Blue Giant—Neptune has a surprisingly active atmosphere powered by an internal heat source rather than the Sun. The Great Dark Spot at latitude 22°S was an anticyclonic storm that drifted westward, completing one circuit every 18.3 hours. At 42°S was "Scooter," a bright cloud that overtook the Great Dark Spot roughly once every three Neptune days. Further south was the "Small Dark Spot 2," highlighted by bright cloud in the center. This almost matched the planet's rotation period, reappearing every 16 hours. (NASA–JPL)

10 NEPTUNE: THE LAST GIANT

The Eighth Planet

With the discovery of Uranus, the Solar System once again seemed complete. A search through old records going back to the seventeenth century revealed that the first Astronomer Royal, John Flamsteed, had observed Uranus on no fewer than six occasions from 1690 onwards, without realizing that it was a planet rather than a star. Armed with this invaluable positional information that extended over more than one Uranian year, it seemed to be relatively straightforward to determine the planet's orbit.

Meanwhile, German astronomer Johann Bode—the first person to suggest the name Uranus for the seventh planet—had revived interest in a mysterious arithmetic progression that seemed to match the relative distances of the planets from the Sun. (This so-called "Bode's Law" had first been described by Johann Titius in 1766.)

Beginning with the series 0, 3, 6, 12, 24, in which each successive number after 3 was double the one before it, the next steps were to add 4 and divide each number by 10. For no apparent reason, this mathematical game provided a good approximation to the actual distances of the planets in astronomical units. (One astronomical unit or AU is the mean distance of the Earth from the Sun.)

The progression came back into prominence when Uranus was shown to be 19.2 AU from the Sun, very close to the Law's prediction of 19.6. Further apparent confirmation came in 1801 when the asteroid Ceres was found between Mars and Jupiter, almost exactly where the "Law" predicted.

Unfortunately, Uranus refused to behave properly.

Until 1822 it moved faster than predicted, while after 1822 it seemed to drag its heels. Scientists were baffled. Attempts to avoid errors in the calculations by using only observations made since its discovery in 1781 failed to resolve the problem. Was it possible that another planet was pulling on Uranus and causing the strange behavior?

A brilliant young Cambridge undergraduate named John Couch Adams decided to find out. Refusing to accept that it was a virtual impossibility to predict the location of an unseen planet, he began to think about how to calculate where the new world might be found, based on its presumed influence on Uranus and Bode's Law.

By the time he graduated in 1843, the precocious student felt confident enough to inform James Challis, Professor of Astronomy at Cambridge University, about his part-time work. In February 1844, Challis wrote to George Airy, Astronomer Royal and Director of the Royal Greenwich Observatory, requesting some additional information for his "young friend." Airy wasted no time in replying, and by mid-1845, Adams informed Challis that he had completed his calculations.

Adams's solution was actually within two degrees of arc of Neptune's actual position. If a search had begun there and then, Adams would probably have gone down in history as the unchallenged discoverer of the eighth planet from the Sun. Instead, as a result of bad luck and incompetence, nothing was done. The young man's attempts to speak to Airy were thwarted on two occasions when the great man was not at home, and then when he was dining in the middle of the afternoon. A letter summarizing his work met with a disappointing response, prompting Adams to ignore Airy's request for clarification.

Johann Galle—Johann Galle (1812–1910) was born in Saxony, Prussia. He worked at the Berlin Observatory until 1851, when he became director of the Breslau Observatory (now in Wroclaw, Poland). In 1846, Galle received a letter from Le Verrier, in which he requested a search for a planet beyond Uranus. Within one hour of beginning their search, Galle and his junior colleague, Heinrich d'Arrest (1822–1875), had located Neptune close to the predicted location. Eight years earlier, Galle had been the first to recognize Saturn's inner Crêpe Ring. He also suggested a method of measuring the scale of the Solar System by observing the parallax of asteroids, first applying his method to the asteroid Flora in 1873. (RAS)

The Influence of Neptune on Uranus 1781–1846—The discovery of Neptune was made as a result of unexpected perturbations in the orbital motion of Uranus. Until 1822, the recently discovered seventh planet was moving faster than expected against the background stars. After 1822, it slowed down. Adams and Le Verrier recognized that the gravitational pull of another giant planet beyond Uranus could be influencing its orbital motion. As Uranus overtook Neptune, the pull of the unseen planet increased. After opposition in 1822, when the planets were at their closest, Neptune's influence gradually diminished.

Meanwhile, on the other side of the English Channel, a French mathematician, Urbain Le Verrier, had been persuaded to take up the role of planet detective. By June 1846, he had completed his own independent calculations and reached a similar conclusion to Adams. A copy of the paper presented to the Paris Academy of Sciences landed on Airy's desk on June 23 or 24. Finally, on July 9, the Astronomer Royal wrote to Challis, asking him to begin an urgent search.

However, Adams's "friend" was reluctant to begin what he thought to be a laborious wild goose chase. Not until July 29 did the search with the 30-cm (11.8-in) North-umberland refractor get underway. Ignoring the predicted positions of Le Verrier and Adams, he chose to sweep a

huge swath of sky 30 degrees long and 10 degrees wide with the intention of recording and checking the positions of some 3,000 stars. On August 12, the planet was recorded almost exactly where it had been predicted—but no one recognized its true nature.

Le Verrier was also struggling to persuade his colleagues to take action. His third paper presented to the Paris Academy of Sciences on August 31, predicting that the new planet would be about 5 degrees east of the star delta Capricorni, was greeted with general apathy. Fortunately, the frustrated genius remembered a young astronomer,

Johann Galle, whose doctoral dissertation he had read the previous year. Le Verrier wrote to Galle at the Berlin Observatory, requesting that he initiate a planet search. The letter arrived in Berlin on September 23.

With the backing of Johann Encke, the observatory's director, Galle and a graduate student, Heinrich d'Arrest, immediately set out to search Le Verrier's location with a 23-cm (9-in) refractor. As Galle stared through the telescope and called out each star in the field of view, d'Arrest checked them off on a new map of the heavens.* Within one hour they found an object that was not on the chart. Excitedly, they rushed to drag Encke away from his birthday party. The next night, they were able to confirm not only that the newcomer had moved, but that it had a noticeable disk. The discovery had been made within one degree of Le Verrier's predicted position, and an elated Galle immediately wrote to him, "The planet whose position you have pointed out actually exists."

Acrimonious Discovery

Not until October 1, 1846, did the news of the planet's discovery break in the London *Times*. The breakthrough came as a great shock to the British astronomical community, particularly Challis, who had been plodding along for two months, unaware of Le Verrier's ultimate success. As recently as September 29, he had logged a star that seemed to present a disk, but neglected to follow up immediately. Indeed, he had now seen the new planet on four separate occasions, but remained blissfully unaware of his success.

Studies of the new world confirmed that it was the cause of the orbital perturbations of Uranus. Before 1822 the outer planet was pulling on Uranus to speed up its passage; during 1822 they were in opposition; and after that date the outsider's gravity held back Uranus as it strived to race ahead.

---

* By chance they found an excellent new sky map by Carl Bremiker that was only available in Germany.

John Couch Adams—John Couch Adams (1819–1892) was the son of a Cornish tenant farmer. He earned a scholarship to Cambridge University, where he raised funds by tutoring other students. In 1841, while still an undergraduate, Adams set out to analyze the erratic orbital motion of Uranus, calculating the probable position and orbit of an unseen eighth planet. After Neptune's discovery in 1846, there was a lengthy dispute over priority, but Le Verrier and Adams eventually shared the honors. In 1858 he became Lowndean Professor of Astronomy and Geometry at Cambridge. In 1861 he replaced Challis as director of Cambridge Observatory. He refused a knighthood in 1847 and the position of Astronomer Royal when Airy retired in 1881. (RAS)

The episode appeared to be a triumph for theoretical astronomy, in particular Newton's Theory of Gravitation. Unfortunately, it was also a cause of strife and acrimony among the scientists who had participated in the historic search.

With the name of "Uranus" still not universally accepted, it was hardly surprising that the major participants in the search could not agree on what to call the eighth planet. Challis suggested the name "Oceanus" and Galle offered "Janus." Le Verrier sought to prevent a protracted debate by writing to Galle, "The Bureau of Longitudes here has decided upon 'Neptune.' The symbol is

Urbain Le Verrier—Like Adams, Le Verrier (1811–1877) came from a family of modest means. His father sold the family home to pay for him to go to college in Paris, and a year later he won a mathematics prize. After working as a chemist for the Ministry of Tobacco, he became professor of astronomy at the École Polytechnique in 1837. Eight years later, he was asked to study the "Uranus problem," eventually calculating the position of an unknown planet that was influencing Uranus. In 1859, he proposed that Mercury's orbit was being modified by another unseen planet, which he called Vulcan. Unfortunately, Vulcan was never found. (RAS)

to be a trident. The name 'Janus' would imply that this is the last planet in the Solar System, which we have no reason at all to believe."

In fact, the Bureau of Longitudes had no authority to name the new planet, and apparently had not even given an official opinion. This piece of fiction seems to have been Le Verrier's way of expressing his preference without becoming involved in a political wrangle. Unfortunately, within a few days he had changed his mind and persuaded François Arago, Director of the Paris Observatory, to argue that the planet should be named "Le Verrier."

This blatant piece of nationalistic propaganda and egotistical bravado attracted wide criticism and had little chance of success. Eventually, by common consent, it was agreed that the planet should be named Neptune, after the Roman god of the oceans.

This debate was quite harmonious compared with the acrimonious argument over who should claim the honor for its discovery. Since Le Verrier was completely unaware that an Englishman was also striving to calculate the location of Uranus' great attractor, he was taken aback when Airy and Challis revealed that Adams had produced an almost identical solution. Outraged by the British counterclaim, François Arago drafted a fiery response which stated that, "Mr Adams has no right to figure in the history of the discovery of the planet Le Verrier, neither by a detailed citation, nor by the slightest allusion."

As each side sought to establish its superior claim, the one redeeming feature was the refusal of the two leading characters in the saga to become involved. When Le Verrier and Adams met for the first time at the British Association meeting in Oxford in June 1847, they soon established a rapport and became firm friends. Today, both men stand side by side in the pantheon of fame.

One further twist to the story came in 1980, when Charles Kowal and Stillman Drake were searching through the notebooks of Galileo Galilei, the first telescopic observer of the heavens. They found a diagram of Jupiter's moons, dated December 28, 1612, on which he had marked a star that did not exist. In fact, he had almost certainly made the first observation of Neptune—234 years before Galle! On January 28, 1613, he even made a note that the "star" and one of its neighbors "seemed further apart." Astonishingly, Galileo had in his grasp the opportunity to discover the eighth planet from the Sun almost 170 years before the seventh planet was found by Herschel.

Although it is often brighter than eighth magnitude, Neptune is too faint to be seen with the naked eye. With an apparent diameter of little more than 3 arc seconds, it is 600 times smaller than a Full Moon. However, with a magnification of over 100, the planet clearly shows a bluish disk.

Not surprisingly, a number of prediscovery observations were uncovered, including two from the French astronomer,

Joseph de Lalande, in May 1795. These were of great assistance in determining Neptune's orbit. It soon became apparent that, although Le Verrier and Adams had been so successful in fixing its position, both had failed to predict an accurate distance. The fallibility of Bode's Law had also been demonstrated: Neptune was located about 30 AU (4.5 billion km or 2.8 billion mi) from the Sun, rather than the 38.8 AU anticipated by Bode's believers.

At such a vast distance, the planet crawls around the Sun at just 5.5 km/s (12,375 mph), taking almost 165 years to complete one circuit. Not until 2010 will Neptune return to the same part of the sky where it first swam into the view of Johann Galle.

The limited number of observations obviously added to the difficulty in determining Neptune's exact position. Only months before the Voyager 2 flyby of 1989, astronomers were concerned that they only knew its location to within 5,000 km (3,125 mi). Indeed, in the late 1970s, Thomas Van Flandern and Robert Harrington (US Naval Observatory) believed that Neptune was misbehaving, just as Uranus had done 150 years earlier. This led them to predict the presence of another giant planet beyond Pluto, but no sign of this remote world has ever been found.

In From the Cold

From its apparent size and distance, it was immediately clear that Neptune was the fourth giant to orbit the Sun, so it was hardly surprising that the planet was able to influence the orbit of Uranus so markedly. Unfortunately, precise measurements of its size were not possible using ground-based telescopes hampered by atmospheric turbulence. An apparent diameter of between 2 and 3 arc seconds gave a ballpark figure of about 50,000 km (31,250 mi)—about the same as Uranus.

As astronomers struggled to refine this number, there was disagreement over which of the worlds could claim precedence. This debate was important because it had implications for the relative densities of the two blue giants.

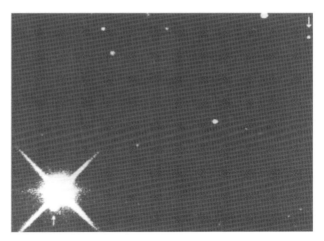

Neptune with Triton and Nereid—Before Voyager 2 flew past Neptune in 1989, only two satellites of Neptune had been observed. Triton (marked by the arrow at lower left) was discovered by William Lassell in October 1946, only 17 days after the planet was found. The other moon, Nereid (arrowed at top right), was found by Gerard Kuiper in 1949 by studying photographs taken with the 208-cm (82-in) McDonald reflector. Note the overexposure of Neptune in order to show the much fainter moons. (Mt Wilson Observatory)

Today, the accepted figure for the equatorial diameter of Neptune is 49,528 km (30,775 mi)—about 1,600 km (1,000 mi) smaller than Uranus.*

Fortunately, the discovery of moons around both planets made it relatively straightforward to calculate the masses of the two planets. It turned out that Neptune is noticeably more massive than its neighbor, weighing as much as 17 Earths, compared with 14.5 Earths for Uranus. With an overall density 1.6 times that of water (compared with 1.3 for Uranus), it was clear that the neighbors were more cousins than twins.

---

* Like the other giant planets, Neptune's rapid rotation causes a considerable bulge at the equator, so that its polar diameter is about 1,000 km (625 ml) smaller than its equatorial diameter.

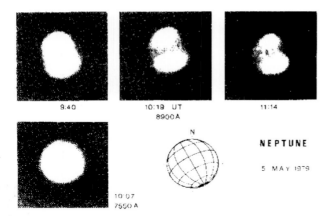

9:40
10:19 UT
8900 A
11:14

N

NEPTUNE

5 MAY 1979

10:07
7550 A

Ground-based Images of Neptune—A series of four photographs taken in 1979 by Brad Smith, Harold Reitsema, and Stephen Larson, using the 154-cm (60-in) reflector at Catalina Observatory. The images were taken in near-infrared light. The dark patches close to the equator and on the edge of the disk show where light is absorbed by methane gas. Images such as these enabled astronomers to estimate that Neptune's rotation period was between 17 and 18 hours. (University of Arizona)

Ground-based observations made it clear that Neptune has a deep atmosphere, presumably composed mainly of hydrogen and helium. Spectroscopic studies indicated the presence of methane, which absorbs red light—thus explaining the planet's blue color. Its tremendous distance and tiny apparent size meant that it was not possible to be certain of any weather activity in the form of bands or spots, although there was evidence of a variable, high-altitude haze of ice crystals.

Models of Neptune's interior were developed to take into account the planet's size and formation in the icy outer reaches of the solar nebula. These suggested that both Uranus and Neptune had rocky cores about 16,000 km (10,000 mi) in diameter, surrounded by an 8,000-km (5,000-mi) layer of liquid "ices," probably water, methane, and ammonia. This was topped by an atmosphere that was 9,000 km (5,625 mi) deep and rich in hydrogen and helium.

However, scientists' understanding of how substances behave under immense pressure at temperatures of 2,000–4,000°C remained very limited, opening the door to some interesting speculation. One unorthodox theory, put forward by Marvin Ross in 1981, proposed that methane in the mantle was split into carbon and hydrogen, allowing the carbon to become compressed into tiny diamond crystals. These might sink toward the planet's center, surrounding the core with a diamond-rich layer thousands of kilometers deep.

At the outer limits of the Solar System—Neptune is sometimes the furthest planet from the Sun—the surface temperature of the planet should be around –230°C, since the planet receives 900 times less solar radiation than the Earth. Under such conditions, there should be insufficient warmth to power atmospheric convection and storm systems.

However, infrared observations made in the mid-1970s revealed a temperature of "only" –216°C at the cloud tops. This meant that Neptune was as warm as Uranus, despite the fact that it is one-third as far again from the Sun. Apparently, Neptune was emitting 2.4 times as much energy as it received from the Sun.

Absorption of ultraviolet radiation by methane in the atmosphere was not a sufficient explanation, since this also occurs on Uranus, so the only conclusion to be drawn was that Neptune is being warmed by an internal source. The nature of this heat source was a mystery, especially since the internal structures of Uranus and Neptune were thought to be almost identical. One early suggestion was tidal heating by Neptune's nearby, massive satellite Triton, but this was subsequently discounted. A more likely scenario was slow contraction of the planet and separation of the icy constituents in its mantle.

If the nature of Neptune was often a subject of controversy, there was also considerable uncertainty over the number of satellites and whether the planet possessed a system of rings.

Astronomers did not have long to wait for the first satellite discovery. At the request of John Herschel, William Lassell—a brewer and distinguished amateur astronomer from Lancashire, England—began to scour the sky for Neptunian satellites. Lassell met with almost immediate success when, on October 10, 1846—only two and a half weeks after Neptune was recognized—he discovered a nearby moon. More than 60 years passed before the name Triton was generally accepted. A much smaller and more remote moon, Nereid, was found during a photographic search by Gerard Kuiper in 1949.

Meanwhile, Lassell became convinced that his excellent 61-cm (24-in) reflector had also revealed a faint ring. Support for this view came from a number of observers, including James Challis, who reported "a distinct impression" that the planet had a ring like that of Saturn. Others who searched in vain for this apparition, were more skeptical. Finally, in 1852, Lassell himself was forced to abandon his claim when he discovered that the so-called "ring" changed position when the tube of his telescope was rotated. One hundred and thirty-seven years passed before Voyager 2 proved that Neptune really does possess a ring system, sadly unconnected with Lassell's phantom.

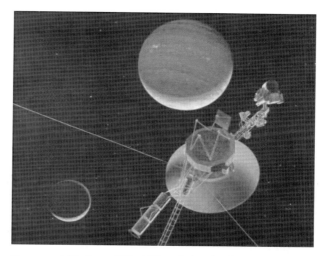

Voyager at Neptune—This 1981 painting by Don Davis shows Voyager 2 as it looks back at a crescent Neptune and its largest moon, Triton, seven hours after closest approach to the giant planet. Note the absence of rings or ring arcs, since at this time there was no indication that they existed. A later version produced prior to the flyby included the newly discovered ring arcs. (NASA–JPL)

Voyager's Grand Finale

After the successful completion of its third planetary encounter in January 1986, NASA's Voyager 2 spacecraft set course for Neptune on the final leg of its Grand Tour of the gas giants.

Having already endured 10 years of interplanetary travel, the spacecraft was suffering from partial deafness and a sticky scan platform, but still remarkably robust. The mission team now had three and a half years to prepare their ship and the ground equipment for the ultimate test, a high-speed flyby of a dark world located at 30 AU, i.e. 30 times Earth's distance from the Sun. As little as possible would be left to chance for this once-in-a-lifetime experience.

In order to ensure that Voyager's weak signals were safely gathered in, all three of NASA's 64-m (211-ft) Deep Space Network antennas were enlarged to 70 m (231 ft) and made as efficient as possible. Two 34-m (112-ft) antennas were also available at Canberra, Goldstone, and Madrid.

Additional listening stations were enlisted to track Voyager, including the Parkes radio telescope in Australia and another dish in Usuda, Japan. Telemetry data was also to be retrieved by the 27 antennas of the Very Large Array near Socorro, New Mexico.

The spacecraft's elderly computers were also reprogrammed to accommodate light levels 900 times darker than on Voyager's home planet. Typical camera exposures at Neptune were to last for 15 seconds or longer, compared with 5 seconds at Uranus. However, exposures of up to 61 seconds were now possible, with additional extensions in multiples of 48 seconds.

In order to minimize blurring, the spacecraft was made much steadier, reducing camera wobble to 30 times slower than the motion of the hour hand on a clock. Compensation for the high speeds involved in the encounter was provided by programming the spacecraft to keep the camera pointed at the target without turning too far to

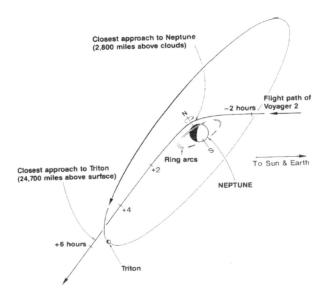

Closest approach to Neptune
(2,800 miles above clouds)

Flight path of
Voyager 2

−2 hours

N

Ring arcs

To Sun & Earth

+2

NEPTUNE

Closest approach to Triton
(24,700 miles above surface)

S

+4

+6 hours

Triton

The Voyager 2 Flyby of Neptune—Voyager 2 passed only 4,500 km (2,800 mi) above Neptune's northern hemisphere. The planet's gravity bent its trajectory south of the ecliptic, so that it passed only 38,500 km (24,060 mi) from the center of Triton less than six hours after the Neptune encounter. The extreme course deviation resulted in Voyager 2 heading out of the Solar System in a southerly direction. (NASA–JPL)

interrupt the flow of data back to Earth. Voyager was now able to perform simultaneous gyro-controlled turns for each axis at a rate of 120 degrees per hour.

Since there were no more destinations on Voyager's itinerary, planners had much more freedom to choose the optimum flight path than during previous encounters. However, since scientists around the world were clamoring for priority to be given to their particular observations, hard choices had to be made in order to satisfy as many scientific priorities as possible, without undue risk to the spacecraft.

The main constraint was the aiming point—exactly where Voyager would have to make its closest approach to Neptune in order to gain a grandstand view of Triton. On the scheduled arrival date (August 25, 1989) Triton would be well south of Neptune's orbital plane, so Voyager would have to sweep low over the north pole, allowing the planet's powerful gravity to bend its path southward. In the event, Voyager safely passed only 4,500 km (2,800 mi) above Neptune's northern hemisphere—a feat likened by the mission team to sinking a 3,360-km (2,260-mi) golf putt.

By May 1988, images sent back from Voyager were already outstripping anything obtained by ground-based telescopes. Then, in January 1989, Voyager's camera discovered a huge dark spot in the planet's southern hemisphere. Excitement continued to mount as the spacecraft closed on Neptune at a rate of 1,450,000 km (900,000 mi) per day.

The first success in the ongoing search for new moons came in early July, when a 400-km (250-mi) wide satellite was found inside the orbit of Triton. Another five small moons were found even closer to Neptune's cloud tops in the weeks that followed.

One of the most intriguing puzzles that scientists expected Voyager to solve was the apparent presence of partial rings around Neptune. Not until August 11 did two ring arcs, 50,000 km (31,250 mi) and 10,000 km (6,250 mi) long respectively, appear in a narrow-angle image. Subsequent imaging and occultation observations showed that the arcs were actually elongated clumps within a set of narrow, dark ringlets. Voyager crossed the plane of the rings twice, about 60 minutes before and 85 minutes after its closest approach to Neptune.

Meanwhile, the spacecraft crossed the bow shock—the boundary where the solar wind ploughed into Neptune's magnetosphere—about 9½ hours before flying over the north pole. It eventually exited the magnetosphere 38 hours later, but multiple bow-shock crossings were subsequently experienced almost 4 million km (2.5 million mi) from the planet.

Voyager's grazing approach curved its trajectory about 48 degrees south of the ecliptic, enabling it to pass only 38,500 km (24,060 mi) from the center of Triton less than six hours after the Neptune encounter. The extreme course deviation resulted in Voyager 2 heading out of the Solar

System in a southerly direction at a speed of just over 53,400 km/h (33,100 mph). There was no possibility of redirecting it toward Pluto—the only way to change course sufficiently was to target the spacecraft *inside* Neptune's atmosphere.

In many ways, Triton was the star of the show. Voyager switched its attention to the mysterious moon 110 minutes after leaving Neptune behind. Over the next three hours the illuminated southern hemisphere of this unique icy world grew ever larger, revealing a strange panorama of pink ice, cantaloupe patterns, and dark streaks.

Voyager's farewell view of Neptune and Triton was taken on August 28, when the narrow-angle camera captured two pale crescents hanging in a black sky. Interstellar space beckoned.

Neptune Unveiled

After almost 150 years of speculation based on unreliable data from ground-based instruments, Neptune finally came in from the cold on August 25, 1989. Although the sky blue disk was no surprise, astronomers were not expecting a Jupiter-like vista of bands and storms. However, months before Voyager 2 arrived in Neptune's vicinity its narrow-angle camera was picking up tantalizing evidence of atmospheric variability, notably a huge dark spot hovering south of the equator. Neptune was obviously very different from bland Uranus.

Global views showed a number of zones parallel to the equator, though with less contrast and structure than those of Jupiter. Images taken at different wavelengths indicated that the darker regions were deeper in the atmosphere, while the highest ice clouds appeared as white streaks that stretched around the planet from west to east.

The most amazing cloud image, returned only two hours before Voyager's closest approach, showed streamers of silvery cirrus cloud casting shadows onto the blue cloud deck 100 km (62 mi) below. A thin blanket of haze reflected sunlight at the edge of the planet's disk but otherwise allowed the weak sunlight to penetrate to the underlying methane layer.

Models of the atmosphere based on these observations suggest that Neptune has several distinct cloud layers. The uppermost level consists of methane hazes and bright clouds of methane ice. Beneath this, where temperature and pressure increase, should be layers of hydrocarbons such as ethylene and acetylene, and a layer of hydrogen sulfide and ammonia (possibly combined to form ammonium hydrosulfide). Water ice clouds may exist at even greater depths.

Time lapse movies pieced together from continuous imaging during Voyager 2's long approach provided important evidence about wind speeds at different latitudes and the nature of the various cloud features.

The southern hemisphere was dominated by the "Great Dark Spot," an anticyclonic storm that appeared to resemble Jupiter's Great Red Spot in terms of circulation, latitude, and size relative to its home planet. However, Neptune's storm seemed to be much more unstable, as it stretched and contracted while rocking a little from side to side over an eight-day cycle. The spot was also drifting toward the equator at 15 degrees per year, a migration that, if continued, would ensure its doom.

Measuring about 15,000 × 7,000 km (9,320 × 4,350 mi), the oval spot was large enough to swallow planet Earth. Streaks of ice cloud were particularly noticeable to the east (the lee side) of the spot and around its southern edge, but wisps of cirrus also appeared above the "eye" itself and to the west, like a bow wave ahead of a giant ship. Bradford Smith, head of the Voyager imaging team, speculated that the Great Dark Spot acted as an obstacle over which the methane-rich air was forced to rise. As the gas cooled, methane condensed out to form white ice clouds.

The Great Dark Spot moved around Neptune once every 18.3 hours—slower than the planet's rotation. Strong winds were driving the storm westward at a rate of 325 m/s (730 mph).

Two smaller cloud features enabled scientists to determine wind speeds at higher latitudes. One of these was an isolated patch of wispy white cloud nicknamed "Scooter"

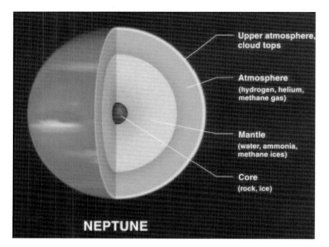

Neptune's Internal Structure—The atmosphere of Neptune is similar to that of Uranus. It consists mainly of hydrogen, helium, and methane. Below it is thought to be a liquid hydrogen layer including helium and methane. The mantle contains liquid hydrogen compounds, oxygen, and nitrogen. It is believed that the planet's core, which is larger in proportion to its size than those of its giant cousins, is a mixture of rock and ice. Neptune's average density is the greatest of all the gaseous planets. (LPI)

Labels on figure:
- Upper atmosphere, cloud tops
- Atmosphere (hydrogen, helium, methane gas)
- Mantle (water, ammonia, methane ices)
- Core (rock, ice)
- NEPTUNE

showed that Neptune could generate the strongest jet streams anywhere in the Solar System, with equatorial winds reaching up to 2,000 km/h (1,250 mph). Voyager's studies of the surprisingly active atmosphere showed various layers of different hazes, beneath which was a band of methane haze where the temperature measured about –210°C. Visible through this haze was an optically thick layer of unknown composition—possibly hydrogen sulfide. Deeper cloud layers not observed by Voyager were thought to exist, possibly composed of ammonium hydrosulfide, ammonia, and water.

Voyager confirmed that Neptune's internal heat is much more significant than solar energy in powering the winds and storms—although the actual amount of heat released by the planet is much lower than for Jupiter or Saturn.

Neptune's magnetic field was 25 times stronger than Earth's, though at its visible surface the field was actually weaker than we experience on our small planet. Perhaps the biggest surprise was the lopsided nature of the magnetic field, which was remarkably similar to that of Uranus. Voyager found that it was tilted 47 degrees to the rotation axis and offset from the planet's center. The field is thought to be generated by currents within the planet's icy mantle. As Neptune spins, this results in a magnetic field that appears to gyrate wildly in the solar wind. Substantial amounts of nitrogen were apparently fed into the magnetosphere from Triton's atmosphere.

Triton

Little was known about Triton before the arrival of Voyager 2. Even its size was uncertain, with speculation about its diameter ranging from 2,800 to 3,600 km (1,700 to 2,200 mi). In other words, Triton could be larger than all of the moons in the Solar System apart from Ganymede, Titan, Callisto and possibly Io—or it might be significantly smaller than Earth's Moon.

One of the most intriguing revelations came in 1983, when a team led by Dale Cruikshank (University of

because it overtook the Great Dark Spot roughly once every three Neptune days. Scooter changed shape from round to square and then triangular before Voyager's electronic eyes as it moved around the planet in just under 16 hours. Observations indicated that it was floating below the main deck of methane haze.

At 55°S was the "Small Dark Spot 2" or "D2," which outwardly resembled its larger brother. However, unlike the Great Dark Spot, D2 developed a bright core and appeared to rotate in a clockwise direction, indicating that it was a cyclone or region of low atmospheric pressure. Gases should, therefore, be rising in the cloudy center of the spot and descending in the dark oval around it. Voyager's cameras showed individual cloud structures no more than 20 km (12.5 mi) across in its center. D2 almost matched the planet's rotation period, reappearing roughly every 16 hours.

Despite its frigid location far from the Sun, Voyager

Triton—This Voyager 2 mosaic of Triton shows a surface covered in nitrogen ice. Pinkish deposits near the south polar cap (bottom) are believed to contain methane ice, which reacts under sunlight. Dark streaks are believed to be dust deposited from huge geyser-like plumes. The bluish-green band near the equator may consist of relatively fresh nitrogen frost deposits. The greenish area includes what is called the cantaloupe terrain, whose origin is unknown, and some "cryovolcanic" landscapes, apparently produced by icy-cold liquids (now frozen) that erupted from Triton's interior. (USGS)

A Frozen Lava Lake?—A composite view showing Neptune on Triton's horizon. Neptune's south pole is to the left. In the foreground is a computer-generated view of a circular walled plain, Ruach Planitia, as it would appear from 45 km (28 mi) above the surface. The terraces indicate multiple episodes of "cryovolcanic" flooding. Relief has been exaggerated roughly 30 times and the actual range of the relief is about 1 km (0.6 mi). As a result of Triton's motion relative to Neptune, the planet would appear to move along the horizon, eventually rising and setting at high latitudes. (USGS)

Hawaii) studied infrared spectra of Triton and discovered the presence of frozen methane on its surface. By studying different regions of the moon as it orbited Neptune, they concluded that the methane ice was not uniformly distributed, but scattered in continent-sized patches. It seemed that Triton was very different from the water ice satellites of Jupiter, Saturn, and Uranus.

Methane alone would be sufficient to ensure a sparse atmosphere, but if gaseous nitrogen was also present, there was a distinct possibility that the atmosphere could be more substantial than that of Mars. With temperatures possibly within 50 degrees of absolute zero, gases would also turn to ice, causing scientists to speculate that Triton could be the only place in the Solar System to exhibit frozen lakes of nitrogen.

Voyager 2 succeeded beyond everyone's wildest dreams in unveiling the mysteries of Triton. As the satellite loomed ever larger in Voyager's images, it soon became clear that Triton was at the lower end of the size scale, with a fairly modest diameter of 2,706 km (1,683 mi).

With a surface temperature of −235°C, it was coldest place in the Solar System. This was due to a combination of remoteness from the Sun and the mirror-like reflective properties of the moon's icy surface. However, the images also showed that Triton was far from a dead, frigid world. Indeed, it was unlike any object ever seen.

The moon's southern hemisphere was largely blanketed by a bright coating of nitrogen frost, mixed with ices made of methane, carbon dioxide, and carbon monoxide. Its pink tinge was thought to be caused by the interaction of cosmic rays with the methane, producing a natural overlay of complex hydrocarbons. Close to the equator and extending all the way around Triton was a darker region, which was tinted blue by the scattering of sunlight reflected from surface frost—probably fairly fresh deposits of nitrogen or water ice.

One of the most obvious characteristics was the virtual absence of impact craters—a certain sign of resurfacing. The largest impact crater seen was a hollow, 27 km (17 mi) wide, and later named Mazomba. It seemed clear that Triton's youthful appearance was produced by cryovolcanism, the large-scale eruption of liquid or "slushy" ices onto the surface.

Much of the near-equatorial zone displayed a unique pattern of "dimples" that bore a remarkable resemblance to the skin of a cantaloupe melon. Possibly created by material upwelling from below, this pockmarked landscape was thought to be the oldest area on Triton. Criss-crossing the cantaloupe terrain was a series of narrow, interconnecting ridges that were reminiscent of those associated with the fractures on Europa. As was the case with Jupiter's moon, it seemed likely that ices had been pushing up through cracks in the frozen surface.

Other evidence of geological activity included circular, walled plains several hundred kilometers across. These seemed to be the result of periodic flooding by cryovolcanic ices, followed by freezing and then collapse. Less clear-cut were the "maculae," dark patches that resembled paw marks surrounded by lighter material.

Yet more evidence of ongoing activity came in the form of numerous dark streaks. Often hundreds of kilometers long, they were generally aligned from the south pole toward the northeast. To scientists' amazement, Voyager even captured images of at least four dust-laden nitrogen plumes. These were seen to rise vertically to a height of about 8 km (5 mi) before horizontal winds carried the dark material away and deposited it on the surface.

More evidence of a world in never-ending transition was the sculpted edge of the ice cap, between the equator and 30°S. The polar frosts were sublimating (turning to gas), despite the apparent lack of warmth.

The defrosting effect was the result of Triton's unique orbital and rotational motion, which creates a 688-year climate cycle. When the Sun is overhead at the equator, gases migrate to the poles before freezing out. However, at the time of the Voyager flyby, Triton was enjoying a decades-long southern summer. The Sun was overhead near 50°S and shining directly on the polar cap. Over time, this causes the southern ice cap to disappear, but when the gases reach its dark, northern counterpart, they freeze back onto the surface, creating a fresh coat of exotic frost.

The vaporization of nitrogen ice explained the existence of a tenuous, nitrogen-rich atmosphere on such a small world.* Although it was more than 70,000 times thinner than the air we breathe, the atmosphere on Triton could generate winds capable of carrying material long distances. Other images showed individual clouds and a thin layer of haze particles or photochemical smog. Some of the clouds extended about 100 km (63 mi) along Triton's limb.

Small Moons and Rings

Despite numerous observations over a period of 150 years, only two satellites of Neptune—Triton and Nereid—were known before the Voyager 2 flyby. However, based on past experience with the other giant planets, expectations were high that Neptune's list of companions would multiply as a result of the spacecraft's encounter. The scientists were not disappointed.

Voyager's first discovery swept into view in early July, early enough to schedule an imaging sequence during the flyby. 1989 N1 (later named Proteus) was found about 117,650 km (73,500 mi) from Neptune and, with a diameter of more than 400 km (250 mi), it replaced Nereid as the planet's second largest moon. Proteus had an odd, box-like

---

* Triton is one of only three objects in the Solar System known to have a nitrogen-dominated atmosphere (the others are Earth and Saturn's moon, Titan).

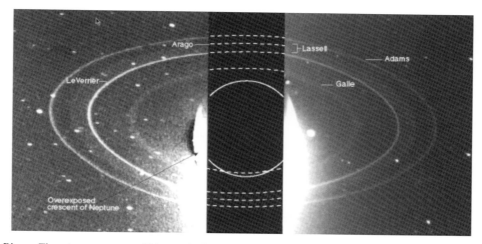

Faint Rings—These two exposures of Neptune's rings were taken by Voyager 2's wide-angle camera on August 26, 1989, from a distance of 280,000 km (175,000 mi). The two main rings (Adams and Le Verrier) appear complete over the region imaged. The time between exposures was 1 hour 27 minutes. The images do not show the ring arcs in the Adams ring—they were on the opposite side of the planet for each exposure.

Also visible is the inner faint ring (Galle) at about 42,000 km (25,000 mi) from the center of Neptune, and Lassell, the broad, faint band that extends smoothly from the Le Verrier ring toward the Adams ring. These long-exposure images were taken while the rings were back-lit by the Sun, allowing fainter, dusty regions to be seen. In order to show the faint rings, crescent Neptune has been overexposed. The two gaps in the outer ring (top left) are due to computer processing. (NASA–JPL)

shape, disfigured by a large impact crater. With a slightly higher mass, its gravity would have been sufficient to pull itself into a sphere.

Five more small moons were found even closer to Neptune's cloud tops in the following weeks. They were all dark and followed circular orbits close to Neptune's equator. Apart from Proteus, they all scooted around the planet in less than one Neptunian day. They were also arranged in order of size, with the smallest (Naiad) closest to Neptune and the largest (Proteus) furthest away.

Perhaps the most significant aspect of the five inner satellites was their potential interaction with the planet's narrow rings. Putting aside Lassell's phantom ring, observers first began to suspect the presence of rings around Neptune in 1981, when they re-examined observations obtained during an occultation 13 years earlier. The

data showed a marked dip in the brightness of a background star that occurred about three minutes after it emerged from behind Neptune.

The same year, James Elliot, who had successfully observed the faint rings of Uranus, noted another short-lived fading episode, but he could not be sure whether it was caused by a ring or a satellite. Eight years later, the Voyager data confirmed that Elliot had, indeed, found a moon (Larissa).

Realizing the potential of this powerful technique, astronomers eagerly grabbed any opportunities over the next few years to study stellar occultations. However, they were largely disappointed. In more than 100 occultation studies, the presence of possible ring material was only observed 11 times.

Even more frustrating was the inability to detect similar

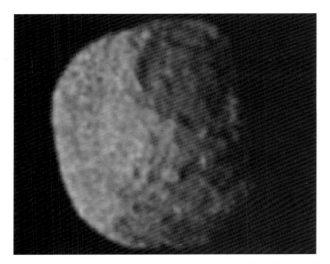

Proteus—Proteus (1989 N1) was the first new satellite discovered by Voyager 2 in July 1989. More than 400 km (250 mi) across, it is the second largest moon of Neptune after Triton. The satellite has an odd box-like shape, disfigured by a large impact crater. If the moon's mass was just a little higher, its gravity would cause it to reshape itself into a sphere. Proteus follows a circular, highly inclined orbit 117,600 km (73,500 mi) from Neptune. This image was acquired on August 25, 1989, from a range of 146,000 km (91,000 mi). (NASA–JPL)

dimming on either side of the planet. In the months leading up to the Voyager encounter, the leading theory was that Neptune possessed at least three partial rings, although no one could satisfactorily explain how these "arcs" could have been created or maintained.

Despite the gloomy conditions during the Neptune flyby, Voyager 2's cameras soon solved the mystery. Narrow, elongated arcs of material do, indeed, exist above Neptune's cloud tops, but only as distinct sections of complete rings.

Voyager discovered five continuous rings between 41,900 km (26,035 mi) and 62,933 km (39,100 mi) from the center of Neptune. They were named after some of the leading historical characters involved in the planet's discovery. In order of distance from the planet, the rings are Galle, Le Verrier, Lassell, Arago, and Adams. There was also some evidence for a faint, partial ring sharing the orbit of Galatea—about 1,000 km inside the Adams ring.

Three of the rings were extremely narrow—less than 100 km (62 mi) in width—with only Galle and Le Verrier measuring several thousand kilometers across. The Adams ring was of particular interest. Not only was it the densest and slimmest member of the ring system, but it was also found to contain four arcs that were noticeably brighter than the remainder of the 50-km (30-mi) wide ribbon.

The arcs made up about one-tenth of the Adams ring's circumference, sufficient to explain the occasional occultation observations that had so puzzled scientists. Leading the procession of bright beads was an arc called Courage, with Liberté, Egalité (subdivided into two separate sections), and Fraternité in its wake.

Voyager's cameras had shown that three small shepherd satellites—Naiad (1989 N8), Thalassa (1989 N7), and Despina (1989 N5)—orbit Neptune between the innermost ring (Galle) and Le Verrier. Galatea (1989 N6) patrolled between the Arago and Adams rings, while Larissa (1989 N3) policed the outer edge of the Adams ring. It did not take long for scientists to show that the gravitational influence of these satellites was actively preventing any ring particles from escaping. The rings were confined by sentries on either side. Less easy to explain were the weird arcs in the Adams ring. If left alone to evolve, the arcs should spread out into a uniform ring very quickly—perhaps in no more than one year. Clearly, some unusual process was holding them together.

Several years passed before scientists were able to show that the close presence of Galatea, which followed a slightly inclined orbit just inside the Adams ring, would be enough to stabilize the clumps of dust that made up the four arcs.

Post-Voyager Observations

Although the Voyager flyby remains our most important source of information about the eighth planet, there have

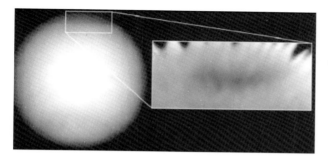

The Northern Dark Spot—By 1994, Hubble Space Telescope observations had shown that the Great Dark Spot seen by Voyager had disappeared. However, another comparable dark spot had appeared in the northern hemisphere. Unfortunately, the new feature appeared near the limb of the planet and was difficult to observe. Like its predecessor, it was outlined by high-altitude clouds of methane ice. The image was taken on November 2, 1994, when Neptune was 4.5 billion km (2.8 billion mi) from Earth. (H. Hammel [Massachusetts Institute of Technology] and NASA)

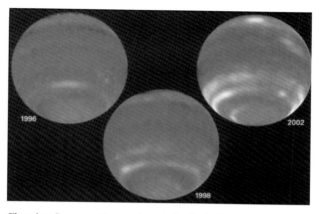

Changing Seasons—Neptune's axis is tilted 29 degrees from the ecliptic, so, like Earth, it experiences marked seasons. However, as the gas giant takes 165 times longer to complete one orbit of the Sun, seasonal changes are slow to appear—each season lasts about 41 years. However, over a period of seven years, the Hubble Space Telescope revealed a marked brightening and broadening of cloud bands in the southern hemisphere. (L. Sromovsky, P. Fry [Univ. of Wisconsin]/NASA/ESA)

been some very interesting and valuable observations, both by the Hubble Space Telescope (HST) and large ground-based instruments, in recent years.

An ongoing observational campaign with the HST has provided fascinating insights into the remarkably fluid cloud formations on Neptune. The first major change was noticed in 1994, when Heidi Hammel and others showed that Neptune's most prominent feature, the Great Dark Spot, had completely disappeared. Clearly, comparisons made with Jupiter's venerable Great Red Spot had been very wide of the mark. "We weren't surprised the other spot disappeared," said Hammel. "It was kind of 'floppy' because it changed shape as atmospheric circulation carried it around the planet."

Neptune's disk did not remain bland for long. Within a year, Hammel's team had discovered a new dark spot in the northern hemisphere that was a near mirror image of its southern counterpart. There was speculation that, like its predecessor, the newcomer might be a hole in Neptune's methane clouds that offered a window to lower levels of the atmosphere. Once again, the storm system was accompanied by bright, high altitude clouds of methane ice,

presumably formed by cooling of gas that was rising and flowing over the spot.

Scientists also suggested that Neptune's atmospheric dynamism was largely being driven by the planet's internal heat source, rather than solar energy. A slight change in the temperature difference between the top and bottom of the cloud layers could trigger rapid, large-scale changes in atmospheric circulation.

More recent changes have been monitored by scientists from the University of Wisconsin–Madison. Their observations showed a distinct increase in the area and brightness of banded cloud features, particularly in the planet's southern hemisphere, between 1997 and 2003. These findings were consistent with observations made by G.W. Lockwood at Lowell Observatory, which showed that Neptune has been gradually getting brighter since 1980, and studies made with the Keck telescope in Hawaii.

The changes were especially prominent at infrared wavelengths, which are much more sensitive to high-altitude clouds. Scientists believe the changes are a response to seasonal variations in sunlight. Neptune's 29-degree tilt means that the amount of sunlight striking the northern and southern hemispheres varies considerably throughout the planet's year. At the present time, it is summer in the south.

"We would expect heating in the hemisphere getting the most sunlight. This in turn could force rising motions, condensation and increased cloud cover," said Wisconsin scientist Lawrence Sromovsky. "Neptune's nearly constant brightness at low latitudes gives us confidence that what we are seeing is indeed seasonal change as those changes would be minimal near the equator and most evident at high latitudes where the seasons tend to be more pronounced." If this is the case, the planet is likely to continue brightening for another 20 years. Meanwhile, other advances in our understanding of the interiors of both Uranus and Neptune have been made in recent years.

In 1999, results from ESA's Infrared Space Observatory (ISO) were announced by French and German scientists. ISO found that Uranus and Neptune have three times more deuterium (heavy hydrogen) in their atmospheres than Jupiter or Saturn. This is consistent with ices making up more than half the mass of the planetary pair, a much higher fraction than found in their larger cousins.

However, the team questioned whether incoming comets could have made up the bulk of the icy planets. Since comets typically contain five times as much deuterium as Uranus and Neptune, the data suggested that the ice must have come from some other, mysterious, source in the solar nebula.

The team went on to suggest that all four giant planets may have started with similar icy cores, each roughly 10 times the mass of the Earth. However, the outer regions of the solar nebula were relatively sparse feeding grounds, so Jupiter was able to gather more material and swell in size much more effectively, gathering about 40 times as much gas as Uranus and Neptune.

Other recent studies support theories that the magnetic fields of the blue giants are generated in a thin outer layer of the icy mantle, rather than in the lower mantle regions where convection does not take place. This thin outer layer

acts as the planet's dynamo, combining with the spin of the planet to produce tangled magnetic field lines. According to Sabine Stanley and Jeremy Bloxham of Harvard University, this would explain not only the highly tilted magnetic poles of Uranus and Neptune, but also the fact that the field lines loop in and out in shifting patches.

More Moons

Despite its remoteness, Neptune has not been ignored or forgotten. Observations with the Hubble Space Telescope and the new generation of large ground-based instruments have enabled astronomers not only to continue monitoring the Neptune system, but also to make new discoveries.

One of the most intriguing results was announced in 1998, when James Elliot's team from the Massachusetts Institute of Technology announced that Triton seemed to have heated up significantly since Voyager 2's visit nine years earlier. "Since 1989, at least, Triton has been undergoing a period of global warming—percentage-wise, it's a very large increase," said Elliot.

The scientists used one of Hubble's three Fine Guidance Sensors (used to keep the telescope pointed at a celestial target by monitoring the brightness of guide stars) to measure Triton's atmospheric density when it passed in front of a star known as "Tr180" in November 1997. The guidance sensor measured the star's gradual decrease in brightness as its light traveled through slightly thicker layers of the satellite's sparse atmosphere. This led them to conclude that Triton's atmosphere had doubled in density. Because of the unusually strong link between the temperature of Triton's surface ice and its atmospheric pressure, the scientists inferred that a modest warming of $2°C$ had taken place over nine years. Triton was now enjoying a balmy temperature of $-234°C$ (about $39°$ above absolute zero).

The warming trend was thought to be driven by seasonal changes in its polar ice caps. As Triton's southern hemisphere receives more direct sunlight during its extreme southern summer (a season that occurs every few hundred

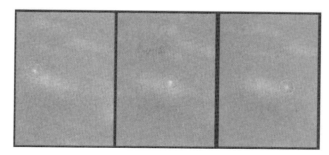

A New Moon—The moon temporarily designated S/2002 N1 is shown here in images from the Cerro Tololo Inter-American Observatory's 4-m (13.2-ft) Blanco telescope. S/2002 N1 is the innermost of the five small moons discovered by ground-based telescopes in 2002–2003. About 50 km (30 mi) across, it orbits more 16.6 million km (10.3 million mi) from Neptune and takes over five years to complete one circuit of the planet. The orbit is also highly inclined and elliptical. (Matt Holman, Harvard-Smithsonian Center for Astrophysics)

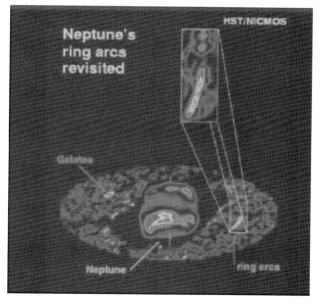

Evolving Ring Arcs—Modern near-infrared cameras and telescopes have enabled astronomers to view Neptune's faint inner moons and ring arcs. In particular, observations made with the NICMOS instrument on the Hubble Space Telescope have provided new information about the orbital periods of the inner moons and subtle changes in the rings. NICMOS data indicates that the ring arc named Liberté has edged forward, toward Courage. (Christophe Dumas, NASA/JPL)

years), the nitrogen ice is vaporizing, causing a considerable leap in atmospheric pressure.

Triton may be the dominant member of Neptune's retinue, but the giant planet has been found to have many more minor companions. No fewer than five small moons were discovered far beyond both Triton and Nereid in 2002 and 2003, boosting the number of known satellites to 13. This was something of a surprise, since it was thought that cataclysmic events related to the capture of Triton would have dispersed any outer satellites.

Four of the moons were found by a team led by Matthew Holman (Harvard–Smithsonian Center for Astrophysics) and John Kavelaars (National Research Council of Canada), while the other was found by Scott Sheppard and his team at the University of Hawaii.

Using the 4-m (158-in) Blanco telescope at Cerro Tololo in Chile and the 3.6-m (142-in) Canada–France–Hawaii Telescope they took multiple exposures of the sky surrounding Neptune. After tracking the motion of the planet across the sky, they then combined many digital images to boost the signal of any faint objects. Since they tracked the planet's motion, stars showed up in the final

combined image as streaks of light, while the moons appeared as points of light.

The satellites are all exceedingly faint and very small, 30–50 km (20–30 mi) across. They are located a long way from Neptune, at distances between 16 and 25 million km (10 and 16 million mi). Indeed, 2003 N1 is the furthest known satellite of any planet, taking about 26 years to travel once around Neptune. Two of the moons have normal prograde orbits and the others have backward or retrograde orbits. Their elliptical, inclined orbits indicate that they must all have been captured by Neptune.

Meanwhile, observations with Hubble and the Canada–

France–Hawaii Telescope have also thrown doubt over the long-term stability of Neptune's ring arcs. Until now, the standard explanation was that the arcs are confined by the gravity of Galatea. However, images taken since 1998 show that the arcs are not where they should be if Galatea is the sole controlling influence.

Since Galatea seems to be responsible for only part of the arc confinement, astronomers are now seeking alternative explanations. One possibility is that tiny, unseen moonlets are helping to prevent the "beads" from spreading out and disappearing.

Precise measurements of the arcs also indicate that subtle changes have taken place since Voyager's visit. The most noticeable of these is a deterioration and forward shift of about two degrees by Liberté—the only arc to have altered its position relative to the others. Also detected was a decrease in intensity of Liberté and a slight widening of Egalité. One explanation is that some material has been transferred from Liberté toward Egalité.

Another interesting development has been the discovery of Neptune's Trojan asteroids. The first, known as 2001 QR322, was found in deep digital images taken on August 21, 2001, with the Blanco Telescope at Cerro Tololo by Marc Buie, Robert Millis, and Lawrence Wasserman of Lowell Observatory.

Subsequent observations confirmed that the object behaves just like the 1,600 Trojan asteroids that occupy Jupiter's orbit. 2001 QR322 remains, on average, about 60 degrees ahead of Neptune—a position that it can continue to occupy almost indefinitely. Approximately 230 km (140 mi) in diameter, it, like Neptune, requires about 165 years to complete each circuit of the Sun.

What about the future? No missions to Neptune are on the drawing board, although NASA has begun to study a nuclear-powered Neptune polar orbiter that might fly a few decades from now. Such a spacecraft could also conduct a detailed study of Triton and release instrumented probes deep into the planet's atmosphere.

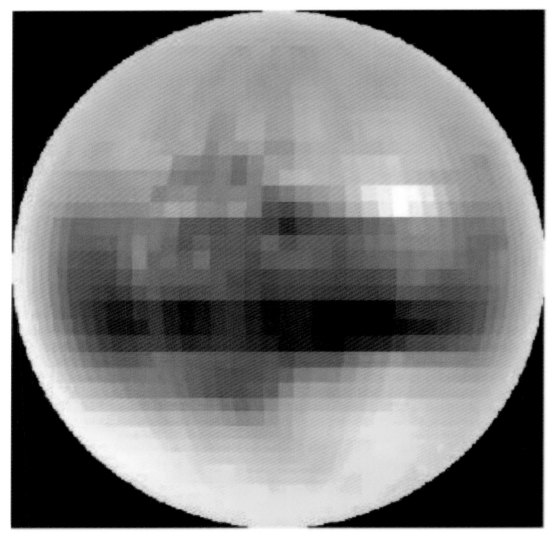

Shades of Brown—Pluto displays more large-scale contrast than any planet except Earth. This map shows the true colors of Pluto and the highest surface resolution so far available. It was created by tracking brightness changes of Pluto during times when it was being partially eclipsed by its moon Charon. Pluto's pink–brown color may be caused by the action of cosmic rays on frozen methane deposits. (Eliot Young [SwRI] et al., NASA)

11 PLUTO: KING OF THE KUIPER BELT

Astronomers have devoted centuries of study and observation to explaining and predicting the movements of the heavenly bodies and they dislike any behavior which does not fit in with their theories or forecasts. Once the period of consternation involving the wanderings of Uranus came to an end with the discovery of Neptune in 1846, equilibrium was restored and the Solar System once more seemed to be complete. Then, after about 50 years of peace and stability, observers noticed that Uranus seemed to be straying again, if only very slowly and very slightly.

Percival Lowell, most famous for his theories and observations concerning Martian canals, believed there might be another unknown planet beyond Neptune. Using the supposed tiny irregularities in the orbit of Uranus, he began calculating the possible position of this "Planet X." Impatient to start, a search was initiated at his observatory near Flagstaff, Arizona, as early as 1905, though his final results were not published until 1914. Lowell predicted that Planet X would be about six times the mass of Earth, and about 1.6 billion km (1 billion mi) beyond Neptune.

When the eccentric businessman died in 1916, he left an endowment of over a million dollars to keep his observatory operating. However, the will was contested by Lowell's widow, and the court battle to settle the estate—which eventually consumed most of the endowment—dragged on until 1927. Although the Flagstaff search was temporarily abandoned, a former supporter of Lowell, named William Pickering, had also taken up the quest.

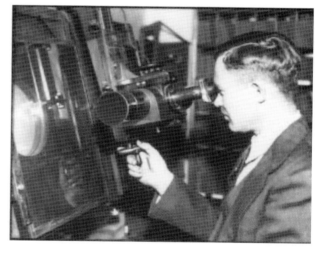

Clyde Tombaugh—Clyde Tombaugh (1906–1997) discovered Pluto on February 18, 1930. Brought up on a farm in Kansas, he was a keen amateur astronomer who built his own telescopes. He was later hired to search for a ninth planet at Lowell Observatory in Flagstaff, Arizona. Here he is comparing ("blinking") photographic plates in order to detect faint objects moving among the "fixed" stars. During this program, he discovered numerous clusters of stars and galaxies, hundreds of asteroids, two comets, and one nova. He later worked on missile-tracking telescopes at White Sands Proving Ground. (Lowell Observatory)

Pickering published his own predictions for the missing world, which he contrarily dubbed "Planet O." In 1919, he and Milton Humason began another search at Mt Wilson Observatory, making use of time-exposure photography that enabled astronomers to study photographs of the sky

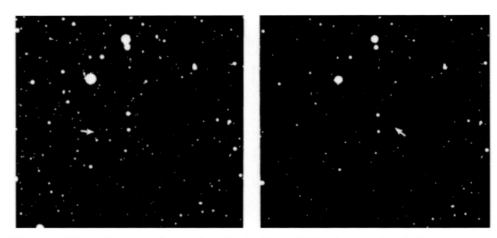

The Discovery of Pluto—These two photos, taken with the 33-cm (13-in) refractor at Lowell Observatory, are small sections of the plates on which Pluto was discovered. They show the motion of Pluto—barely visible against the "fixed" background stars—between January 23 and 29, 1930. By rapid "blinking" of the two images in a special microscope, observatory assistant Clyde Tombaugh was able to detect the movement of the tiny planet. (Lowell Observatory)

at their leisure and to compare plates of the same area taken at different times. The "fixed" stars would stay in the same positions while any planet or other nearby body would shift its position slightly and, hopefully, be detected. Sadly, Planet X (or O) failed to appear once again.

In 1929, the spotlight returned to Lowell Observatory. Under the direction of Vesto M. Slipher, a young assistant named Clyde Tombaugh was given the onerous task of comparing photographic plates taken with a new, specially designed wide-field search telescope. The 33-cm (13-in) refractor had been purchased with the aid of a gift from A. Lawrence Lowell, the president of Harvard and Percival Lowell's brother.

Day after day, Tombaugh diligently set up the plates in a blink comparator which enabled him to switch quickly between two views of the same area of sky. However, it was like looking for a needle in a haystack—a typical photographic plate could contain over 150,000 stars, while those taken of the Milky Way could contain close to a million. It took Tombaugh up to a week to examine each pair of exposures.

Then, on February 18, 1930, the persistent young man struck gold when he noticed a tiny star-like object that shifted position against the background of stars on two negatives. Careful study of its apparent motion over the next few weeks confirmed that this was no ordinary asteroid, and on March 13, 1930, the discovery of the ninth planet was officially announced. Planet X had been found within six degrees of the position predicted by Lowell.

Although the 24-year-old farmer's son became a celebrity overnight, the only person to discover a planet in the twentieth century still had to struggle for academic qualifications while continuing his search for distant planets. Eventually, he achieved his dream by becoming a highly respected professional astronomer. Fifty years after his momentous discovery, he was honored by having asteroid 1604, which he had discovered during his search, named after him.

What name should be given to Planet X? Lowell Observatory was flooded with suggestions from all over the world, but the winning name came from Venetia Burney, a young schoolgirl from Oxford, England. She thought that because the planet was so far away from the Sun, in its own dark realm, it should be named Pluto, after the Roman god of the underworld.

The excitement over the discovery of the "great attractor" that was influencing both Neptune and Uranus soon gave way to a general disappointment and scratching of heads. Pluto seemed to be far too small to be Lowell's Planet X, a world capable of modifying the paths of its distant companions.

The first fairly reliable measurements of its size were made by Gerard Kuiper in 1950. Using the new 5-m (200-in) Palomar reflecting telescope, Kuiper estimated that Pluto's diameter was a modest 5,860 km (3,600 mi), which placed it midway between Mercury and Mars.

Unfortunately, studies of the perturbations of Neptune and Uranus suggested that Planet X should have a mass about 90% that of the Earth. In other words, for little Pluto to have caused major disturbances to its neighbors, it must be almost 53 times denser than water—or more than nine times denser than our planet! For this to be true, Pluto must be composed almost entirely of rare, heavy elements such as osmium, gold, platinum, tungsten, and uranium.

Perhaps Pluto was not Lowell's Planet X at all. Instead, it seemed increasingly likely that its discovery might rank as one of the greatest coincidences in the history of science. Whatever the truth of its discovery, there was no doubt that Pluto was a dark, desolate world dimly illuminated by a tiny, distant star, the Sun. Any indigenous life forms would never know of the existence of Earth and the other inner planets.

A World in Deep Freeze

Fortunately, the newcomer on the edge of the Solar System had been observed a number of times before it was recognized as a planet, the first such record going back as far as 1846. This enabled its orbit to be calculated with reasonable accuracy, and a very strange one it turned out to be.

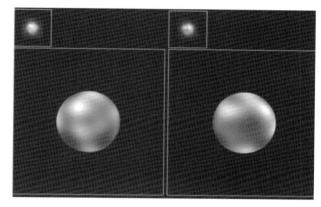

A World of Contrast—In 1994, ESA's Faint Object Camera on the Hubble Space Telescope obtained the first high-quality visible and ultraviolet images of surface features on Pluto. The dark and light regions included a "ragged" northern polar cap bisected by a dark strip, a permanent bright spot, a cluster of dark spots, and a bright linear marking. The left image is centered at 289°E, while the right image is centered at 112°E. The south pole was not visible. (Alan Stern [SWRI], Marc Buie [Lowell Observatory], NASA, and ESA)

Pluto was found to have the most elongated path of any planet. Apart from being highly eccentric, its orbit is also inclined to the ecliptic at an angle greater than 17 degrees. Indeed, its path is more like that of a comet than a *bona fide* planet.

The tremendous distance Pluto has to cover and the slow speed at which it moves—only 4½ km/s (3 mi/s)—means that one orbit lasts no less than 248 years. Since its discovery, sluggish Pluto has traveled less than one-third of its path around the Sun.

For most of this prolonged voyage through deep space, Pluto is the furthest of the nine planets, eventually receding to the unimaginable distance of 7.4 billion km (4.5 billion mi) from the Sun. In contrast, for a brief spell of 20 years it moves inside the orbit of Neptune, with a perihelion distance of 4.5 billion km (2.75 billion mi). This planetary swap last occurred between 1979 and 1999.

Despite appearances, there is no possibility of a collision between the two planets. One reason is that Pluto is always well above or below its giant cousin. Their orbital cycles also ensure that they stay far apart. Pluto completes exactly two circuits of the Sun for every three orbits by Neptune, so they are never in the same vicinity when Pluto crosses its neighbor's orbit. In fact, Pluto comes closer to Uranus than it ever does to Neptune.

Not surprisingly, Pluto's remoteness and small size have caused problems for astronomers ever since it was identified. (Even when on the innermost leg of its orbit, 18,000 Plutos would have to be lined up to match the apparent diameter of a Full Moon.) Until the introduction of the Hubble Space Telescope and 8-m (26-ft) class telescopes equipped with adaptive optics in the 1990s, Pluto was visible only as a tiny speck of light under even the best viewing conditions.

The first challenge was to refine its size, mass, and density. After Kuiper's initial measurement in 1950, follow-up studies indicated that Pluto was even smaller than he had suggested. In particular, Pluto's bright surface and ability to act like a mirror indicated that it could be only 2,500 km (1,560 mi) across. If true, Pluto would be relegated to the minor league of planetary bodies, even outranked by Earth's Moon.* (In 1980, Alex Dessler and Chris Russell jokingly remarked that if Pluto's size continued to shrink at the current rate, it would eventually disappear altogether!)

As time went by, observers gradually began to piece together a fair picture of conditions on this incredibly distant little world. In 1955, regular changes in the planet's brightness were detected, indicating a period of rotation of 6 days 9 hours and a highly tilted axis.

Pluto's surface temperature was found to range from –218 to –238°C during the course of its orbit, as the planet's distance from the Sun increases from 30 to 50 AU. At aphelion it receives only 1/2,500 of the light that reaches Earth. In these frigid conditions, almost all gases will freeze,

so it was not too surprising when infrared measurements made in 1976 showed that Pluto is probably covered by a thick layer of methane ice.

Then came the serendipitous discovery of a satellite, which revolutionized Pluto studies. Assuming that the satellite was orbiting above Pluto's equator, this confirmed that the planet must be rotating on its side, much like Uranus. Indeed, Pluto's axis is tilted about 120 degrees, even more than that of Uranus, with the result that Pluto spins in the opposite direction to most of the planets.

Using Charon's orbital period and its distance from the planet, it was possible to calculate the combined mass of the system. It immediately became clear that Pluto was even more of a featherweight than previously suggested, with only 0.002% the mass of our Earth. Even the Moon is 10 times heavier. Surface gravity is so low that a human visitor foolish enough to visit Pluto would tip the scales at only 4% of his or her weight at home.

Clearly, Pluto is far too small and insubstantial to have caused the supposed perturbations of Uranus and Neptune that were the basis of Lowell's and Pickering's calculations. It seems that they arrived at the right answer but used completely the wrong calculations.

The discovery of Charon also confirmed that Pluto's overall density is low, about twice that of water. The planet is probably 50–70% rock mixed with water and other ices. Its icy surface is likely to be almost entirely (98%) frozen nitrogen, with some methane plus traces of carbon monoxide and other hydrocarbons.

Maps based on Hubble images and Pluto–Charon eclipses show a patchwork of bright areas of frost or ice. These are interspersed with darker terrain, possibly coated with organic material turned brown by cosmic rays and ultraviolet radiation. Its nitrogen-rich atmosphere is extremely tenuous, with a surface pressure of only a few microbars—or about 1/100,000 that of Earth's surface pressure.

Unfortunately, the gases freeze onto the surface as Pluto moves away from the Sun, remaining in that state for two centuries. Desperate to observe Pluto during its brief summer bloom, scientists have been clamoring to send a spacecraft to study the planet by 2020.

* Measurements made by the Hubble Space Telescope in the 1990s give a diameter of about 2,320 km (1,440 ml), accurate to within about 1%.

Charon and Other Companions

The existence of a large satellite close to Pluto came as a major surprise. The discovery was made by James Christy, an astronomer at the US Naval Observatory, on June 22, 1978. During a routine study of photographic plates taken to refine Pluto's orbit, Christy was concerned to note that many of the 18 plates had been labeled "defective." A closer look showed that Pluto seemed distorted, as if the camera had been out of focus. However, the nearby stars looked perfectly normal.

Since the "blob" on Pluto's limb seemed to shift position, the solution clearly involved the planet rather than the instrumentation. It gradually dawned on Christy that Pluto must have a large moon with an orbital period of about one week, the same as Pluto's period of rotation. Checks of previous photos confirmed the discovery.

Christy quickly suggested the name Charon, in honor of his wife Charlene. However, aware that this was unlikely to impress the International Astronomical Union, he undertook a search of mythological names. To his surprise and delight, he found that in Roman mythology Charon was a boatman who ferried the souls of the dead across the River Styx into Hades, the realm of Pluto, god of the underworld. The name could hardly have been more appropriate.

Charon was soon found to be unique among planetary satellites. Like a geostationary satellite above Earth's equator, Charon keeps pace with the planet's rotation. As it orbits Pluto once every 6 days 9 hours 17 minutes, the satellite appears to hover over the same point on the surface. For an inhabitant of Pluto, the moon would always be above the horizon in one location, but it would never grace the skies above the opposite hemisphere.

Charon is also very close to Pluto, following a circular orbit at a distance of only 19,640 km (12,200 mi)—about

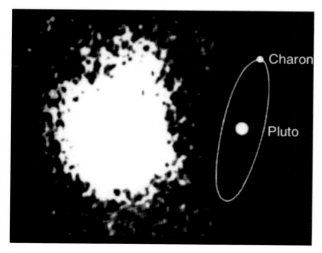

The Discovery of Charon—Pluto's moon, Charon, was discovered in 1978 by James Christy. This photograph taken with the US Naval Observatory's 1.55-m (60-in) telescope in Flagstaff, Arizona shows a "blob" (top) that was seen to move around Pluto about once a week. Charon's orbital period is identical to Pluto's rotation period, so it is only visible from one hemisphere. Like the Moon, Charon is tidally locked, always keeping the same face toward the planet. Charon travels in a roughly north–south direction, indicating that Pluto's spin axis is tilted 120 degrees to the plane of its orbit. (US Naval Observatory)

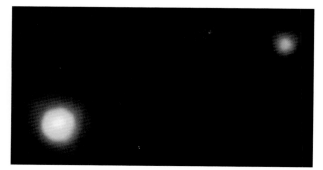

Pluto and Charon—The best picture so far of Pluto and its moon, Charon, was taken by ESA's Faint Object Camera on the Hubble Space Telescope. The image was taken on February 21, 1994, when the planet was 4.4 billion km (2.6 billion mi) from Earth. Pluto's diameter is about 2,320 km (1,440 mi). Charon is comparatively large, with a diameter of 1,270 km (790 mi). The two worlds are 19,640 km (12,200 mi) apart. (R. Albrecht, ESA/ESO Space Telescope European Coordinating Facility; NASA)

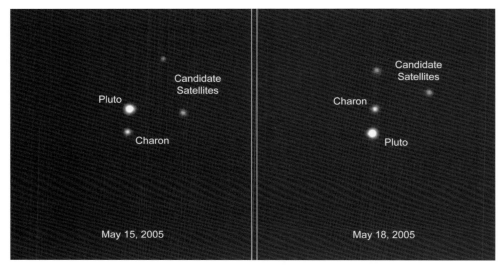

Pluto Sats—These Hubble Space Telescope images, taken by the Advanced Camera for Surveys, show Pluto, its large moon Charon, and two new satellites. (NASA, ESA, H. Weaver [JHU/APL], A. Stern [SwR1], and the HST Pluto Companion Search Team)

eight Pluto diameters. However, during Pluto's travels around the Sun, the viewing angle from Earth changes dramatically. At the time of its discovery, the whole of Charon's orbit was clearly visible in ground-based telescopes. Then, between 1985 and 1991, the orbit appeared almost edge-on, as viewed from the Earth. This meant that the satellite passed in front of and behind the planet, resulting in a series of mutual transits, eclipses, and occultations.

This special alignment—which occurs only twice during each 248-year orbit—offered astronomers a unique opportunity to determine the sizes of both Pluto and Charon, in addition to measuring variations in the brightness of each world. The eclipse observations confirmed that the moon was surprisingly big, with a diameter of about 1,270 km (roughly half that of Pluto). In fact, Charon was much larger than any other satellite in relation to its primary, persuading many astronomers to classify the Pluto–Charon system as a double planet.

The latest calculations indicate that Charon's mass is about one-eighth that of Pluto. Given that it probably has a density about twice that of water, Charon must contain relatively little rock and may be similar to the icy moons of Saturn and Uranus.

Since its discovery, some of the secrets of Charon have been revealed by the Hubble Space Telescope and ground-based observations. Curiously, its color is neutral gray and the surface is much darker than Pluto, implying that both worlds have a different surface composition and structure. There is little variation in brightness, and spectroscopic studies show that Charon's surface is probably coated by water ice, rather than the nitrogen ice found on Pluto. Other, as yet unidentified, substances have also been detected in recent years. The moon is almost certainly too small to hold onto an atmosphere worthy of the name.

One possible explanation for their different appearance is that lighter molecules on Charon's surface, such as methane, have escaped the moon's weak gravity. It seems possible that some of this material may have eventually reached Pluto and enriched its surface over time.

In May 2005, during a search for further companions of Pluto, a team led by Harold Weaver and Alan Stern used the Hubble Space Telescope to image two more moon-like objects. Provisionally designated S/2005 P1 and P2, the new satellites are much smaller and fainter than Charon— probably around 70–80 km in diameter. Both follow circular, counterclockwise orbits close to Pluto's equator. P1's orbital distance is about 64,700 km, while P2's is about 49,500 km.

The presence of Charon and Pluto's two smaller companions has raised some awkward questions about the origin of the system. Early theories attempted to link Pluto's small size, mass, density, and strange orbit with the equally unusual satellites of Neptune, especially Triton, with its highly inclined, retrograde orbit.

Pluto certainly behaved and looked more like a satellite than a major planet. This led R.A. Lyttleton and Gerard Kuiper to suggest that Pluto may be an escaped satellite of Neptune, which was thrown into its present orbit by a near collision with Triton. A later version of the theory introduced an unknown tenth planet which disrupted Neptune's satellites, including Pluto. The suggested outcome was that Pluto was torn out of orbit round Neptune, a piece of Pluto was ripped away to form Charon, and the intruder was flung out to the depths of space where it still lurks, awaiting discovery.

Today, Pluto and its moons are thought to have originated in a giant impact, much like the Earth–Moon system. This collision early in the planet's history would account for the system's unusual orbital arrangement, the axial tilt, and the large size of Charon in relation to Pluto.

The planet itself is thought to have formed some 4 billion years ago as a member of the Kuiper Belt. Pluto is now regarded as the largest known remnant of the billions of icy planetesimals that formed from the solar nebula—the building blocks for the planets themselves.

Pluto's Rivals and the Kuiper Belt

Over the years, there have been numerous predictions of planets as yet unknown lurking beyond even the twilight world of Pluto. Clyde Tombaugh, himself, continued to search for another planet for 14 years after he found Pluto, and satisfied himself that no such world exists.

The apparent emptiness of the outer Solar System puzzled many astronomers. Some suggested that the regions beyond Neptune may have been swept clean by the gravitational influence of the giant planets. Others believed that the missing objects were simply too faint to be detected.

The first suggestion that a reservoir of comets should exist beyond the planets was put forward by British astronomer Kenneth Edgeworth in 1943. He noted that, at such great distances from the Sun, collisions between planetesimals were so infrequent that only small bodies could form. In 1951, the Dutch-American astronomer Gerard Kuiper stated his belief that enormous numbers of icy bodies could populate the region beyond Neptune, perhaps as far out as 120 AU (three times Pluto's average distance from the Sun).

During the 1980s, various researchers suggested that this flattened belt could be the source of short-period comets that occasionally enter the inner Solar System. These intruders are influenced by the gravity of Jupiter, either being ejected from the System altogether or becoming periodic comets. One key piece of evidence was that these comets travel around the Sun in the same direction as the planets, following orbits that are only slightly inclined to the plane of the ecliptic.

Unfortunately, no one could prove that these dormant comets actually exist. The breakthrough came on August 30, 1992. After five years of searching the ecliptic with the 2.2-m (86-in) University of Hawaii telescope, David Jewitt and Jane Luu discovered a point of light drifting slowly among the background stars. Subsequent observations suggested that the newcomer, known as 1992 QB1,

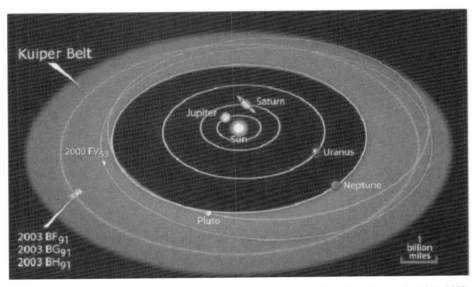

The Edgeworth–Kuiper Belt—The Edgeworth–Kuiper Belt is named after Kenneth Edgeworth (1880–1972) and Gerard Kuiper (1905–1973), who proposed that enormous numbers of icy bodies exist in the region of space beyond Neptune. Today, the name is usually shortened to simply "Kuiper Belt." After many years of searching, the first object was found in 1992. Since then, more than 1,000 have been discovered, mostly by ground-based observatories. The orbits of four recently found members are shown on this diagram. (NASA, A. Field [STScI])

measured about 250 km (155 mi) across and took 296 years to complete one circuit of the Sun. It was the first object in the Solar System known to spend its entire life beyond Neptune.

Within a year or so, another half dozen objects had been discovered in similar orbits. The long-hypothesized Edgeworth–Kuiper Belt (now generally known as the "Kuiper Belt") had finally emerged from the darkness. Today, more than 1,000 occupants of the Belt have been found, though the total population probably approaches 100 million.

The Kuiper Belt is almost certainly home to many more pieces of primordial space debris than the asteroid belt between Mars and Jupiter. Estimates suggest that the total mass of the Kuiper Belt between 30 and 50 AU is about one-tenth that of Earth (or about eight Moon masses). There may be more than 10,000 Kuiper Belt Objects

(KBOs) with diameters greater than 100 km (62 mi), compared to a mere 200 asteroids of that size.

Most of the objects so far discovered have been 100–500 km (60–300 mi) across, though much smaller and much larger members have been sighted. About 1 in 100 of the KBOs is a binary—like a mini version of the Pluto–Charon system.

With the introduction of robotic survey telescopes and modern optics, astronomers have been able to peer further out into the Kuiper Belt and various objects that rival Pluto have been found. The largest of these is 2003 UB313, which was discovered by Mike Brown (Caltech), Chad Trujillo (Gemini Observatory), and David Rabinowitz (Yale University).

Thought to have a diameter of about 2,400 km (1,490 mi), slightly larger than Pluto, the object was first

The "Tenth Planet"—On July 29, 2005, Mike Brown (California Institute of Technology) and colleagues announced the discovery of an object that is larger than Pluto. At the time, the "tenth planet" was on the outermost part of its elongated orbit, 97 AU (astronomical units) from the Sun, making it the most distant known object in the Solar System. First photographed with the Samuel Oschin Telescope at the Palomar Observatory, California, on October 31, 2003, it was rediscovered on January 8, 2005, and later given the temporary name of 2003 UB313. A satellite was discovered in September 2005. (Mike Brown, Caltech/Chad Trujillo, Gemini Observatory/David Rabinowitz, Yale University)

photographed with the 1.2-m (48-in) Samuel Oschin Telescope on October 31, 2003. When it was rediscovered on January 8, 2005, the team were able to study its motion and calculate an orbit.

2003 UB313 has an orbital period of 557 years, and currently lies close to its maximum possible distance from the Sun (aphelion), about 14,500 million km (9,070 million mi or 97 astronomical units) away. Its eccentric orbit brings it to within 5,236 million km (3,272 million mi or 35 AU) of the Sun at perihelion. (Pluto's distance varies between 29 and 49.5 AU.) Unlike the terrestrial planets and gas giants, which all travel roughly in the same plane, 2003 UB313's orbit is tilted at an angle of about 44 degrees to the ecliptic.

Other notable discoveries include 2005 FY9, discovered by Brown's team, and 2003 EL61, discovered by Jose-Luis Ortiz (Sierra Nevada Observatory, Spain), which seem to be about 70% of Pluto's diameter. 2003 EL61 has two tiny satellites. The larger moon completes one orbit every 49 days at a distance of about 49,500 km (30,760 mi).

However, size calculations must be regarded as preliminary until accurate measurements of the KBO albedos (how much light they reflect) can be obtained. A warning light appeared in 2004 when the Spitzer Space Telescope found that a large KBO known as 2002 AW197 was markedly smaller and more reflective than expected. If the object is very dark, with an albedo of only 4%, it must be about 1,500 km (932 mi) in diameter. However, Spitzer's infrared data showed that it reflects 18% of incoming sunlight. This unexpected mirror-like quality means that it can have the same brightness in the sky with a diameter of only 700 km (435 mi).

Observations have also shown that some of the KBOs are unusually red. This is probably caused by a mantle of organic compounds that have been transformed into a dark, tar-like substance by the bombardment of cosmic rays and ultraviolet radiation over billions of years. Others are gray due to a coating of carbon-rich materials.

Pluto's New Horizons

Pluto remains the only planet never visited by spacecraft. This is about to change, following the launch of NASA's New Horizons mission on January 19, 2006.

Aware that Pluto is now heading away from the Sun on its long trek toward the obscurity of the Edgeworth–Kuiper Belt, scientists have been pressing for such a mission for

Pluto New Horizons—This artist's concept shows the New Horizons spacecraft during its encounter with Pluto (foreground) and Charon. The craft carries cameras, a radio science experiment, ultraviolet and infrared spectrometers, and space plasma experiments to map Pluto and Charon, study their surface compositions and temperatures, and examine Pluto's atmosphere. Its most prominent feature is a 2.1-m (8-ft) dish antenna, used for communication with Earth from a distance of 7.5 billion km (4.7 billion mi). (JHU–APL/SwRI)

many years. In particular, they wanted a mission to reach Pluto before any residual atmosphere froze and disappeared from sight, leaving only a coating of exotic ices.

Many concepts were studied and rejected, until, in the late 1990s, NASA selected a mission called Pluto–Kuiper Express. Unfortunately, as the cost of the proposed mission soared toward $800 million, the agency took fright and canceled it in the autumn of 2000.

Inundated with protests from many quarters, including members of Congress, NASA succumbed to the pressure by opening the door to proposals from universities, research labs, and aerospace companies for a lower cost approach to exploring Pluto. The downside of this innovative approach was that the agency reserved the right to reject all of them if they exceeded the $500-million funding limit and held no promise of exploring Pluto before 2020.

The winner was a mission known as New Horizons, proposed by a team under the leadership of the Southwest Research Institute and the Applied Physics Laboratory at Johns Hopkins University. The nuclear-powered spacecraft carries more instruments and should send back 10 times as much scientific data for around $250 million less than the canceled Pluto–Kuiper Express.

The launch date was critical. New Horizons had to get

under way in January 2006 in order to receive a slingshot from Jupiter. A departure after February 2, 2006 would drastically extend the journey time, delaying a Pluto flyby until 2019–2020, when it might be too late to study Pluto's summer atmosphere. The viewing conditions would also be less favorable at that time—because of Pluto's extreme axial tilt and its inclined orbit, much of the planet's southern hemisphere would be hidden in darkness.

By beginning the voyage during the January window, New Horizons will be able to reach distant Pluto in less than 10 years, courtesy of a Jupiter gravity assist. During a reconnaissance of the Jupiter system in early 2007, it will conduct an intensive, four-month study of the giant planet and its retinue of moons.

The spacecraft will eventually fly past Pluto and Charon in the summer of 2015. Observations of the "double planet" system will begin six months before closest approach. Once the distance is down to 100 million km (62.5 million mi)— about 75 days before close encounter—New Horizons will begin to send back images superior to anything yet obtained.

As the two worlds loom larger, the instruments will be able to map their surfaces in increasing detail and study any clouds or other variable phenomena, such as possible

First To The Kuiper Belt—After its Pluto–Charon flyby, the nuclear-powered New Horizons spacecraft will be directed toward one or more Kuiper Belt Objects (KBOs). In this view, the Sun, more than 6.7 billion km (4.1 billion mi) away, appears as a bright star embedded in the glow of the zodiacal dust cloud. Jupiter and Neptune are visible as orange and blue "stars" to the right of the Sun. The artist has also included other KBOs in order to give the impression of an extensive disk of icy worlds beyond Neptune. (JHU-APL/SwRI)

cryovolcanism. On the day of closest approach, New Horizons should sweep within a few thousand kilometers of both Pluto and Charon, obtaining the first close-up observations of this unique pairing. After passing Pluto, it will turn to image the planet's night side, illuminated by dim moonshine from Charon. Measurements of the behavior of radio beams traveling between Earth and the spacecraft will also reveal the vertical structure of the sparse atmosphere.

Mission planners hope that the probe will then go on to explore one or more Kuiper Belt Objects at close quarters over the next six years. The number of flybys will depend on the available targets close to the flight path, the amount of hydrazine propellant remaining after the Pluto flyby, and the amount of power still being generated by the single radioisotope thermoelectric generator.*

The New Horizons spacecraft has a mass of 478 kg—heavier than Pioneers 10 or 11 but lighter than the Voyagers. Most of the spacecraft's subsystems, including its computers and propulsion control system, are based on designs used in other APL spacecraft.

It carries four sets of scientific instruments. A high-resolution camera will image selected regions, revealing objects down to 50 m (165 ft) in size. Studies of the composition and temperature of Pluto's ices and atmosphere will be undertaken by the infrared imaging spectrometer, which will also be able to map the surface at a resolution of 1 km (0.6 mi).

A radio science instrument will probe the structure of Pluto's tenuous atmosphere and gauge the surface temperatures on both day and night sides by measuring the intensity of the microwave radiation reaching the 2.5-m (8.25-ft) wide dish antenna. A suite of charged particle detectors will sample gases escaping from Pluto's atmosphere and determine their escape rate.

If the New Horizons mission is successful, humanity's preliminary exploration of the major worlds in the Solar System will be complete—an adventure that began back in the 1960s with the first primitive probes to Venus and Mars. If it fails to achieve its objective, the next favorable opportunity will not arrive until the twenty-third century, when the feeble warmth of the Sun once more awakens Pluto from its frozen slumber.

---

* RTG power output generally decreases by 3 to 5 watts per year. A security shutdown at Los Alamos in 2004 meant that the RTG will carry less plutonium-238 than originally planned. This may limit the science that can be done when New Horizons encounters any Kuiper Belt Objects.

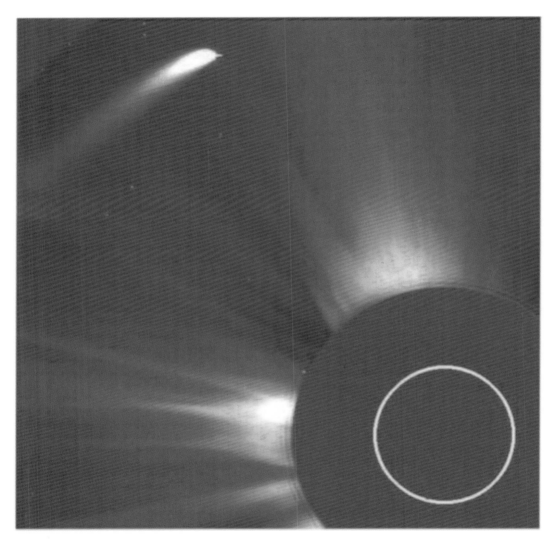

Comet Machholz—The orbits of some comets are modified by close encounters with the planets, particularly Jupiter. These periodic comets may reappear every five or six years. This image of 96P/Machholz (top) was taken on January 8, 2002 by the LASCO C2 instrument on the ESA–NASA SOHO spacecraft. The mask in the LASCO coronagraph hides the brilliance of the Sun, the size of which is shown by the white circle. Comet Machholz has a 63-month orbit, approaching to within 18 ½ million km (11 ½ million mi) of the Sun. (SOHO/LASCO, ESA, NASA)

12 COMETS: COSMIC ICEBERGS

Signs and Portents

For centuries, comets have inspired awe and wonder. Many ancient civilizations saw them as portents of death and disaster, omens of great social and political upheavals. Shrouded in thin, luminous veils with tails streaming behind them, these "long-haired stars" were termed "comets" by the ancient Greeks (the Greek word *kome* meant "hair"). Even today, explanations involving spirits and supernatural occurrences prevail among primitive people living in isolated communities. However, ancient philosophers and scientists tried to provide a more scientific explanation for these unpredictable phenomena.

Aristotle (384–322 BC) argued that inflammable gases escaping from clefts in rocks would collect in the upper layers of the sublunar world ("world under the Moon") and ignite. Rapid release of such gases produced a shooting star, whereas a slow seepage gave rise to a comet.

The apparent link between heat and comets was further reinforced by various philosophers, who said that comets produce heat that creates storms and natural disasters. Pliny the Elder (c. 23–79 AD) listed 12 different cometary phenomena and assigned one natural disaster to each class.

With the rise of Christianity, comets came to be regarded as signs from God. The appearance of a bright comet in 1066 (now known to be Halley's comet) was linked by many to the death of King Harold and the Norman conquest of England—hence its portrayal in the Bayeux Tapestry.

The decisive step toward overturning the view that comets are atmospheric phenomena was taken in 1577 by Danish astronomer, Tycho Brahe. Careful observations of a passing comet over a period of two and a half months enabled him to establish that the mysterious object must be located beyond the Moon's orbit.

By this time, astronomers were also aware that comet tails always pointed away from the Sun, although the reason for this strange behavior was unknown. Tycho's former assistant, Johannes Kepler (1571–1630), astutely observed that such tails must be caused by solar rays penetrating the comet and carrying part of the comet away with them—an explanation not too far from our modern understanding. Kepler also suggested there were more comets in the skies than fish in the seas.

On the other hand, despite his pioneering work on explaining planetary orbits, Kepler showed little interest in the paths of comets. He simply assumed that they moved in straight lines, lighting up the skies for a few weeks or months before they disappeared from view, never to be seen again.

The next breakthrough came in 1705, when British astronomer Edmond Halley (1656–1742) began to calculate the orbits of 24 comets. He noticed that the orbit of a bright comet observed in 1682 resembled the orbits of other comets recorded by Johannes Kepler in 1607 and by Peter Apianus in 1531. Halley concluded that the only reasonable explanation was that the same comet had reappeared over a period of 75–76 years. The slight variations in the timing of each return were accounted for by small tugs on the comet by the giant planets. Although he did not live to see the proof, his revolutionary theory was proved to be correct when, as he predicted, the comet duly returned in December 1758. The first periodic comet to be recognized was named 1P/Halley in his honor. (Today, comets are named after the person or persons who first discovered them.) More recent

Halley's Comet in 1066—Comets were often regarded as bad omens or warnings of major upheavals in ancient and medieval societies. Here the 1066 apparition of Halley's comet (top right) is shown in the famous Bayeux Tapestry, which records the Norman invasion of England by William the Conqueror. Note the representation of the round coma and the tails streaming behind it. (ESA)

trawls through ancient records have shown that the famous comet was recorded by the Chinese as long ago as 240 BC.

Searching for comets became a popular pastime, and it is said that Charles Messier—nicknamed the "Ferret of Comets" by the King of France—was even more upset over the discovery of a comet by a rival than he was over his wife's death.

As time went by and more orbits were calculated, it became clear that comets were difficult to classify. At one extreme there are the Jupiter family comets, which typically take no more than 20 years to circle the Sun. Most of these have orbits that have been drastically altered by the gravitational pull of the largest planet. The shortest period

belongs to comet Encke, which travels around the Sun every 3.3 years.

About 20 known comets, including Halley, follow a more leisurely route, traveling beyond the orbit of Neptune before returning to the inner Solar System. Although these short-period comets have also been influenced by encounters with the giant planets, their orbits are more random, often tilted steeply to the plane of the planetary orbits (the ecliptic). Many of these—including Halley's comet—travel in a retrograde (backward) direction.

Comets with orbits of less than 200 years are thought to originate in the Kuiper Belt, a doughnut-shaped region that begins at the orbit of Neptune and extends to at least 50 Sun–Earth distances. They were probably ejected to their present location billions of years ago by gravitational interactions with Uranus and Neptune. The first Kuiper Belt Object was discovered in 1992 and many hundreds are now known (see Chapter 11).

Finally, there is a huge population of comets that follow highly elliptical or near-parabolic orbits that carry them many billions of kilometers from the Sun. These comets come without warning from any direction and may take many thousands of years to return. Some may never return at all.

By studying the paths of long-period comets, Dutch astronomer Jan Oort (1900–1992) realized that there must be a huge spherical reservoir containing billions of dormant comets on the outer reaches of the Solar System, 10,000 to 100,000 times Earth's distance from the Sun. Very occasionally, these ice worlds are dislodged and sent inwards, towards the Sun, by collisions or gravitational tugs from a giant planet or passing star.

Dirty Snowballs

Until the arrival of the Space Age, no one knew what a comet nucleus was really like. Ground-based telescopes are unable to resolve cometary nuclei because of their small size—often just a few kilometers across. In addition, for

Halley's Comet in 1910—This image of comet Halley was processed at the Kitt Peak National Observatory from an original black-and-white plate taken at Lowell Observatory in 1910. The false colors indicate varying levels of brightness. A broad dust tail is visible above a narrow ion (gas) tail. Halley is currently heading outward, and will reach its furthest point from the Sun (aphelion) in December 2023. It will then begin its long return journey, arriving in the inner Solar System in 2062. (Lowell Observatory/NOAO/AURA/NSF)

Comet Hale-Bopp—Comet Hale-Bopp was one of the great comets of recent years. It first appeared as a "little fuzzball" in the telescopes of two amateur astronomers, Alan Hale and Thomas Bopp, on July 23, 1995. Over the next two years, as it approached the Sun, it became extremely bright and active, developing an 8-degree-long ion tail (blue) and a yellowish, 2-degree dust tail. Hale-Bopp's nucleus was estimated to be 40 km (25 mi) in diameter, huge compared with most comets that reach the inner Solar System. (Peter Stättmayer)

most of the comet's orbit, the nucleus is so cold and dark that it is impossible to see or study from the Earth. When the icy heart of the comet does move close enough to us for detailed observation, it is obscured from view by a shroud of gas and dust, known as the coma.

For two and a half centuries, astronomers stared through their telescopes at a startling array of comets and wondered what caused their transformation. The nineteenth century, in particular, was blessed with many "great" comets whose tails stretched far across the night sky.

Studies of several close encounters with planets clearly showed that comets were of very low mass and insubstantial. Comet tails could stretch for millions of kilometers through space, but, despite their enormous length, they were so diaphanous that stars could be seen through them.

On a number of occasions, such as the return of comet Halley in 1910, the Earth passed unscathed through the wispy tails—despite scare stories about "the end of the world" in certain publications.

Other important clues came from studies of meteor showers. In 1866–1867, astronomers were able to determine the paths of the particles that entered Earth's atmosphere during four meteor showers. In each case, the orbits were found to coincide with those of known periodic comets.

The link between meteors and comets was further supported by observations that one of the four comets, Biela, had split into two during its 1846 apparition. The next time it came around, in 1852, the two segments were about 2 million km (1.25 million mi) apart, but still following the same orbit. They were never seen again. In their place, great meteor storms were seen in 1872 and 1885—presumably debris from the disintegration of the comet. Since then, other comets have been seen to break apart, and many more meteor swarms have been linked to periodic comets. Comet Halley, for example, is associated with two annual showers, the Eta Aquarids in May and the Orionids in October.

Occasionally, spectacular meteor storms light up the sky when the Earth ploughs through dense clouds of dust left by a comet that recently passed through that part of the sky. Such was the case in November 1966 when a peak rate of 120,000 shooting stars an hour was recorded over the USA, soon after the passage of comet 55P/Tempel-Tuttle.

The plethora of fiery trails created by particles burning up in the atmosphere led to the misconception that comets must be flying swarms of sand and dust. However, the advent of two powerful new techniques—spectroscopy and photography—in the second half of the nineteenth century showed that these unpredictable wanderers were more complex than previously thought. Bright bands in the spectrum of reflected light indicated the presence of compounds made of hydrogen, carbon, oxygen, and nitrogen, as well as the silicate particles that produced the meteor swarms.

Then, in 1950, American astronomer Fred Whipple (1906–2004) turned cometary science on its head. Whipple knew that some comets have orbited the Sun a thousand times or more. If they were nothing but a large pile of sand mixed with hydrocarbons, they would have broken apart. Whipple's answer was that comet cores are "icy conglomerates" made of water ice and dust mixed with ammonia, methane, and carbon dioxide—what the press later called "dirty snowballs." As the snowball approached the Sun and became warmer, its outer ices began to vaporize, releasing large amounts of dust and gas. The result was the formation of an all-enveloping coma, along with the two characteristic tails if dust and gas production was high enough. Whipple's model also made it possible to understand the variations in comet orbits. By taking into account the thrust from gaseous jets that erupt from the nucleus, astronomers could predict the motions of active comets far more accurately.

Subsequent observations from the ground, suborbital sounding rockets and Earth-orbiting spacecraft such as the International Ultraviolet Explorer allowed this model to be refined. It is now understood that the tiny dust particles in the coma are pushed away by sunlight, forming a yellowish tail, which is usually broad and stubby. Meanwhile, the gases are turned into a "swarm" of electrically charged particles which are carried away with the solar wind, forming an ion tail (also called a gas or plasma tail). This bluish tail is usually narrow and straight, sometimes streaming away from the nucleus for many millions of kilometers.* Gusts in the solar wind can cause the ion tail to swing back and forth, to develop curls, swirls, and temporary knots that sometimes break away and then regrow.

The composition of the gas in the coma is sometimes indicated by the greenish glow of cyanogen (CN) and carbon when illuminated by sunlight. This is called "resonant fluorescence." Other compounds of carbon, hydrogen, and nitrogen have been found. Ultraviolet images from spacecraft have also shown that the visible coma is surrounded by a huge, sparse cloud of hydrogen.

Until recently, the Great March Comet of 1843 held the record for the longest plasma tail, which stretched 330 million km (206 million mi)—more than two Sun–Earth distances—across the sky. Then, in 1998, analysis of data from the Ulysses spacecraft showed that it had passed through the plasma tail of comet Hyakutake at the remarkable distance of 570 million km (360 million mi) from the nucleus.

Outbursts, Collisions, and Sungrazers

The most familiar representatives of the comet tribe are the regular visitors to our skies, such as Encke and Halley. More than 150 periodic comets are currently known, but—with a few exceptions such as Halley—they rarely make the headlines. Indeed, very few comets become visible to the naked eye, let alone develop a tail.

---

* A third tail, made of neutral sodium atoms, was discovered in comet Hale-Bopp in 1997.

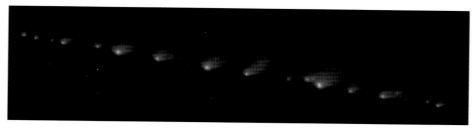

The "String of Pearls"—A number of comets have been seen to disintegrate as they approach the Sun or giant planets. This Hubble Space Telescope image shows 22 separate pieces of comet Shoemaker–Levy 9 after its nucleus shattered during a close approach to Jupiter in 1992. The "string of pearls" was made up of one remnant around 3 km (2 mi) across and more than 20 smaller fragments of between 0.5 and 2 km (0.3 and 1.2 mi). One after the other, these mini nuclei smashed into Jupiter between July 16 and 22, 1994. (NASA/STScI)

In the cold depths of the Solar System, beyond the orbit of Jupiter, the small, dark nuclei are extremely hard to detect. These dormant comets only make their presence felt if they experience an unusual gaseous eruption.

Repeated observations of comet Halley as it moved away from the Sun revealed a steady decrease of activity. When it reached the distance of Saturn, the tail and coma had disappeared completely, leaving only the "dirty snowball" nucleus. This inactivity was rudely interrupted in 1991, when a gigantic explosion enveloped it for several months with an extensive, expanding cloud of dust. The cause of this eruption remains unknown. Perhaps it was caused by a collision with an unknown piece of rock or a sudden release of pent-up gas.

This was not the first example of a cosmic iceberg suddenly making its presence known. Perhaps the most unusual of these was an object known as Chiron, which spends its life between the orbits of Jupiter and Uranus. Discovered by Charles Kowal in 1977, Chiron was classified as a rocky asteroid, 180 km (115 mi) in diameter. Then, in 1988, two years after its closest passage to the Sun, it unexpectedly doubled in brightness. Over the coming months, astronomers observed the development of a fuzzy coma, and by 1991 the detection of cyanogen gas made it clear that Chiron had reverted to a comet. Since then, many more of these objects, known as Centaurs, have been found beyond the orbit of Jupiter. It is now realized that they are giant comet nuclei that have strayed from the Kuiper Belt beyond Neptune. Like Chiron, most of them would take on the characteristics of normal comets if they were moved closer to the Sun.

Although they are often regarded as pristine objects that have changed little over billions of years, this is obviously not true of the periodic comets. During each circuit of the Sun, several meters of surface material are vaporized or blown into space. As the object shrinks, a "crust" of black, organic material coats much of the outer skin, restricting outgassing activity to a few jets.

This eventually causes gas and dust production rates to plummet in periodic comets that have evolved during dozens of solar passages. After perhaps 10,000 years, a member of the Jupiter family is likely to fade into a dark, dead remnant of its former glory. It may continue to circle the Sun as an asteroid look-alike, break into smaller pieces, or completely disintegrate into dust.

Some of these events can be quite spectacular. In August 2000, the Hubble Space Telescope and ground-based instruments captured graphic images showing the sudden break-up of comet Linear (C/1999 S4) as it approached the Sun.

Planets may also interfere with the welfare of comets. The most dramatic example of this was Shoemaker–Levy 9, a comet that was discovered in a two-year orbit around Jupiter in March 1993. Calculations showed that the comet

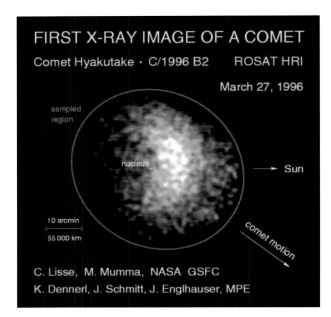

FIRST X-RAY IMAGE OF A COMET

Comet Hyakutake · C/1996 B2   ROSAT HRI

March 27, 1996

sampled region

nucleus

Sun

10 arcmin

55 000 km

comet motion

C. Lisse, M. Mumma, NASA GSFC
K. Dennerl, J. Schmitt, J. Englhauser, MPE

Comet X-rays—One of the most unexpected results of the ROSAT X-ray astronomy mission was the discovery of very bright X-ray and extreme-ultraviolet emission from comets. This X-ray image of comet Hyakutake was obtained by ROSAT on March 27, 1996. Since X-rays are normally produced in hot, high-energy environments, this discovery showed that there must be previously unsuspected processes taking place in the comet, probably due to the influence of the Sun's radiation and/or the solar wind. (NASA-GFC/MPE)

had been orbiting the giant planet unnoticed for several decades. Then, in July 1992, the nucleus had been disrupted by a close approach to Jupiter. The ease with which the nucleus had been pulled apart by Jupiter was a clear demonstration of the low-strength, possibly porous, nature of the comet's core. All that remained was a line of 20 or so glowing fragments, strung out like a string of pearls. And these were on a collision course with Jupiter. One by one, these mini-nuclei smashed into Jupiter between 16 and 22 July 1994, resulting in huge explosions that left bruise-like marks in the planet's clouds for many months.

Similar impacts with planets have taken place throughout the history of the Solar System. In its first billion years or so, the Sun's realm was like a cosmic shooting alley, with asteroids and comets flying in all directions. Inevitably, many of these left their mark on the fledgling planets, including the young Earth.

Many astronomers believe that comet ice was one of the main sources of the water that filled Earth's oceans. Comets may also have provided the water ice that—against the odds—is thought to exist in the permanently shaded craters on Mercury and the Moon.

Even today, comets occasionally collide with Earth. Traveling at speeds of up to 260,000 km/h (164,000 mph), a large comet on a collision course with our world would pose a major threat to modern civilization. Fortunately, small impact events are much more likely. One such event occurred on June 30, 1908, when a 50-m (165-ft) wide object—a comet or asteroid—exploded in mid-air over the Tunguska region of Siberia, flattening trees for hundreds of kilometers in all directions. If the explosion had taken place further west, over London or Paris, an entire city could have been obliterated.

With the introduction of personal computers, internet links and the ESA–NASA SOHO spacecraft, astronomers have recently found a sizeable population of small, sungrazing comets. Most of these follow similar paths, so they are thought to have come from the break-up of a single, giant parent long ago. By mid-2005, the number of comets discovered in SOHO views of the Sun was rapidly approaching 1,000. More than 75% of these were found by amateur comet hunters who searched the freely available SOHO images on the internet.

The Armada to Halley

Halley's comet is unique. Not only is it the largest and brightest of all the comets that follow a predictable timetable for their flights through the inner Solar System, but for well over two millennia it has also played a part in

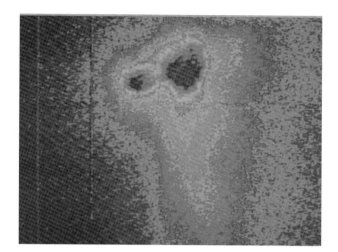

Vega 2 Closes on Halley—The Soviet Vega spacecraft studied comet Halley after dropping off landers and balloons at Venus. This Vega 2 image shows the dust-obscured nucleus just two seconds before closest approach to the comet's sunlit side on March 9, 1986. The spacecraft was then 8,030 km (5,000 mi) from the comet. The Sun is shining from lower left. Apart from the peanut-shaped nucleus (blue), the image shows bright jets of gas and dust (bottom) that pointed toward the Sun. The Vegas's TV system produced about 1,500 images. Images such as these enabled ESA's Giotto spacecraft to make an even closer flyby a few days later. (IKI)

Giotto—The European Space Agency's first deep space mission was named Giotto, after the Italian artist Giotto di Bondone. Launched on July 2, 1985, it flew past the nucleus of comet Halley at a distance of less than 600 km (373 mi) on March 13, 1986. Although its camera was disabled by a collision with a large dust particle, Giotto survived to become the first spacecraft to visit two comets when it passed within 200 km (124 mi) of comet Grigg–Skjellerup on July 10, 1992. (ESA)

human affairs. Not surprisingly, the cosmic iceberg's first apparition since the advent of the Space Age led excited scientists to propose various missions that would unveil its secrets. Unfortunately, the timing of the 1986 apparition could not have been worse for US astronomers. Hit by costly delays in the Shuttle program and tight budgets, NASA had to abandon its plan to send a spacecraft to fly alongside the icy nucleus. The world's greatest space power had to settle for a long-range reconnaissance by spacecraft such as the Pioneer Venus orbiter, and a cut-price flyby of a much smaller comet by a spacecraft that was originally designed for observations of the solar wind.

In 1981, it was proposed that NASA's International Sun–Earth Explorer 3 should be redirected toward a comet. After 15 maneuvers through the Earth's magnetotail and five lunar flybys to get the spacecraft out of the Earth–Moon system, the spacecraft was sent on its way toward comet Giacobini-Zinner.

Renamed the International Cometary Explorer (ICE), the spacecraft was aimed to fly behind the small periodic comet so that its fields and particles instruments could sample the tail. On September 11, 1975, ICE made history when it became the first spacecraft to encounter a comet during a high-speed flyby at a distance of 7,862 km (4,885 mi). It found a region of interacting cometary and solar wind ions (charged particles), with a gas tail about 25,000 km (15,625 mi) wide. Water and carbon monoxide ions were also identified, which supported the "dirty snowball" theory. Although the spacecraft did not carry any protection against comet particles, it survived the encounter unscathed, since the dust impact rate peaked at only about one per second—lower than expected. ICE went on to observe Halley's comet from a distance of 28 million km (17 million mi) in March 1986.

Meanwhile, Europe, the Soviet Union, and Japan pressed ahead with a joint program of Halley observations. The adventure began in December 1984 with the launches of two huge Soviet Vega spacecraft that were designed to take advantage of opportunities to explore Venus en route to the speeding comet. (Since there is no "G" in the Cyrillic alphabet, Vega was the nearest abbreviation for Venus–Halley.)

After releasing landers that descended to the surface of Venus, the Vegas headed for Halley. Acting as pathfinders for Europe's Giotto spacecraft, they provided invaluable information on the size, shape, and location of the dust-shrouded nucleus. Vega 1 made its closest approach to the comet on March 6, 1986, at a distance of 8,890 km (5,520 mi). Three days later, its twin flew past at a distance of 8,030 km (5,000 mi).

The early images from Vega 1 showed two bright areas that were initially interpreted as a double nucleus, but later turned out to be distinct jets of gas and dust. The potato-shaped nucleus was about 16 km (10 mi) long with a rotation period of about 53 hours. It was surprisingly dark, with a surface temperature of 27–127°C, much warmer than expected for an icy body. It seemed that the comet had a thin insulating layer covering its frozen heart.

Japan's first deep space missions, known as Sakigake (Pioneer) and Suisei (Comet), were launched in January and August 1985 respectively. Sakigake went on to obtain some valuable observations of the comet's interactions with the solar wind, flying to within 7 million km (4.35 million mi) of the nucleus on March 11, 1986. Suisei came much closer, flying past at a distance of 151,000 km (93,825 mi) on March 8, 1986. It carried an ultraviolet camera and a particle experiment that detected water, carbon monoxide, and carbon dioxide ions from the comet. During closest approach, Suisei was hit by two dust particles which slightly changed its spin axis and spin period, but it survived to detect two major and four minor outbursts from the comet.

The flagship of the international armada was the European Space Agency's Giotto, named after the Florentine artist Giotto di Bondone, who included the 1301 apparition of comet Halley in his painting of the Wise Men visiting the infant Jesus. Launched on July 2, 1985, Giotto flew past the nucleus at a distance of less than 600 km (373 mi) on March 13, 1986. This grazing encounter enabled it to image the nucleus in dramatic detail. Unfortunately, the photo opportunity was terminated only seven seconds before closest approach when a 1-mm dust particle sent the spacecraft spinning. Monitor screens went blank as contact with the Earth was temporarily lost. TV audiences and anxious Giotto team members feared the worst, but, to everyone's amazement, occasional bursts of information soon began to filter through. Giotto was still alive.

Over the next 30 minutes, the sturdy spacecraft's thrusters stabilized its motion and contact was fully restored. By then Giotto had skimmed past the nucleus and was heading back into interplanetary space. It continued to return scientific data for another 24 hours on the outward journey, and the last dust impact was detected 49 minutes after closest approach. After a long hibernation, Giotto again made history by completing a second flyby on July 10, 1992, this time passing only 200 km (124 mi) from a less active periodic comet called Grigg-Skjellerup.

Halley Makes an Impact

The European Space Agency could hardly have chosen a more ambitious, high-profile project for its first deep space mission—a close encounter with the most famous comet of all. For the first time, human eyes would see what the nucleus of a comet really looked like. After an eight-month journey from Earth, Giotto closed in on the fast-moving nucleus. Taking advantage of the locational information provided by the two Soviet Vegas, the spacecraft was targeted to fly past on the sunward side at a distance of about 540 km (340 mi).

The historic rendezvous began on March 12, 1986, when

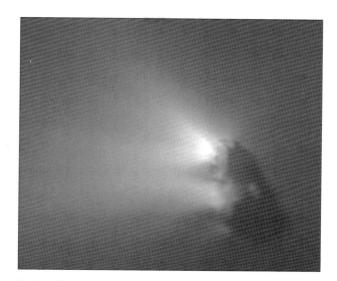

Halley's Nucleus—The first detailed view of a comet's nucleus was taken by ESA's Giotto spacecraft from a distance of about 18,000 km (11,250 mi) on March 13, 1986. Images showed a dark, peanut-shaped nucleus that measured about 8 × 8 × 16 km (5 × 5 × 10 mi). The sunlit side (left) was illuminated by bright jets of gas and dust. The surface was very rugged, with sizeable valleys and depressions. A mountain on the dark side can be seen catching the Sun's rays. The rotation axis of the nucleus was approximately horizontal in this view. (ESA)

the spacecraft's instruments first detected hydrogen ions 7.8 million km (4.9 million mi) from the comet. 22 hours later, Giotto crossed the bow shock of the solar wind (the region where a shock wave is created as the supersonic solar particles slow to subsonic speed) and entered the dusty coma. At this point the camera was switched to tracking mode to follow the brightest object (presumed to be the nucleus) in its field of view and began to send the first, fuzzy images back to Earth.

Excitement rose at the European Space Operations Center in Darmstadt, Germany, as the stream of pictures and data were received. Located in adjacent rooms, each of the 10 experiment teams scrutinized the latest information and struggled to come up with a preliminary analysis of the multicolored pictures.

The first of 12,000 dust impacts was recorded 122 minutes before closest approach, but the rate rose sharply as the spacecraft passed through a jet of material that streamed away from the nucleus. The danger posed by this debris-spewing object was well understood by the mission team, who had provided Giotto with a "bullet-proof vest" in the form of a front bumper shield. Nevertheless, with only eight seconds to go before closest approach, the spacecraft was sent spinning by a collision with a "large" particle about the same size as a grain of sand.

Screens went blank and scientists feared the worst, but after 30 minutes of anxious nail biting, the hardy spacecraft recovered stability. It was later found that the camera was no longer usable after its baffle was damaged as a result of the pummeling. Despite this setback, Giotto was able to transmit images to within 1,372 km (850 mi) of the nucleus. The 100-m (330-ft) resolution pictures showed a potato-shaped object that measured about 8 × 8 × 16 km (5 × 5

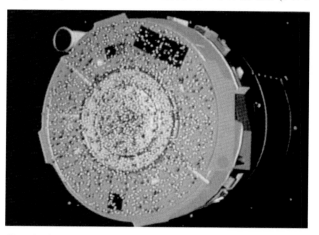

Giotto's Greatest Hits—This computer-generated view shows the impact sites of the dust particles (white dots) that blasted Giotto's front end during its flyby of comet Halley. The black patches are the dust impact sensors. The camera was disabled when the baffle at upper left was largely torn away as a result of the pummeling. One impact with a 1-gram particle, just 8 seconds before closest approach, sent the spacecraft spinning and almost brought the mission to an end. (University of Kent)

× 10 mi). Three bright jets were blasting away from active regions on the sunlit side, but only about 10% of the surface was active. These and their smaller counterparts were throwing about three tonnes of material into space every second, resulting in a strange, wobbling rotation of the nucleus.

The surface of the nucleus was very dark—blacker than coal—which suggested a thick covering of dust or organic material. In fact, it was one of the darkest objects in the Solar System. The nucleus was also very irregular, with a large central depression bordered by numerous hills and depressions. The "fluffy," porous texture meant that its density was only 0.3 g/cm^3 (one-third the density of water). Water accounted for about 80% by volume of all of the material thrown out by the comet. There were also substantial amounts of carbon monoxide, carbon dioxide, methane, and ammonia. Traces of other hydrocarbons, iron, and sodium were also found.

Most of the dust was similar in size to specks of cigarette smoke. Two major types of dust particle were found. One was dominated by the light "CHON" elements—carbon, hydrogen, oxygen and nitrogen. The other was rich in mineral-forming elements—sodium, magnesium, silicon, iron, and calcium.

All the comet's light elements (except nitrogen) were found to be in the same relative abundance as in the Sun, which confirmed that the comet had formed 4.5 billion years ago from ices condensing onto grains of interstellar dust. Since then, it has remained almost unaltered in the cold, outer regions of the Solar System. "The mission forced us to revisit our long-standing image of a comet nucleus as a 'dirty snowball'," said Giotto scientist, Uwe Keller. "The pictures showed that it was more like a lump of icy sludge. The solid part of the nucleus is much larger than the icy part."

Despite the loss of the camera and some of the other instruments, the battle-scarred spacecraft was deemed sufficiently healthy to attempt an unprecedented second comet encounter. On July 10, 1992, Giotto passed within 200 km (124 mi) of the much less active periodic comet 26P/Grigg-Skjellerup. Flyby conditions were very different from those during the Halley encounter. Since Grigg-Skjellerup would approach Giotto at an angle of 68 degrees instead of head-on, the bumper shield afforded no protection. However, with a much slower relative approach speed of 14 km/s (9 mi/s)—as opposed to 68 km/s (42 mi/s) for Halley—and a dust production rate about 1/200 that of Halley, the Grigg-Skjellerup encounter was expected to cause very little damage to the spacecraft. And so it proved. Only three sizeable particles were detected, none of which was large enough to threaten the spacecraft.

Close Encounters

In the mid-1990s, NASA introduced the New Millennium program in order to flight-test new technologies for future space missions. The first mission was Deep Space 1, a spacecraft built to test a dozen new technologies, including an ion engine. A secondary objective of Deep Space 1 was to investigate at least one asteroid and one comet. After launch on October 24, 1998, the revolutionary ion engine gently accelerated Deep Space 1 toward asteroid 9969 Braille. Its primary mission was then extended to include a flyby of comet 19P/Borrelly on September 22, 2001.

Deep Space 1 made amends for its disappointing images of Braille by returning the most detailed pictures ever obtained of a comet's nucleus. Passing by at a distance of about 2,200 km (1,400 mi) the spacecraft found that Borrelly's nucleus measures 8 × 4 km (5 × 2.5 mi) and is apparently broken into two pieces. One part of the comet appeared to be "canted" about 15 degrees with respect to the other segment. At the junction of the two was a series of parallel ridges, possibly caused by compression.

The onboard camera was also able to pick out a series of flat-topped, steep-sided hills in the center of the nucleus, near the most active regions. These mesas were topped by a thick insulating layer of dust, whereas the underlying, ice-rich comet material was exposed on their steep sides. Scientists theorized that the ices were sublimating (vaporizing) from the sides of the mesas, undercutting the thick

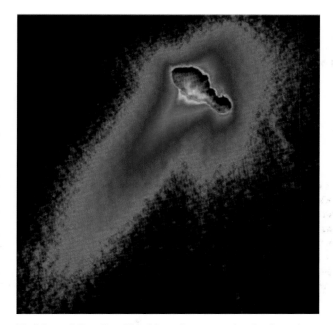

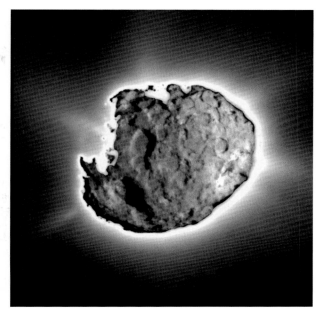

Blackhearted Borrelly—This false-color, composite view is made up of images taken by NASA's Deep Space 1 spacecraft when it flew by comet Borrelly on September 22, 2001. Apart from the dark nucleus, it shows details in the dust jets (purple and blue) and the coma of dust and gas. The potato-shaped nucleus was found to be the blackest object in the Solar System, reflecting less than 3% of the sunlight that it receives. It is about 8 km (5 mi) long and 4 km (2.5 mi) wide. (NASA–JPL)

Comet Wild 2—This composite image of comet Wild 2 was taken during Stardust's flyby on January 2, 2004. The unusual surface displayed several large craters and depressed regions, cliffs, and pinnacles. Shaped like a thick hamburger, Wild 2 is about 5 km (3.1 mi) in diameter. The picture comprises a short exposure image that shows tremendous surface detail overlain on a long-exposure image taken 10 seconds later that reveals at least five jets of dust and gas. (NASA–JPL)

insulating layer and causing sections of it to collapse into the valley floor.

Even more surprises came from NASA's Stardust spacecraft, which was launched toward another periodic comet, 81P/Wild 2, on February 7, 1999. As its name suggests, Stardust was mainly designed to collect samples of dust from interstellar space and from the comet's coma. It would then bring the preserved particles back to Earth for analysis. However, Stardust was also equipped to take advantage of a rare opportunity to examine the nucleus from close range, and the scientists were not disappointed. After a seven-year trek that included a gravity assist from

Earth to slingshot the spacecraft beyond the orbit of Mars, the spacecraft swept past Wild 2 at a distance of 236 km (147 mi) on January 2, 2004. This close encounter enabled Stardust's navigation camera to obtain images that surpassed even those of Deep Space 1. And to everyone's surprise, Wild 2 was completely different from its distant cousins, Borrelly and Halley.

Although the low-density, hamburger-shaped nucleus was only 5 km (3 mi) across, its surface was strong enough to support cliffs, pinnacles, and mesas over 100 m (328 ft) high. The images even showed a 100-m (325-ft) overhang at the top of a cliff. Most noticeable of all were the large circular craters, some measuring 1.5 km (1 mi) wide, and 150 m (500 ft) deep. Some of these displayed a round central pit surrounded by ragged, ejected material, while others had a flat floor and straight sides. One large hollow, called Left Foot, was 1 km (0.6 mi) across—one-fifth of the diameter of the nucleus. Mission scientists concluded that the unique character of Wild 2 could be accounted for by its recent introduction to the inner Solar System. The comet has spent all but a few decades of its 4.5-billion-year life in the frigid outer reaches where its original, impact-cratered surface was preserved. Rather than being chipped off a larger body by an impact or resurfaced by numerous solar encounters, Wild 2 probably remains much the same as when it originally formed from the dust and gas of the presolar disk. Somehow, the surface of the nucleus has been hardened during its long period out in the cold.

Another big surprise was the abundance and behavior of jets of particles shooting up from the comet's surface. Stardust discovered more than two dozen jets, including some that resembled blasts of water from a powerful garden hose. This resulted in quite a wild ride as the spacecraft flew through three huge jets and was bombarded with about a million particles per second. Out of about 10,000 impacts detected on Stardust's shielding, the vast majority of the particles were smaller than a pinhead. However, 12 of these, some larger than a bullet, penetrated the top layer of the spacecraft's protective shield and threatened its survival. Scientists believe that such unusual "swarms" might result from large chunks of material breaking free of the comet's surface and disintegrating into clouds of much smaller particles.

Stardust returned to Earth carrying millions of primordial dust particles trapped in a "catcher's mitt" composed of a frothy material called aerogel. The re-entry capsule descended by parachute onto the Utah desert with its precious cargo on January 15, 2006. Apart from the fluffy grains captured by high-flying aircraft, this provided the first opportunity for scientists to examine cometary material in the laboratory.

Deep Impact

Although scientists were able to learn a great deal from analysis of cometary coma material and remote sensing of nuclei, many questions remained concerning the internal structure and composition of these icy cores. In order to fill this knowledge gap, NASA launched its Deep Impact spacecraft toward comet 9P/Tempel 1 on January 12, 2005. Like all the other comets so far visited by spacecraft, Tempel 1 is a periodic visitor to the inner Solar System. Discovered in 1867 by German lithographer-turned-astronomer Ernst Wilhelm Tempel, it circles the Sun once every 5½ years, passing close enough to Earth on every other orbit to be easily reached and observed. That made it a perfect target. Instead of simply looking at the nucleus or attempting to sample the swarms of lightweight particles surrounding it, Deep Impact was designed to take a much more explosive, brute-force approach. It would blast a

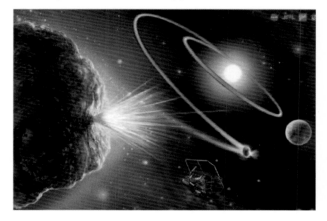

Deep Impact—Mankind's first attempt to look inside a comet took place on July 4, 2005. After a 431-million-km (268-million-mi) journey, NASA's Deep Impact spacecraft despatched a probe that collided with the nucleus of comet Tempel 1 at a speed of 10 km/s (6.3 mi/s). The 373-kg (820-lb) impactor exploded on arrival, excavating a stadium-sized crater and ejecting a huge cloud of ice and dust debris. (NASA–JPL)

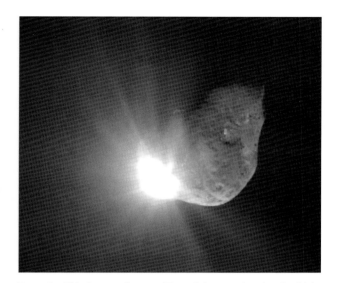

Impact!—This image of comet Tempel 1 was taken by the high-resolution camera on the Deep Impact flyby craft 67 seconds after arrival of the impactor probe. Scattered light from the collision saturated the camera's detector, creating the bright flash. Linear spokes of light radiate from the impact site, while reflected sunlight illuminates most of the comet surface. The nucleus reveals ridges, scalloped edges, and possible impact craters that formed long ago. (NASA/JPL–Caltech/UMD)

crater in the nucleus by firing a 373-kg (820-lb) copper projectile the size of a bathtub into its midriff.

Deep Impact consisted of a flyby spacecraft and a small impactor that would be targeted for a planned high-speed collision on July 4. The mother craft carried three instruments to observe the effects of the collision—a high resolution camera and infrared spectrometer, together with a medium resolution camera. A near duplicate of the latter camera was placed on the impactor to record the final moments before it was run over by comet Tempel 1 at a speed of about 37,000 km/h (23,000 mph). After a trip of six months and 431 million km (268 million mi), the two-part spacecraft closed on the fast-moving target, which was near perihelion, its closest approach to the Sun. Meanwhile, a battery of instruments on ground-based telescopes and

scientific spacecraft focused on the active nucleus in preparation for the collision. Expectations were raised in mid-June when two major outbursts caused a marked brightening of the coma around the nucleus.

On July 3, the impactor was released, enabling the mother ship to maneuver for a grandstand view of the collision. As it closed with Tempel 1, the impactor's camera sent back increasingly detailed pictures of the nucleus. Finally, Deep Impact provided its own Independence Day fireworks with a spectacular, explosive display as the missile smashed into the comet with an energy equivalent to 4.8 tons of TNT. Remarkable pictures of the cosmic collision were sent back to Earth from the impactor, despite the fact that it took two particle hits prior to impact. These particle collisions slewed the spacecraft's camera for a few moments before the attitude control system could get it back on track.

A sequence of increasingly detailed pictures was sent back until three seconds before the impactor ploughed into the nucleus. The comet's icy heart was revealed as a remarkably rounded object, 5 km (3.1 mi) long and 7 km (4.3 mi) wide, marked by ridges, smooth (presumably younger) patches, and numerous craters that were probably the result of ancient impacts of a nonhuman origin. Although the nucleus was generally darker than charcoal, there were scattered bright patches that appeared to represent very smooth and reflective material. The reason for the rougher, higher landscape and the smoother "lowlands" left scientists scratching their heads for an explanation. The penetrator finally committed suicide, striking the comet's surface at an angle of about 25 degrees. It immediately vaporized in an incandescent flash, releasing large quantities of water, carbon dioxide and organic compounds. It then burrowed its way into the surface before a shock wave threw out a huge cloud of dust.

Meanwhile, the instruments on the flyby craft had a bird's eye view as Deep Impact swept by just 480 km (300 mi) from the nucleus. The images portrayed in graphic detail how the cloud of icy debris erupted from the comet's interior. The fireball of vaporized impactor and comet material shot skyward, expanding away from the impact site at about 5 km/s (11,200 mph). As it spread out and dispersed, the cameras were able to capture the creation of the first man-made impact crater on another world. Preliminary analysis suggested that a stadium-sized indentation, 50–250 m (165–820 ft) across, was excavated in a matter of seconds.

Scientists estimated that several tens of thousands of tons of material exploded into space. Instruments on Deep Impact and elsewhere indicated that the immense cloud was largely composed of extremely fine material, more like talcum powder than beach sand. This led scientists to suggest that the comet has a fluffy constituency and is held together more by gravity than by the cohesiveness of its materials.

Studies with Earth-orbiting satellites such as Swift and XMM-Newton showed that comet Tempel 1 became increasingly bright at X-ray and ultraviolet wavelengths after the collision. However, the UV light peaked within a day of impact whereas the X-ray output rose for about five days. The X-rays were created by the newly liberated material as it rose through the comet's thin atmosphere and was affected by the high-energy solar wind from the Sun.

Ground-based observations showed the comet to be two to three times brighter in infrared light than the day before the impact. The coma of gas and dust was also much larger and lopsided, with the rapid growth of a new, bright, fan-like structure. More detailed studies showed that the direction in which the ejected dust particles traveled away from the nucleus seemed to depend on the particle size. One jet structure, made up of smaller dust particles, was seen pointing toward the north, whereas a jet composed of larger particles was rotated by about 45 degrees toward the northeast.

Rosetta

Even before ESA's Giotto spacecraft had completed its historic encounter with comet Halley, European scientists were anticipating an even more ambitious follow-up. Their

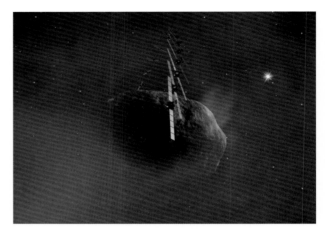

The Rosetta Orbiter—After a 10-year trek, ESA's Rosetta spacecraft will rendezvous with the periodic comet 67P/Churyumov–Gerasimenko in spring 2014, eventually entering orbit about 25 km (16 mi) above the surface. For the next 18 months it will map and study the surface in great detail as the comet plunges headlong toward the inner Solar System. For the first time, scientists will see at close range the dramatic changes that take place as the warmth of the Sun transforms the icy nucleus into a cauldron of activity. (ESA/Medialab)

desire was to collaborate with their counterparts from NASA to build a comet sample return mission. Unfortunately, the joint expedition proved to be too complex and too expensive and was rejected. However, the science community persevered, and in 1993 ESA made the Rosetta comet chaser one of its Cornerstone science projects.

The mission was named after the Rosetta Stone, a slab of basalt that was found in Egypt by Napoleon's invading army in 1799. What made the Stone unique was the carved script in three languages that opened the way to deciphering ancient Egyptian hieroglyphics. Just as the Rosetta Stone provided the key to understanding an ancient civilization, ESA's Rosetta would unlock the mysteries of comets, the oldest, most primitive building blocks in the Solar System.

In order to fulfill this ambition, the spacecraft would be launched in January 2003, then follow a circuitous route to the orbit of Jupiter and eventually rendezvous with periodic

Comet Touchdown—The little Rosetta lander, known as Philae, is scheduled to land on the nucleus of comet Churyumov–Gerasimenko in November 2014. The 100-kg (220-lb) lander will anchor itself to the surface with harpoons and screw-in feet before drilling into the crust, sampling the dark, icy material and sending back pictures and other information on the pristine, alien environment. (ESA/Medialab)

comet 46P/Wirtanen in 2011. After entering orbit around the nucleus, it would release a lander onto the surface. Meanwhile, the orbiter would spend two years in close proximity to the comet's core as it plunged toward the warmer inner reaches of the Sun's domain.

ESA assembled the largest, most complex scientific spacecraft that the agency had ever conceived. In addition to the 3-tonne orbiter, Rosetta also carried a 100-kg (220-lb) lander—later named Philae after a separate, bilingual obelisk that was found on an island in the river Nile. Altogether, the spacecraft was equipped with no fewer than 20 experiments—11 on the orbiter and nine on the lander—in order to survey the comet in unprecedented detail.

Despite a tight schedule, the spacecraft was ready for launch when an unexpected obstacle stopped the countdown in its tracks. During its maiden flight, an advanced version of Rosetta's launch vehicle, the Ariane 5, suffered a catastrophic failure. Nervous ESA officials postponed Rosetta's lift-off.

Unfortunately, this meant that an encounter with comet Wirtanen was no longer possible. Not only would another target have to be found, but a new flight profile would have to be calculated. After evaluating all possibilities within a reasonable time frame, ESA scientists plumped for 67P/Churyumov–Gerasimenko, a slightly larger and more active comet than Wirtanen.

Rosetta finally got underway on March 2, 2004. Over the next 10 years, it will slowly spiral outward, taking advantage of three Earth gravity assists (2005, 2007, and 2009) and a close flyby of Mars (2007) in order to reach the remote target. The long trek will also be punctuated by two fleeting visits to members of the main asteroid belt. By mid-2011, when it is about 800 million km (500 million mi) from the Sun, Rosetta will break the distance record for a solar-cell-powered spacecraft. It will then ignite its main engine for a major maneuver that will place it on an intercept path with the comet, but a further three years will pass before Churyumov–Gerasimenko looms large in images taken by Rosetta's wide-angle camera.

Rosetta will be reactivated in January 2014, as it enters a six-month approach phase, closing in slowly on the comet's nucleus. Since the comet will still be far from the Sun, it should be cold and dormant. By August, the spacecraft will enter orbit around the nucleus at an altitude of some 25 km (16 mi).

Sometimes edging to within just a few kilometers of the giant flying iceberg, it will spend the next three months conducting detailed mapping in preparation for the landing attempt. Once a suitable site on the sunlit side has been selected, Philae will be released from an altitude of about 1 km (0.6 mi). After deploying its three legs, it will make a controlled descent to the pristine surface.

Pulled inward by the tiny gravity of the nucleus, it will touch down at walking speed. Even so, there will be a danger that the little craft will tip over or bounce back into space, so the legs have been designed to damp out most of the energy from the impact and adjust to an uneven surface. Immediately on arrival, Philae will anchor itself to the surface with two harpoons and screw its feet into the comet's crust.

The lander is expected to operate for at least a few weeks, sending back via the orbiter a treasure trove of incredibly detailed pictures and information about the comet's atmosphere and surface conditions. It will also drill at least 20 cm (8 in) into the dark surface to collect samples that can be heated and analyzed in tiny onboard ovens. Meanwhile, the Rosetta mother ship will continue its observations of the comet's nucleus for over a year, providing scientists with a ringside seat to the dramatic awakening of the comet's activity during its headlong plunge toward the inner Solar System. The orbiter will catalog every eruption of gas and dust as Churyumov–Gerasimenko's volatiles vaporize in the warmth of the Sun.

Apart from the usual cameras, spectrometers, and particle experiments, Rosetta also carries a sounding instrument that should revolutionize our understanding of the nucleus. CONSERT will probe the comet's interior by transmitting and receiving radio waves that are reflected and scattered as they pass through the nucleus. If all goes according to plan, this remarkable 4,000-day mission of exploration is expected to end in December 2015, after the comet has begun its retreat to the more frigid vicinity of Jupiter.

APPENDICES

MOON

Name	Country	Launch date	Purpose	Results
Pioneer 0	USA	Aug. 17, 1958	Orbiter	Launch failure
[Unnamed]	USSR	Sep. 23, 1958	Impact	Launch failure
Pioneer 1	USA	Oct. 11, 1958	Orbiter	Failed to reach escape velocity
[Unnamed]	USSR	Oct. 11, 1958	Impact	Failed to reach escape velocity
Pioneer 2	USA	Nov. 8, 1958	Orbiter	Failed to reach escape velocity
[Unnamed]	USSR	Dec. 4, 1958	Impact	Launch failure
Pioneer 3	USA	Dec. 6, 1958	Flyby	Failed to reach escape velocity
Luna 1	USSR	Jan. 2, 1959	Impact	Lunar flyby at distance of 6,400 km (3,796 mi) on Jan. 4, 1959
Pioneer 4	USA	Mar. 3, 1959	Flyby	Lunar flyby Mar. 4, 1959 at distance of 59,545 km (37,000 mi)
[Unnamed]	USSR	Jun. 18, 1959	Impact	Launch failure
Luna 2	USSR	Sep. 12, 1959	Impact	Lunar impact
Luna 3	USSR	Oct. 4, 1959	Circumlunar	Flyby at distance of 7,900 km (4,910 mi) on Oct. 6
Pioneer P-3	USA	Nov. 26, 1959	Orbiter	Launch failure
[Unnamed]	USSR	Apr. 15, 1960	Circumlunar	Failed to reach escape velocity
[Unnamed]	USSR	Apr. 19, 1960	Circumlunar	Launch failure
Pioneer P-30	USA	Sep. 25, 1960	Orbiter	Launch failure
Pioneer P-31	USA	Dec. 15, 1960	Orbiter	Launch failure
Ranger 3	USA	Jan. 26, 1962	Impact	Flyby on Jan. 28, at distance of 36,793 km (22,862 mi)
Ranger 4	USA	Apr. 23, 1962	Impact	Impact on lunar farside Apr. 26, no data
Ranger 5	USA	Oct. 18, 1962	Impact	Flyby on Oct. 21, at distance of 724 km (450 mi)
[Unnamed]	USSR	Jan. 4, 1963	Lander	Failed to leave Earth orbit
[Unnamed]	USSR	Feb. 3, 1963	Lander	Launch failure
Luna 4	USSR	Apr. 2, 1963	Lander	Flyby on Apr. 6, at distance of 8,500 km (5,280 mi)
Kosmos 21	USSR	Nov. 11, 1963	Flyby	Failed to leave Earth orbit

MOON contd.

Name	Country	Launch date	Purpose	Results
Ranger 6	USA	Jan. 30, 1964	Impact	Impact Feb. 2, but no data returned
[Unnamed]	USSR	Mar. 21, 1964	Lander	Failed to reach Earth orbit
[Unnamed]	USSR	Apr. 20, 1964	Lander	Launch failure
Ranger 7	USA	Jul. 28, 1964	Impact	Impact Jul. 31; returned 4,316 pictures
Ranger 8	USA	Feb. 17, 1965	Impact	Impact Feb. 20; returned 7,137 pictures
Ranger 9	USA	Mar. 21, 1965	Impact	Impact Mar. 24; returned 5,814 pictures
[Unnamed]	USSR	Apr. 10, 1965	Lander	Failed to reach Earth orbit
Luna 5	USSR	May 9, 1965	Lander	Crashed on lunar surface
Luna 6	USSR	Jun. 8, 1965	Lander	Flyby on Jun. 11, at distance of 161,000 km (100,625 mi)
Zond 3	USSR	Jul. 18, 1965	Flyby	Flyby Jul. 20, at distance of 9,220 km (5,730 mi)
Luna 7	USSR	Oct. 4, 1965	Lander	Crashed on lunar surface
Luna 8	USSR	Dec. 3, 1965	Lander	Crashed on lunar surface
Luna 9	USSR	Jan. 31, 1966	Lander	Landed on Ocean of Storms Feb. 3; operated until Feb. 6
Kosmos 111	USSR	Mar. 1, 1966	Orbiter	Failed to leave Earth orbit
Luna 10	USSR	Mar. 31, 1966	Orbiter	Arrived Apr. 3; operated until May 30
Surveyor 1	USA	May 30, 1966	Lander	Landed on Ocean of Storms Jun. 2; operated until Jan. 7, 1967
Explorer 33	USA	Jul. 1, 1966	Orbiter	Entered eccentric Earth orbit
Lunar Orbiter 1	USA	Aug. 10, 1966	Orbiter	Arrived Aug. 14; deliberately crashed on Moon on Oct. 29
Luna 11	USSR	Aug. 24, 1966	Orbiter	Arrived Aug. 27; operated until Oct. 1
Surveyor 2	USA	Sep. 20, 1966	Lander	Crashed on lunar surface
Luna 12	USSR	Oct. 22, 1966	Orbiter	Arrived Oct. 25; operated until Jan. 19, 1967
Lunar Orbiter 2	USA	Nov. 6, 1966	Orbiter	Arrived Nov. 10; deliberately crashed on Moon on Oct. 11, 1967
Luna 13	USSR	Dec. 21, 1966	Lander	Landed on Ocean of Storms Dec. 24; operated until Dec. 28
Lunar Orbiter 3	USA	Feb. 5, 1967	Orbiter	Arrived Feb. 7; deliberately crashed on lunar surface Oct. 9
Surveyor 3	USA	Apr. 17, 1967	Lander	Landed on Ocean of Storms Apr. 20; operated until May 4
Lunar Orbiter 4	USA	May 4, 1967	Orbiter	Arrived May 8; operated until Jul. 17
Surveyor 4	USA	Jul. 14, 1967	Lander	Contact lost during descent
Lunar Orbiter 5	USA	Aug. 1, 1967	Orbiter	Arrived Aug. 5; deliberately crashed on Moon Jan. 31, 1968

Name	Country	Launch date	Purpose	Results
Surveyor 5	USA	Sep. 8, 1967	Lander	Landed on Sea of Tranquility Sep. 11; operated until Dec. 16
Surveyor 6	USA	Nov. 7, 1967	Lander	Landed on Central Bay Nov. 10; operated until Nov. 24
Surveyor 7	USA	Jan. 7, 1968	Lander	Landed near Tycho Jan. 10; operated until Feb. 21
Luna 14	USSR	Apr. 7, 1968	Orbiter	Arrived Apr. 10; operated until Jun./Jul.
Zond 5	USSR	Sep. 14, 1968	Circumlunar	Flyby 18 Sep. at distance of 1,950 km (1,210 mi)
Zond 6	USSR	Nov. 10, 1968	Circumlunar	Flyby 14 Nov. at distance of 2,420 km (1,500 mi)
Apollo 8	USA	Dec. 21, 1968	Orbiter	Arrived Dec. 24; returned to Earth Dec. 25 Crew: Borman, Lovell, Anders
[Unnamed]	USSR	Feb. 19, 1969	Lander/rover	Launch failure
[Unnamed]	USSR	Jun. 14, 1969	Sample return	Failed to reach Earth orbit
Apollo 10	USA	May 18, 1969	Orbiter	Arrived May 21; returned to Earth May 24 Crew: Stafford, Young, Cernan
Luna 15	USSR	Jul. 13, 1969	Sample return	Crashed on lunar surface
Apollo 11	USA	Jul. 16, 1969	Orbiter/lander/ sample return	Arrived Jul. 19; returned to Earth Jul. 21, Landed on Sea of Tranquility Jul. 20, Crew: Armstrong, Aldrin, Collins
Zond 7	USSR	Aug. 7, 1969	Circumlunar	Flyby Aug. 10, at distance of 1,200 km (750 mi)
Kosmos 300	USSR	Sep. 23, 1969	Sample return	Failed to leave Earth orbit
Kosmos 305	USSR	Oct. 22, 1969	Sample return	Failed to reach Earth orbit
Apollo 12	USA	Nov. 14, 1969	Orbiter/lander/ sample return	Arrived Nov. 18; returned to Earth Nov. 21, Landed on Ocean of Storms Nov. 19. Crew: Conrad, Bean, Gordon
[Unnamed]	USSR	Feb. 6, 1970	Sample return	Failed to reach Earth orbit
Apollo 13	USA	Apr. 11, 1970	Orbiter/lander/ sample return	Flyby Apr. 14, Crew: Lovell, Haise, Swigert
Luna 16	USSR	Sep. 12, 1970	Sample return	Landed on Sea of Fertility Sep. 20; returned to Earth Sep. 21, with 101 g (3.5 oz) of soil
Zond 8	USSR	Oct. 20, 1970	Circumlunar	Flyby Oct. 24, at distance of 1,200 km (750 mi)
Luna 17/ Lunokhod 1	USSR	Nov. 10, 1970	Lander/rover	Landed Nov. 17; rover operated until Sep. 14, 1971, traveling 10.54 km (6.5 mi)
Apollo 14	USA	Jan. 31, 1971	Orbiter/lander/ sample return	Arrived Feb. 4; returned to Earth Feb. 6; landed at Fra Mauro Feb. 5. Crew: Shepard, Roosa, Mitchell
Apollo 15	USA	Jul. 26, 1971	Orbiter/ subsatellite/ lander/rover/ sample return	Arrived Jul. 29; returned to Earth Aug. 4; landed at Hadley-Apennine Jul. 30; released subsatellite Aug. 4. Crew: Scott, Irwin, Worden
Luna 18	USSR	Sep. 2, 1971	Sample return	Contact lost during descent

MOON contd.

Name	Country	Launch date	Purpose	Results
Luna 19	USSR	Sep. 28, 1971	Orbiter	Arrived Oct. 2; operated until Oct. 1972
Luna 20	USSR	Feb. 14, 1972	Sample return	Landed on Sea of Fertility Feb. 21; returned to Earth Feb. 22, with 55 g (2 oz) of soil
Apollo 16	USA	Apr. 16, 1972	Orbiter/ subsatellite/ lander/rover/ sample return	Arrived Apr. 19; returned to Earth Apr. 24; landed at Descartes Apr. 20; released subsatellite Apr. 24. Crew: Young, Duke, Mattingley
Apollo 17	USA	Dec. 7, 1972	Orbiter/lander/ rover/sample return	Arrived Dec. 10; returned to Earth Dec. 14; landed at Taurus–Littrow Dec. 11. Crew: Cernan, Schmitt, Evans
Luna 21/ Lunokhod 2	USSR	Jan. 8, 1973	Lander/rover	Landed Jan. 15; rover operated until May 9, traveling 37 km (23 mi)
Explorer 49	USA	Jun. 10, 1973	Orbiter	Arrived Jun. 15; no lunar data—conducted studies of radio emissions from distant sources. Operated until Jun. 1975
Luna 22	USSR	May 29, 1974	Orbiter	Arrived Jun. 2; operated until Sep. 2, 1975
Luna 23	USSR	Oct. 28, 1974	Sample return	Damaged during landing on Nov. 6, Contact lost Nov. 9
Luna 24	USSR	Aug. 9, 1976	Sample return	Landed on Sea of Crises Aug. 18; returned to Earth Aug. 19, with 170 g of soil
[Unnamed]	USSR	Oct. 16, 1975	Sample return	Failed to reach Earth orbit
Galileo	USA	Oct. 18, 1989	Venus, Earth and asteroid flybys, Jupiter orbiter/probe	Moon flyby Dec. 8, 1990 at distance of 565,000 km (350,000 mi); Moon flyby Dec. 8, 1992, at distance of 110,000 km (68,350 mi)
Hiten/Hagomoro	Japan	Jan. 24, 1990	Circumlunar/ orbiter	First flyby Mar. 19, at distance of 16,742 km; orbiter released but contact lost
Clementine	USA	Jan. 25, 1994	Lunar orbiter/ asteroid flyby	Arrived at Moon Feb. 19; operated until May 3
Cassini–Huygens	USA/ESA	Oct. 15, 1997	Venus, Earth and asteroid flybys, Saturn orbiter, Titan probe/lander	Moon flyby Aug. 18, 1999 at distance of 377,000 km (235,625 mi)
Lunar Prospector	USA	Jan. 7, 1998	Orbiter	Arrived Jan. 11; deliberately crashed on Moon Jul. 31, 1999

Nozomi	Japan	Jul. 3, 1998	Earth–Moon flybys, Mars orbiter	Moon flyby Sep. 24, 1998 at distance of 2,809 km (1,745 mi); 2nd flyby Dec. 18, 1998 at distance of 1,003 km (623 mi)
Smart-1	ESA	Sep. 27, 2003	Orbiter	Arrived Nov. 15, 2004; operational
Lunar-A	Japan	(?)	Orbiter/2 penetrators	Planned; launch uncertain
Selene	Japan	2007 (?)	Orbiter/2 subsatellites	Planned
Chandrayaan-1	India	Sep. 2007 (?)	Orbiter/ penetrators	Planned
Chang'e 1	China	2007 (?)	Orbiter	Planned
Lunar Reconnaissance Orbiter	USA	2008 (?)	Orbiter/ impactor	Planned

MERCURY

Name	Country	Launch date	Purpose	Results
Mariner 10	USA	Nov. 3, 1973	Venus/Mercury flyby	Venus flyby Feb. 5, 1974; Mercury flybys Mar. 29, 1974, Sep. 21, 1974, Mar. 16, 1975
MESSENGER	USA	Aug. 3, 2004	Venus flyby, Mercury flyby/ orbiter	En route; arrival due Jan. 15, 2008
BepiColombo	ESA/Japan	Sept. 2012 (?)	2 orbiters	Planned

VENUS

Name	Country	Launch date	Purpose	Results
Sputnik 7	USSR	Feb. 4, 1961	Impact	Did not reach Earth orbit
Venera 1	USSR	Feb. 12, 1961	Impact	Contact lost Feb. 22, 1961
Mariner 1	USA	Jul. 22, 1962	Flyby	Did not reach Earth orbit
Sputnik 19	USSR	Aug. 25, 1962	Impact	Failed to leave Earth orbit
Mariner 2	USA	Aug. 27, 1962	Flyby	Flyby Dec. 14, 1962
Sputnik 20	USSR	Sep. 1, 1962	Flyby	Failed to leave Earth orbit
Sputnik 21	USSR	Sep. 12, 1962	Flyby	Destroyed in Earth orbit
[Unnamed]	USSR	Feb. 19, 1964	Flyby	Did not reach Earth orbit
[Unnamed]	USSR	Mar. 1, 1964	Flyby	Did not reach Earth orbit

VENUS, contd.

Name	Country	Launch date	Purpose	Results
Kosmos 27	USSR	Mar. 27, 1964	Atmospheric probe	Failed to leave Earth orbit
Zond 1	USSR	Apr. 2, 1964	Atmospheric probe	Contact lost May 25
Venera 2	USSR	Nov. 12, 1965	Flyby	Flyby Feb. 27, 1966, no return of data
Venera 3	USSR	Nov. 16, 1965	Atmospheric probe	Arrived Mar. 1, 1966; no return of data
Kosmos 96	USSR	Nov. 23, 1965	Flyby	Failed to leave Earth orbit
Venera 4	USSR	Jun. 12, 1967	Atmospheric probe	Arrived Oct. 18, 1967; sent back data for 93 min. during descent
Mariner 5	USA	Jun. 14, 1967	Flyby	Flyby Oct. 19
Kosmos 167	USSR	Jun. 17, 1967	Atmospheric probe	Failed to leave Earth orbit
Venera 5	USSR	Jan. 5, 1969	Atmospheric probe	Arrived May 16, 1969; sent back data for 53 min. during descent
Venera 6	USSR	Jan. 10, 1969	Atmospheric probe	Arrived May 17, 1969; sent back data for 51 min. during descent
Venera 7	USSR	Aug. 17, 1970	Atmospheric probe/lander	Landed on Dec. 15, 1970; sent back data from surface for 23 min
Kosmos 359	USSR	Aug. 22, 1970	Atmospheric probe/lander	Failed to leave Earth orbit
Venera 8	USSR	Mar. 27, 1972	Atmospheric probe/lander	Landed on Jul. 22, 1972; sent back data from surface for 50 min
Kosmos 482	USSR	Mar. 31, 1972	Atmospheric probe/lander	Failed to leave Earth orbit
Mariner 10	USA	Nov. 4, 1973	Venus/Mercury flyby	Venus flyby Feb. 5, 1974; Mercury flybys Mar. 29, 1974, Sep. 21, 1974, Mar. 16, 1975
Venera 9	USSR	Jun. 8, 1975	Orbiter/lander	Arrived Oct. 22, 1975; lander sent back data from surface for 53 min, including first photo
Venera 10	USSR	Jun. 14, 1975	Orbiter/lander	Arrived Oct. 25, 1975; lander sent back data from surface for 65 min., including a photo
Pioneer Venus 1	USA	May 20, 1978	Orbiter	Arrived Dec. 4, 1978; operated until Oct. 8, 1992
Pioneer Venus 2	USA	Aug. 8, 1978	Main bus/4 atmospheric probes	Arrived Dec. 9, 1978; day probe sent back data from surface for 67 min
Venera 11	USSR	Sep. 9, 1978	Orbiter/lander	Arrived Dec. 25, 1978; lander sent back data from surface for 95 min

Name	Country	Launch date	Purpose	Results
Venera 12	USSR	Sep. 14, 1978	Orbiter/lander	Arrived Dec. 9, 1978; lander sent back data from surface for 110 min
Venera 13	USSR	Oct. 30, 1981	Flyby/lander	Arrived Mar. 1, 1982. Lander sent back data from surface for 127 min., including 8 color photos
Venera 14	USSR	Nov. 4, 1981	Flyby/lander	Arrived Mar. 3, 1982; lander sent back data from surface for 57 min., including 8 color photos
Venera 15	USSR	Jun. 2, 1983	Orbiter	Arrived Oct. 10, 1983; mapped N. hemisphere with radar until Jul. 10, 1984
Venera 16	USSR	Jun. 7, 1983	Orbiter	Arrived Oct. 14, 1983; mapped N. hemisphere with radar until Jul. 10, 1984
Vega 1	USSR	Dec. 15, 1984	Flyby/lander/ balloon; comet Halley flyby	Venus flyby and landing Jun. 11, 1985; lander sent back 56 min. of data; balloon operated until Jun. 13
Vega 2	USSR	Dec. 21, 1984	Flyby/lander/ balloon; comet Halley flyby	Venus flyby and landing Jun. 15, 1985; lander sent back 57 min. of data; balloon operated until Jun. 17
Magellan	USA	May 4, 1989	Orbiter	Arrived Aug. 10, 1990. Mapped planet with radar, operational until Oct. 12, 1994
Galileo	USA	Oct. 18, 1989	Venus, Earth and asteroid flybys, Jupiter orbiter/probe	Venus flyby Feb. 10, 1990
Cassini–Huygens	USA/ESA	Oct. 15, 1997	Venus, Earth and asteroid flybys, Saturn orbiter, Titan probe/lander	Venus flybys Apr. 26, 1998 and Jun. 24, 1999
MESSENGER	USA	Aug. 3, 2004	Venus flyby/ Mercury orbiter	En route; Venus flybys Oct. 2006 and Jun. 2007
Venus Express	ESA	Nov. 9, 2005	Orbiter	Arrived Apr. 11, 2006
Planet-C	Japan	Feb. 2007 (?)	Orbiter	Arrival due Sep. 2009

MARS

Name	Country	Launch date	Purpose	Results
[Unnamed]	USSR	Oct. 10, 1960	Flyby	Did not reach Earth orbit
[Unnamed]	USSR	Oct. 14, 1960	Flyby	Did not reach Earth orbit
[Unnamed]	USSR	Oct. 24, 1962	Flyby	Achieved Earth orbit only
Mars 1	USSR	Nov. 1, 1962	Flyby	Radio failed at 106 million km (65.9 million mi)
[Unnamed]	USSR	Nov. 4, 1962	Flyby	Achieved Earth orbit only
Mariner 3	USA	Nov. 5, 1964	Flyby	Launcher shroud failed to jettison
Mariner 4	USA	Nov. 28, 1964	Flyby	First successful Mars flyby Jul. 14, 1965; returned 21 photos
Zond 2	USSR	Nov. 30, 1964	Flyby	Passed Mars but radio failed; returned no planetary data
Mariner 6	USA	Feb. 24, 1969	Flyby	Mars flyby Jul. 31, 1969; returned 75 photos
Mariner 7	USA	Mar. 27, 1969	Flyby	Mars flyby Aug. 5, 1969; returned 126 photos
[Unnamed]	USSR	Mar. 27, 1969	Orbiter	Did not reach Earth orbit
[Unnamed]	USSR	Apr. 2, 1969	Orbiter	Did not reach Earth orbit
Mariner 8	USA	May 8, 1971	Orbiter	Failed during launch
Kosmos 419	USSR	May 10, 1971	Orbiter	Achieved Earth orbit only
Mars 2	USSR	May 19, 1971	Orbiter/lander	Arrived Nov. 27, 1971; no useful data; lander burned up due to steep entry
Mars 3	USSR	May 28, 1971	Orbiter/lander	Arrived Dec. 3, 1971; lander operated on surface for 20 seconds before failing
Mariner 9	USA	May 30, 1971	Orbiter	Arrived Nov. 13, 1971; end of mission Oct. 27, 1972; returned 7,329 photos
Mars 4	USSR	Jul. 21, 1973	Orbiter	Flew past Mars Feb. 10, 1974
Mars 5	USSR	Jul. 25, 1973	Orbiter	Arrived Dec. 2, 1974; lasted a few days; returned 43 photos
Mars 6	USSR	Aug. 5, 1973	Flyby module/lander	Arrived Mar. 12, 1974; lander failed due to fast impact
Mars 7	USSR	Aug. 9, 1973	Flyby module/lander	Arrived Mar. 9, 1974; lander missed the planet
Viking 1	USA	Aug. 20, 1975	Orbiter/lander	Arrived Jun. 19, 1976; orbiter operated until Aug. 7, 1980; lander operated on surface Jul. 20, 1976–Nov. 13, 1982
Viking 2	USA	Sep. 9, 1975	Orbiter/lander	Arrived Aug. 7, 1976; orbiter operated until Jul. 24, 1978; lander operated on surface Sep. 3, 1976–Apr. 11, 1980; Viking orbiters and landers returned 50,000+ photos

Name	Country	Launch date	Purpose	Results
Phobos 2	USSR	Jul. 12, 1988	Mars/Phobos orbiter/lander	Contact lost Mar. 27, 1989 near Phobos; returned photos of Mars and Phobos
Mars Observer	USA	Sep. 25, 1992	Orbiter	Contact lost just before Mars arrival Aug. 21, 1993
Mars Global Surveyor	USA	Nov. 7, 1996	Orbiter	Arrived Sep. 12, 1997; still operational
Mars 96	Russia	Nov. 16, 1996	Orbiter and landers	Did not reach Earth orbit
Mars Pathfinder	USA	Dec. 14, 1996	Lander/rover	Landed Jul. 4, 1997; operated until Sep. 27, 1997
Nozomi	Japan	Jul. 4, 1998	Earth-Moon flybys, Mars orbiter	Mars flyby Dec. 14, 2003 at distance of 1,000 km (620 mi); no Mars data returned
Mars Climate Orbiter	USA	Dec. 11, 1998	Orbiter	Contact lost upon arrival at Mars Sep. 23, 1999
Mars Polar Lander/Deep Space 2	USA	Jan. 3, 1999	Lander/ penetrators	Contact lost on arrival Dec. 3, 1999
Mars Odyssey	USA	Mar. 7, 2001	Orbiter	Arrived Oct. 24, 2001; still operational
Mars Express/ Beagle 2	ESA	Jun. 2, 2003	Orbiter/lander	Orbiter arrived Dec. 25, 2003; still operational; lander lost during entry Dec. 25, 2003
Mars Exploration Rover A (Spirit)	USA	Jun. 10, 2003	Lander/rover	Landed Jan. 4, 2004; still operational
Mars Exploration Rover B (Opportunity)	USA	Jul. 7, 2003	Lander/rover	Landed Jan. 25, 2004; still operational
Mars Reconnaissance Orbiter	USA	Aug. 12, 2005	Orbiter	Arrived Mar. 10, 2006; still operational
Phoenix	USA	Aug. 2007 (?)	Lander	Arrival due May 25, 2008
Mars Science Laboratory	USA	Dec. 2009 (?)	Lander/rover	Arrival due Oct. 2010

ASTEROIDS

Name	Country	Launch date	Purpose	Results
Galileo	USA	Oct. 18, 1989	Venus, Earth and asteroid flybys, Jupiter orbiter/ probe	Flyby of Gaspra Oct. 29, 1991; flyby of Ida and Dactyl Aug. 28, 1993
Clementine	USA	Jan. 25, 1994	Lunar orbiter/ asteroid flyby	Mission to Geographos canceled May 3, 1994 after thruster malfunction
NEAR-Shoemaker	USA	Feb. 17, 1996	Earth and asteroid flybys, Eros orbiter	Mathilde flyby Jun. 27, 1997; Eros flyby Dec. 23, 1998; orbit around Eros Feb. 14, 2000; landed on Eros Feb. 12, 2001
Cassini–Huygens	USA/ESA	Oct. 15, 1997	Venus, Earth and asteroid flybys, Saturn orbiter, Titan probe/lander	Flyby of Masursky Jan. 23, 2000
Deep Space 1	USA	Oct. 24, 1998	Asteroid and comet flybys	Flyby of Braille Jul. 29, 1999
Stardust	USA	Feb. 7, 1999	Earth, asteroid and comet flybys/comet sample return	Flyby of Anne Frank Nov. 2, 2002. Returned capsule to Earth Jan. 15, 2006
Hayabusa	Japan	May 9, 2003	Orbiter/sample return	Arrival at Itokawa Sep. 12, 2005; sample return summer 2007 (?)
Rosetta	ESA	Feb. 24, 2004	Earth, Mars and asteroid flybys; comet orbiter/ lander	Flyby of Steins Sep. 5, 2008; flyby of Lutetia Jul. 10, 2010
Dawn	USA	2007 (?)	Dual orbiter	Arrival at Vesta Oct. 2011, Ceres Aug. 2015

JUPITER

Name	Country	Launch date	Purpose	Results
Pioneer 10	USA	Mar. 2, 1972	Flyby	Jupiter flyby Dec. 4, 1973
Pioneer 11	USA	Apr. 6, 1973	Jupiter and Saturn flybys	Jupiter flyby Sep. 1, 1979
Voyager 2	USA	Aug. 20, 1977	Jupiter, Saturn, Uranus and Neptune flybys	Jupiter flyby Jul. 9, 1979
Voyager 1	USA	Sep. 5, 1977	Jupiter and Saturn flybys	Jupiter flyby Mar. 5, 1979
Galileo	USA	Oct. 18, 1989	Venus, Earth and asteroid flybys, Jupiter orbiter/ atmospheric probe	Orbiter arrived Dec. 8, 1995; operational until atmospheric entry Sep. 21, 2003; Probe entry Dec. 7, 1995.
Ulysses	ESA/USA	Oct. 6, 1990	Solar orbiter/ Jupiter flybys	Jupiter flyby Feb. 8, 1992 at distance of 378,400 km (235,125 mi); distant flyby Feb. 5, 2004
Cassini–Huygens	USA/ESA	Oct. 15, 1997	Venus, Earth and asteroid flybys, Saturn orbiter, Titan probe/ lander	Jupiter flyby Dec. 30, 2000 at distance of 9.7 million km (6 million mi)

SATURN

Name	Country	Launch date	Purpose	Results
Pioneer 11	USA	Apr. 6, 1973	Jupiter and Saturn flybys	Saturn flyby Sep. 1, 1979
Voyager 2	USA	Aug. 20, 1977	Jupiter, Saturn, Uranus and Neptune flybys	Saturn flyby Aug. 22, 1981
Voyager 1	USA	Sep. 5, 1977	Jupiter and Saturn flybys	Saturn flyby Nov. 12, 1980
Cassini–Huygens	USA/ESA	Oct. 15, 1997	Venus, Earth and asteroid flybys, Saturn orbiter, Titan probe/lander	Orbiter arrived Jul. 1, 2004. Still operational Huygens landed on Titan Jan. 14, 2005

URANUS

Name	Country	Launch date	Purpose	Results
Voyager 2	USA	Aug. 20, 1977	Jupiter, Saturn, Uranus and Neptune flybys	Uranus flyby Jan. 24, 1986

NEPTUNE

Name	Country	Launch date	Purpose	Results
Voyager 2	USA	Aug. 20, 1977	Jupiter, Saturn, Uranus and Neptune flybys	Neptune flyby Aug. 25, 1989

PLUTO

Name	Country	Launch date	Purpose	Results
New Horizons	USA	Jan. 19, 2006	Pluto, Kuiper Belt Object flybys	Pluto–Charon flyby due Jul. 2015; KBO flybys 2016–2020

COMETS

Name	Country	Launch date	Purpose	Results
International Cometary Explorer	USA	Aug. 12, 1978	Study of solar wind etc. from L1; comet flyby	Flyby of Giacobini–Zinner Sep. 11, 1985; distant flyby of Halley Mar. 28, 1986
Vega 1	USSR	Dec. 15, 1984	Flyby/lander/balloon; comet Halley flyby	Halley flyby Mar. 6, 1986
Vega 2	USSR	Dec. 21, 1984	Flyby/lander/balloon; comet Halley flyby	Halley flyby Mar. 9, 1986
Sakigake	Japan	Jan. 7, 1985	Comet and Earth flybys	Halley flyby Mar. 11, 1986

Name	Country	Launch date	Purpose	Results
Giotto	ESA	Jul. 2, 1985	Comet and Earth flybys	Halley flyby Mar. 14, 1986; Grigg-Skjellerup flyby Jul. 10, 1992
Suisei	Japan	Aug. 18, 1985	Comet flyby	Halley flyby Mar. 8, 1986
Deep Space 1	USA	Oct. 24, 1998	Asteroid and comet flybys	Flyby of Borrelly Sep. 22, 2001
Stardust	USA	Feb. 7, 1999	Earth, asteroid and comet flybys/comet sample return	Flyby of Wild 2 on Jan. 2, 2004; sample return Jan. 15, 2006
Rosetta	ESA	Feb. 24, 2004	Earth, Mars and asteroid flybys; comet orbiter/ lander	Arrival at Churyumov-Gerasimenko Aug. 2014; landing Nov. 2014
Deep Impact	USA	Jan. 12, 2005	Flyby/impact	Arrival at Tempel 1 on Jul. 4, 2005

Appendix 2: Planetary Data

	Mercury	Venus	Earth	Mars	Jupiter	Saturn	Uranus	Neptune	Pluto
Equatorial diameter (km)	4,879	12,104	12,756	6,792	142,984	120,536	51,118	49,528	2,324
Rotation period	58d 15h 30m	243d 0h 36m (r)	23h 56m	24h 37m	9h 50m	10h 14m	17h 14m (r)	16h 7m	6d 9h 17m (r)
Density (water = 1)	5.43	5.24	5.52	3.91	1.33	0.69	1.29	1.64	2.05
Mass (Earth = 1)	0.055	0.814	1	0.11	317.8	95.2	14.53	17.14	0.0021
Surface gravity (Earth = 1)	0.378	0.903	1	0.38	2.69	1.19	0.79	0.98	0.06
Inclination of equator (deg)	7	3.4	23.5	25.2	3.1	26.7	97.5	29.6	122
Orbital period	87.97d	224.7d	365.25d	687d	11.86y	29.46y	84.01y	164.79y	248.54y
Av. distance from Sun (million km)	57.9	108.2	149.6	227.9	778.3	1,427	2,871	4,497.1	5,913.5
Orbital eccentricity	0.2	0.007	0.0167	0.093	0.048	0.056	0.046	0.0097	0.249
Average orbital velocity (km/s)	47.88	35.02	29.8	24.1	13.06	9.65	6.81	5.43	4.72
Average temp. (°C)	350 (day), −170 (nt)	467	7	−63	−148	−178	−218	−220	−228
Atmosphere	Potassium, sodium	Carbon dioxide	Nitrogen, oxygen	Carbon dioxide	Hydrogen, helium	Hydrogen, helium	Hydrogen, helium	Hydrogen, helium	Nitrogen, methane?
Moons	0	0	1	2	63	47	27	13	3

r = retrograde

Appendix 3: Satellite Data

Earth

Satellite	Discoverer	Year of discovery	Diameter (km)	Visual magnitude	Distance (km)	Orbital period (d)	Orbital ecc.	Orbital inc. (degrees)
Moon	–	–	3,475	–12.7	384,400	27.3	0.055	5.2

Mars

Satellite	Discoverer	Year of discovery	Diameter (km)	Visual magnitude	Distance (km)	Orbital period (d)	Orbital ecc.	Orbital inc. (degrees)
Phobos	A. Hall	1877	13.4 × 11.2 × 9.2	11.4	9,380	0.3	0.015	1.1
Deimos	A. Hall	1877	7.5 × 6.1 × 5.2	12.5	23,460	1.3	0.000	1.8

Jupiter

Satellite	Discoverer	Year of discovery	Diameter (km)	Visual magnitude	Distance (km)	Orbital period (d)	Orbital ecc.	Orbital inc. (degrees)
Metis	S. Synott/Voyager 2	1980	43	17.5	128,100	0.3	0.001	0.0
Adrastea	D. Jewitt, E. Danielson	1979	16	18.7	128,900	0.3	0.002	0.1
Amalthea	E. Barnard	1892	167	14.1	181,400	0.5	0.003	0.4
Thebe	S. Synott/Voyager 1	1980	99	16.0	221,900	0.7	0.018	1.1
Io	Galileo	1610	3,643	5.0	421,800	1.8	0.004	0.0
Europa	Galileo	1610	3,122	5.3	671,100	3.6	0.009	0.5
Ganymede	Galileo	1610	5,262	4.6	1,070,400	7.2	0.001	0.2
Callisto	Galileo	1610	4,821	5.7	1,882,700	16.7	0.007	0.2
Themisto	C. Kowal, E. Roemer	1975	8	21.0	7,507,000	130.0	0.243	43.3
Leda	C. Kowal	1974	20	19.5	11,165,000	240.9	0.164	27.4
Himalia	C. Perrine	1904	170	14.6	11,461,000	250.6	0.162	27.5
Lysithea	S. Nicholson	1938	36	18.3	11,717,000	259.2	0.112	28.3
Elara	C. Perrine	1905	86	16.3	11,741,000	259.6	0.217	26.6
S/2000 J11	S. Sheppard et al.	2000	4	22.4	12,560,000	287.0	0.248	28.3

Jupiter, contd.

Satellite	Discoverer	Year of discovery	Diameter (km)	Visual magnitude	Distance (km)	Orbital period (d)	Orbital ecc.	Orbital inc. (degrees)
S/2003 J12	S. Sheppard et al.	2003	1	23.9	15,912,000	489.5	0.606	151.9
Carpo	S. Sheppard et al.	2003	3	23.2	16,989,000	456.1	0.430	51.4
Euporie	S. Sheppard et al.	2001	4	23.1	19,304,000	550.7 (r)	0.143	145.8
S/2003 J3	S. Sheppard et al.	2003	2	23.8	20,221,000	583.9 (r)	0.197	147.6
S/2003 J18	B. Gladman et al.	2003	2	23.4	20,514,000	596.6 (r)	0.012	146.1
Orthosie	S. Sheppard et al.	2001	4	23.1	20,720,000	622.6 (r)	0.281	145.9
Euanthe	S. Sheppard et al.	2001	6	22.8	20,797,000	620.5 (r)	0.232	148.9
Harpalyke	S. Sheppard et al.	2000	4	22.2	20,858,000	623.3 (r)	0.227	148.6
Praxidike	S. Sheppard et al.	2000	7	21.2	20,907,000	625.4 (r)	0.231	149.0
Thyone	S. Sheppard et al.	2001	8	22.3	20,939,000	627.2 (r)	0.229	148.5
S/2003 J16	B. Gladman et al.	2003	2	23.3	20,963,000	616.4 (r)	0.224	148.5
Iocaste	S. Sheppard et al.	2000	5	21.8	21,061,000	631.5 (r)	0.216	149.4
Mneme	S. Sheppard et al.	2003	2	23.3	21,069,000	620.0 (r)	0.227	148.6
Hermippe	S. Sheppard et al.	2001	8	22.1	21,131,000	633.9 (r)	0.210	150.7
Thelxinoe	S. Sheppard et al.	2004	2	23.5	21,162,000	628.1 (r)	0.221	151.4
Helike	S. Sheppard et al.	2003	4	22.6	21,263,000	634.8 (r)	0.156	154.8
Ananke	S. Nicholson	1951	28	18.8	21,276,000	629.8 (r)	0.244	148.9
S/2003 J15	S. Sheppard et al.	2003	2	23.5	22,627,000	689.8 (r)	0.192	146.5
Eurydome	S. Sheppard et al.	2001	6	22.7	22,865,000	717.3 (r)	0.276	150.3
Arche	S. Sheppard et al.	2002	3	22.8	22,931,000	723.9 (r)	0.259	165
S/2003 J17	B. Gladman et al.	2003	2	23.4	23,001,000	714.5 (r)	0.238	164.9
Pasithee	S. Sheppard et al.	2001	4	23.2	23,004,000	719.4 (r)	0.268	165.1
S/2003 J10	S. Sheppard et al.	2003	2	23.6	23,042,000	716.2 (r)	0.430	165.1
Chaldene	S. Sheppard et al.	2000	4	22.5	23,100,000	723.7 (r)	0.252	165.2
Isonoe	S. Sheppard et al.	2000	4	22.5	23,155,000	726.2 (r)	0.247	165.3
Erinome	S. Sheppard et al.	2000	3	22.8	23,196,000	728.5 (r)	0.267	164.9
Kale	S. Sheppard et al.	2001	4	23.0	23,217,000	729.5 (r)	0.260	165.0
Aitne	S. Sheppard et al.	2001	6	22.7	23,229,000	730.2 (r)	0.264	165.1
Taygete	S. Sheppard et al.	2000	5	21.9	23,280,000	732.4 (r)	0.253	165.3
S/2003 J9	S. Sheppard et al.	2003	1	23.7	23,384,000	733.3 (r)	0.263	165.1
Carme	S. Nicholson	1951	46	17.6	23,404,000	734.2 (r)	0.253	164.9
Sponde	S. Sheppard et al.	2001	4	23.0	23,487,000	748.3 (r)	0.312	151.0
Megaclite	S. Sheppard et al.	2000	5	21.7	23,493,000	752.9 (r)	0.420	152.7
S/2003 J5	S. Sheppard et al.	2003	4	22.4	23,495,000	738.7 (r)	0.210	165.2
S/2003 J19	B. Gladman et al.	2003	2	23.7	23,533,000	740.4 (r)	0.256	165.2
S/2003 J23	S. Sheppard et al.	2003	2	23.6	23,563,000	732.4 (r)	0.271	146.3

Kalyke	S. Sheppard et al.	2000	5	21.8	23,566,000	743.0 (r)	0.246	165.2
S/2003 J14	S. Sheppard et al.	2003	2	23.6	23,614,000	779.2 (r)	0.344	144.5
Pasiphae	P. Melotte	1908	60	17.0	23,624,000	743.6 (r)	0.409	151.4
Eukelade	S. Sheppard et al.	2003	4	22.6	23,661,000	781.6 (r)	0.345	163.4
S/2003 J4	S. Sheppard et al.	2003	2	23.0	23,930,000	755.2 (r)	0.362	149.6
Sinope	S. Nicholson	1914	38	18.1	23,939,000	758.9 (r)	0.250	158.1
Hegemone	S. Sheppard et al.	2003	3	23.2	23,947,000	739.6 (r)	0.328	155.2
Aoede	S. Sheppard et al.	2003	4	22.5	23,981,000	761.5 (r)	0.432	158.3
Kallichore	S. Sheppard et al.	2003	2	23.7	24,043,000	764.7 (r)	0.264	165.5
Autonoe	S. Sheppard et al.	2001	8	22.0	24,046,000	761 (r)	0.317	152.4
Callirrhoe	Spacewatch	1999	9	20.7	24,103,000	758.8 (r)	0.283	147.2
Cyllene	S. Sheppard et al.	2003	2	23.2	24,349,000	751.9 (r)	0.319	149.3
S/2003 J2	S. Sheppard et al.	2003	2	23.2	29,541,000	982.5 (r)	0.380	151.8

Saturn

Satellite	Discoverer	Year of discovery	Diameter (km)	Visual magnitude	Distance (km)	Orbital period (d)	Orbital ecc.	Orbital inc. (degrees)
Pan	M. Showalter/ Voyager 2	1990	20	19.4	133,600	0.575	0.000	0.0
Daphnis	C. Porco et al./Cassini	2005	7	-	136,500	0.59	0.0	0.0
Atlas	R. Terrile/Voyager 1	1980	32	19.0	137,700	0.6	0.001	0.0
Prometheus	S. Collins/Voyager 1	1980	100	15.8	139,400	0.6	0.002	0.0
Pandora	S. Collins/Voyager 1	1980	84	16.4	141,700	0.6	0.004	0.1
Epimetheus	J. Fountain et al./ Voyager 1	1980	119	15.6	151,400	0.7	0.010	0.4
Janus	A. Dollfus	1966	178	14.4	151,500	0.7	0.007	0.2
Mimas	W. Herschel	1789	397	12.8	185,600	0.9	0.021	1.6
Methone	C. Porco et al./Cassini	2004	3	23	194,300	1.0	0.001	0.0
Pallene	C. Porco et al./Cassini	2004	4	23	212,300	1.1	0.004	0.2
Enceladus	W. Herschel	1789	499	11.8	238,100	1.4	0.000	0.0
Calypso	D. Pascu et al.	1980	19	18.7	294,700	1.9	0.001	1.5
Telesto	B. Smith et al./ Voyager 1	1980	24	18.5	294,700	1.9	0.001	1.2
Tethys	G. Cassini	1684	1,060	10.2	294,700	1.9	0.000	0.2
Dione	G. Cassini	1684	1,118	10.4	377,400	2.7	0.000	0.0
Helene	P. Laques, J. Lacacheux	1980	32	18.4	377,400	2.7	0.000	0.2
Polydeuces	C. Porco et al./Cassini	2004	4	23.0	377,400	2.7	0.018	0.2
Rhea	G. Cassini	1672	1,528	9.6	527,100	4.5	0.001	0.3

Saturn, contd.

Satellite	Discoverer	Year of discovery	Diameter (km)	Visual magnitude	Distance (km)	Orbital period (d)	Orbital ecc.	Orbital inc. (degrees)
Titan	C. Huygens	1655	5,150	8.4	1,221,900	16.0	0.029	1.6
Hyperion	W. Bond, W. Lassell	1848	283	14.4	1,464,100	21.3	0.018	0.6
Iapetus	G. Cassini	1671	1,436	11.0	3,560,800	79.3	0.028	7.6
Kiviuq	B. Gladman et al.	2000	14	22.0	11,365,000	449.2	0.334	46.1
Ijiraq	J. Kavelaars et al.	2000	10	22.6	11,442,000	451.5	0.322	46.7
Phoebe	W. Pickering	1898	220	16.4	12,944,300	550.48 (r)	0.164	174.8
Paaliaq	B. Gladman et al.	2000	19	21.3	15,198,000	686.9	0.363	45.1
Skathi	J. Kavelaars et al.	2000	6	23.6	15,641,000	728.2 (r)	0.269	152.6
Albiorix	M. Holman et al.	2000	26	20.5	16,394,000	783.5	0.479	34.0
S/2004 S11	D. Jewitt et al.	2004	6	24.1	16,950,000	822	0.336	41.0
Erriapo	J. Kavelaars et al.	2000	9	23.0	17,604,000	871.2	0.474	34.5
Siarnaq	B. Gladman et al.	2000	32	20.1	18,195,000	895.6	0.296	45.5
Tarvos	J. Kavelaars et al.	2000	13	22.1	18,239,000	926.1	0.536	33.5
S/2004 S13	D. Jewitt et al.	2004	6	24.5	18,450,000	906 (r)	0.273	167.4
S/2004 S17	D. Jewitt et al.	2004	4	25.2	18,600,000	986 (r)	0.259	166.6
Mundilfari	B. Gladman et al.	2000	6	23.8	18,722,000	951.6 (r)	0.208	167.5
S/2004 S15	D. Jewitt et al.	2004	6	24.2	18,750,000	1,008 (r)	0.18	156.9
Narvi	S. Sheppard et al.	2003	8	24.0	19,140,800	988.6 (r)	0.325	135.8
S/2004 S10	D. Jewitt et al.	2004	6	24.4	19,350,000	1,026 (r)	0.241	167.0
Suttungr	B. Gladman et al.	2000	6	23.9	19,465,000	1,016.5 (r)	0.114	175.8
S/2004 S12	D. Jewitt et al.	2004	5	24.8	19,650,000	1,048 (r)	0.401	164.0
S/2004 S18	D. Jewitt et al.	2004	7	23.8	19,650,000	1,052 (r)	0.795	147.4
S/2004 S9	D. Jewitt et al.	2004	5	24.7	19,800,000	1,077 (r)	0.235	157.6
S/2004 S7	D. Jewitt et al.	2004	6	24.8	19,800,000	1,103 (r)	0.401	165.1
S/2004 S14	D. Jewitt et al.	2004	6	24.4	19,950,000	1,081 (r)	0.292	162.7
Thrymr	B. Gladman et al.	2000	6	23.9	20,219,000	1,086.9 (r)	0.485	175.8
S/2004 S16	D. Jewitt et al.	2004	4	25.0	22,200,000	1,271 (r)	0.135	163.0
S/2004 S8	D. Jewitt et al.	2004	6	24.6	22,200,000	1,355 (r)	0.213	168.0
Ymir	B. Gladman et al.	2000	16	21.7	23,300,000	1,315.3 (r)	0.334	173.1

Uranus

Satellite	Discoverer	Year of discovery	Diameter (km)	Visual magnitude	Distance (km)	Orbital period (d)	Orbital ecc.	Orbital inc. (degrees)
Cordelia	R. Terrile/Voyager 2	1986	40	23.6	49,800	0.4	0.000	0.1
Ophelia	R. Terrile/Voyager 2	1986	42	23.3	53,800	0.4	0.01	0.1
Bianca	B. A. Smith/Voyager 2	1986	51	22.5	59,200	0.4	0.001	0.2
Cressida	S. Synott/Voyager 2	1986	80	21.6	61,800	0.5	0.000	0.0
Desdemona	S. Synott/Voyager 2	1986	64	22.0	62,700	0.5	0.000	0.1
Juliet	S. Synott/Voyager 2	1986	94	21.1	64,400	0.5	0.001	0.1
Portia	S. Synott/Voyager 2	1986	135	20.4	66,100	0.5	0.000	0.1
Rosalind	S. Synott/Voyager 2	1986	72	21.8	69,900	0.6	0.000	0.3
Cupid	M. Showalter, J. Lissauer	2003	24	26.0	74,800	0.6	0.000	0.0
Belinda	S. Synott/Voyager 2	1986	81	21.5	75,300	0.6	0.000	0.0
Perdita	E. Karkoschka/ Voyager 2	1999	20	23.6	76,400	0.6	0.000	0.0
Puck	S. Synott/Voyager 2	1985	162	19.8	86,000	0.8	0.000	0.3
Mab	M. Showalter, J. Lissauer	2003	32	26.0	97,734	0.9	0.000	0.0
Miranda	G. Kuiper	1948	472	15.8	129,900	1.4	0.001	4.3
Ariel	W. Lassell	1851	1,158	13.7	190,900	2.5	0.001	0.0
Umbriel	W. Lassell	1851	1,169	14.5	266,000	4.1	0.004	0.1
Titania	W. Herschel	1787	1,578	13.5	436,300	8.7	0.001	0.1
Oberon	W. Herschel	1787	1,523	13.7	583,500	13.5	0.001	0.1
Francisco	M. Holman, J. Kavelaars	2001	22	25.0	4,281,000	266.6 (r)	0.143	147.6
Caliban	B. Gladman et al.	1997	72	22.4	7,231,000	579.5 (r)	0.159	140.9
Stephano	B. Gladman et al.	1999	32	24.1	8,004,000	677.4 (r)	0.23	144.1
Trinculo	M. Holman et al.	2001	18	25.4	8,578,000	759.0 (r)	0.208	167.0
Sycorax	B. Gladman et al.	1997	150	20.8	12,179,000	1288.3 (r)	0.522	159.4
Margaret	D. Jewitt, S. Sheppard	2001	20	25.2	14,688,700	1694.8 (r)	0.783	50.7
Prospero	M. Holman et al.	1999	50	20.8	16,243,000	1977.3 (r)	0.443	152.0
Setebos	J. Kavelaars et al.	1999	47	23.3	17,501,000	2234.8 (r)	0.584	158.2
Ferdinand	M. Holman, B. Gladman	2001	21	25.1	21,000,000	2823.4 (r)	0.426	167.3

Neptune

Satellite	Discoverer	Year of discovery	Diameter (km)	Visual magnitude	Distance (km)	Orbital period (d)	Orbital ecc.	Orbital inc. (degrees)
Naiad	R. Terrile/Voyager 2	1989	58	24.6	48,200	0.3	0.000	4.7
Thalassa	R. Terrile/Voyager 2	1989	80	23.9	50,100	0.3	0.000	0.2
Despina	S. Synott/Voyager 2	1989	148	22.5	52,500	0.3	0.000	0.1
Galatea	S. Synott/Voyager 2	1989	158	22.4	62,000	0.4	0.000	0.1
Larissa	H. Reitsema et al.	1989	192	22.0	73,500	0.6	0.001	0.2
Proteus	S. Synott/Voyager 2	1989	416	20.3	117,600	1.1	0.000	0.0
Triton	W. Lassell	1846	2,707	13.5	354,800	5.9 (r)	0.000	156.8
Nereid	G. Kuiper	1949	340	19.7	5,513,400	360.1	0.751	7.2
S/2002 N1	M. Holman et al.	2002	54	24.2	16,600,000	1,874.4 (r)	0.43	114.9
S/2002 N2	M. Holman et al.	2002	31	25.4	22,300,000	2,914.7	0.27	50.4
S/2002 N3	M. Holman et al.	2002	37	25.0	23,500,000	3,116.7	0.36	35.9
Psamathe	S. Sheppard et al.	2003	36	25.1	47,600,000	9,708.3 (r)	0.49	125.1
S/2002 N4	M. Holman et al.	2002	43	24.7	48,600,000	9,412.5 (r)	0.39	137.4

Pluto

Satellite	Discoverer	Year of discovery	Diameter (km)	Visual magnitude	Distance (km)	Orbital period (d)	Orbital ecc.	Orbital inc. (degrees)
Charon	J. Christy	1978	1,186	17.3	19,410	6.4	0.000	99.1
S/2005 P2	H. Weaver, S. Alan Stern et al.	2005	46?	23.4	48,675	24.86	0.002	96.18
S/2005 P1	H. Weaver, S. Alan Stern et al.	2005	61?	24.4	64,780	38.2	0.005	96.36

r = retrograde

Appendix 4: Planetary Rings

Rings of Jupiter

Name	Distance from center of planet (planet radii)	Distance from center of planet (km)	Width (km)	Thickness (km)
Halo	1.40–1.71	100,000–122,000	22,000	(20,000)
Main	1.71–1.81	122,000–129,000	7,000	Less than 30
Gossamer (inner)	1.81–2.55	129,200–182,000	52,800	?
Gossamer (outer)	2.55–3.15	182,000–224,900	42,900	?

Rings of Saturn

Name	Distance from center of planet (planet radii)	Distance from center of planet (km)	Width (km)	Thickness (km)
D	1.11–1.236	66,900–74,510	7,610	?
C	1.239–1.527	74,658–92,000	17,342	5 m
B	1.527–1.951	92,000–117,580	25,580	5–10 m
A	2.027–2.269	122,170–136,775	14,605	10–30 m
F	2.326	140,180	30–500	?
G	2.82–2.90	170,000–175,000	5,000	100 km
E	3–8	181,000–483,000	302,000	10,000 km

Rings of Uranus

Name	Distance from center of planet (planet radii)	Distance from center of planet (km)	Width (km)	Thickness (km)
1986U2R	1.448–1.545	37,000–39,500	Approx. 2,500	?
6	1.637	41,837	Approx. 1.5	(100)
5	1.652	42,235	Approx. 2	(100)
4	1.666	42,571	Approx. 2.5	(100)
Alpha (α)	1.75	44,718	4–10	(100)
Beta (β)	1.786	45,661	5–11	(100)
Eta (η)	1.834	47,176	1.6	(100)
Gamma (γ)	1.863	47,626	1–4	(100)
Delta (δ)	1.900	48,303	3–7	(100)
Lambda (λ)	1.958	50,024	Approx. 2	(100)
Epsilon (ε)	2.00	51,149	20–96	150
R/2003 U2	2.586–2.735	66,100–69,900	3,800	?
R/2003 U1	3.265–4.03	86,000–103,000	17,000	?

Rings of Neptune

Name	Distance from center of planet (planet radii)	Distance from center of planet (km)	Width (km)	Thickness (km)
Galle	1.692	41,900–43,900	Approx. 2,000	?
Leverrier	2.148	~53,200	Approx. 110	?
Lassell	2.148–2.31	~53,200–57,200	Approx. 4000	?
Arago	2.31	52,200	Less than 100	?
Unnamed (v. faint)	2.501	61,950	?	?
Adams	2.541	62,933	Approx. 50	?

Arcs in Adams ring are called Courage, Liberté, Egalité 1, Egalité 2, and Fraternité.

Appendix 5: The Largest Known Kuiper Belt Objects

Object	Discoverer	Year of discovery	Diameter (km)	Mean distance (AU)	Orbital period (y)	Orbital ecc.	Orbital inc. (degrees)
2003 UB313	M. Brown, C. Trujillo, and D. Rabinowitz	2003	Approx. 2,400	67.7	557	0.44	44.2
Pluto	C. Tombaugh	1930	2,320	39.4	248.54	0.249	17.2
2005 FY9	M. Brown, C. Trujillo, and D. Rabinowitz	2005	Approx. 1,700	45.7	308	0.15	29.0
(90377) Sedna	M. Brown, C. Trujillo, and D. Rabinowitz	2003	Approx. 1,700	502.04	11,249.05	0.849	11.93
2003 EL61	J. Ortiz et al	2003	Approx. 1,600	43.3	285	0.19	28.19
(90482) Orcus	M. Brown, C. Trujillo, and D. Rabinowitz	2004	Approx. 1,500	39.47	247.94	0.218	20.56
(50000) Quaoar	C. Trujillo and M. Brown	2002	Approx. 1,260	43.405	285.97	0.034	7.98
84522	NEAT	2002	Approx. 1,200	55.14	409.16	0.293	35.1
Charon	J. Christy	1978	1,207	Satellite of Pluto			
(20000) Varuna	R. McMillan (Spacewatch)	2000	Approx. 1,060	43.129	283.2	0.051	17.2

N.B. The sizes of most of these objects are very uncertain.

Appendix 6: Lunar and Planetary Firsts

Mercury

First flyby:
Mariner 10 (USA), March 29, 1974

Venus

First successful flyby:
Mariner 2 (USA), December 14, 1962

First impact:
Venera 3 (USSR), March 1, 1966

First successful atmospheric entry:
Venera 4 (USSR), October 18, 1967

First successful soft-landing:
Venera 7 (USSR), December 15, 1970

First surface photos:
Venera 9 (USSR), October 22, 1975

First orbiter:
Venera 9 (USSR), October 22, 1975

First surface color photos:
Venera 13 (USSR), March 1, 1982

First soil analysis:
Venera 13 (USSR), March 1, 1982

First multi-probe atmospheric entry:
Pioneer Venus 2 (USA), December 9, 1978

First balloon flight:
Vega 1 (USSR/France), June 11, 1985

Moon

First impact:
Luna 2 (USSR), September 14, 1959

First flyby:
Luna 3 (USSR), October 6, 1959

First photos of the far side:
Luna 3 (USSR), October 6, 1959

First survivable landing:
Luna 9 (USSR), February 3, 1966

First orbiter:
Luna 10 (USSR), April 3, 1966

First lift-off from the Moon:
Surveyor 6 (USA), November 17, 1967

First robotic sample return mission:
Luna 16 (USSR), September 12–21, 1970

First wheeled vehicle on the Moon:
Lunokhod 1 (USSR), November 17, 1970

First crewed orbital mission:
Apollo 8 (USA), December 21–27, 1968

First human landing:
Apollo 11 (USA), July 20, 1969

Mars

First successful flyby:
Mariner 4 (USA), July 15, 1965

First orbiter:
Mariner 9 (USA), November 14, 1971

First impact:
Mars 2 (USSR), November 27, 1971

First surface photos:
Viking 1 (USA), July 20, 1976

First wheeled vehicle:
Sojourner (USA), July 5, 1997

Asteroids

First flyby (Gaspra):
Galileo (USA), October 29, 1991

First orbiter (Eros):
NEAR Shoemaker (USA), February 14, 2000

First soft landing (Eros):
NEAR Shoemaker (USA), February 12, 2001

Jupiter

First flyby:
Pioneer 10 (USA), December 4, 1973

First atmospheric entry:
Galileo Probe (USA), December 7, 1995

First orbiter:
Galileo Orbiter (USA), December 8, 1995

Saturn

First flyby:
Pioneer 11 (USA), September 1, 1979

First orbiter:
Cassini (USA), June 30, 2004

First landing on Titan:
Huygens (ESA), January 14, 2005

Uranus

First flyby:
Voyager 2 (USA), January 24, 1986

Neptune

First flyby:
Voyager 2 (USA), August 25, 1989

Comets

First flyby (Giacobini-Zinner):
ICE/ISEE-3 (USA), 11 September 1985

First impact (Tempel 1):
Deep Impact (USA), July 4, 2005

Appendix 7: Selected Reading List

General

Deep Space Chronicle: a Chronology of Deep Space and Planetary Probes, 1958-2000 (NASA Monograph in Aerospace History #24), Asif A. Siddiqi, 2002

Beyond the Moon: Golden Age of Planetary Exploration 1971-1978, Robert S. Kraemer, Smithsonian, 2000

Far Travelers: The Exploring Machines, Oran W. Nicks (NASA SP-480), 1985

Exploration of Terrestrial Planets from Spacecraft (2nd Edition), Yuri Surkov, Wiley–Praxis, 1997

Journey Into Space: The First Thirty Years of Space Exploration, Bruce Murray, W. W. Norton & Co., 1989

Journey Beyond Selene, Jeffrey Kluger, Simon & Schuster, 1999

Cosmos, Carl Sagan, MacDonald & Co., 1980

Pale Blue Dot, Carl Sagan, Headline, 1995

The Compact NASA Atlas of the Solar System, Ronald Greeley and Raymond Batson, Cambridge Univ. Press, 2001

Life in the Solar System and Beyond, Barrie Jones, Springer–Praxis, 2004

Exploring the Unknown: Selected Documents in the History of the U.S. Civil Space Program, Volume V, Exploring the Cosmos, John M. Logsdon (ed.) with Amy Paige Snyder, Roger D. Launius, Stephen J. Garber, and Regan Anne Newport (NASA SP-4407), 2001

"Solar System Log," Andrew Wilson, *Jane's*, 1987

Moons & Planets (5th Edition), William K. Hartmann, Thomson, 2005

The Planetary System (3rd Edition), David Morrison and Tobias Owen, Addison Wesley, 2003

The Planets, David McNab and James Younger, Yale Univ. Press, 1999

The New Solar System (4th Edition), J. Kelly Beatty, Carolyn Collins Petersen and Andrew Chaikin, Sky Publishing Corp. / Cambridge Univ. Press, 1999

Uplink–Downlink: A History of the Deep Space Network 1957–1997, Douglas J. Mudgway (NASA SP-2001-4227), 2002

Planetary Geology in the 1980s, Joseph Veverka (NASA SP-467), 1985

"Planetary Oceans," David J. Stephenson, *Sky & Telescope*, November 2002

Mercury

The Voyage of Mariner 10: Mission to Venus and Mercury, James A. Dunne and Eric Burgess (NASA SP-424), 1978

Mercury, F. Vilas, C. R. Chapman, and M.S. Matthews (eds), University of Arizona Press, 1988

Mercury, The Elusive Planet, Robert Strom, Smithsonian Institution Press, 1987

Exploring Mercury, Robert Strom and Ann Sprague, Springer–Praxis, 2003

Atlas of Mercury, Merton E. Davies, Stephen E. Dwornik, Donald E. Gault, and Robert G. Strom (NASA SP-423), 1978.

Flight to Mercury, Bruce Murray and Eric Burgess, Columbia Univ. Press, 1977

"Mercury: The Forgotten Planet," Robert M. Nelson, *Scientific American Special Edition*, 2003

"Mariner 10 Preliminary Science Report," *Science*, 1974

"The Planet Mercury: Mariner 10 Mission," Bruce Murray et. al., *Journal of Geophysical Research*, 1975

"Recalibrated Mariner 10 Color Mosaics: Implications for Mercurian Volcanism," Mark S. Robinson and Paul G. Lucey, *Science*, 1997

"Mercury," Bruce C. Murray, *Scientific American*, September 1975

Venus

Venus Revealed: A New Look Below the Clouds of Our Mysterious Twin Planet, David Harry Grinspoon, Perseus, 1998

The Voyage of Mariner 10: Mission to Venus and Mercury, James A. Dunne and Eric Burgess (NASA SP-424), 1978

Pioneer Venus, Richard Fimmel, Lawrence Colin and Eric Burgess (NASA SP-461), 1983

Atlas of Venus, Peter Cattermole and Patrick Moore, Cambridge Univ. Press, 1997

Magellan: The Unveiling of Venus (JPL-400-345), 1989

The Planet Venus, Mikhail Marov and David Grinspoon, Yale, 1998

"Global Climate Change on Venus," Mark Bullock and David Grinspoon, *Scientific American Special Edition,* 2003

"The Volcanoes and Clouds of Venus," Ronald Prinn, *Scientific American Special, Exploring Space,* 1990.

"Venus," Andrew and Louise Young, *Scientific American,* September 1975

"Overview of VEGA Venus balloon in situ meteorological measurements," R.Z. Sagdeev et al., *Science,* 1986

Moon

Lunar Exploration, Paolo Ulivi, Springer–Praxis, 2004

The Once and Future Moon, Paul D. Spudis, Smithsonian, 1998

Exploring the Moon: The Apollo Expeditions, David M. Harland, Springer–Praxis, 1999

Lunar Impact: A History of Project Ranger, R. Cargill Hall (NASA SP-4210), 1977

Destination Moon: A History of the Lunar Orbiter Program (NASA TM-3487), 1977

The Clementine Atlas of the Moon, Ben Bussey and Paul D. Spudis, Cambridge Univ. Press, 2004

Mapping and Naming the Moon: A History of Lunar Cartography and Nomenclature, Ewen A. Whitaker, Cambridge Univ. Press, 1999

"The Moon," John A. Wood, *Scientific American,* September 1975

Mars

On Mars: Exploration of the Red Planet, 1958–1978, Edward Clinton Ezell and Linda Neuman Ezell (NASA SP-4212), 1984

Mars: The NASA Mission Reports, Robert Godwin (ed.), Apogee, 2000

Mars: The NASA Mission Reports, Vol. 2, Robert Godwin (ed.), Apogee, 2004

Mapping Mars, Oliver Morton, Fourth Estate, 2002

A Traveler's Guide to Mars, William K. Hartmann, Workman, 2003

The Martian Landscape, Viking Lander Imaging Team (NASA SP-425), 1978

Viking Orbiter Views of Mars, Michael Carr et al. (NASA SP-441), 1980

"Uncovering the Secrets of the Red Planet," Paul Raeburn, *National Geographic,* 1998

The Planet Mars: A History of Observation and Discovery, William Sheehan, Univ. of Arizona, 1996

Water on Mars, Michael Carr, Oxford Univ. Press, 1995

"The Climate of Mars," Robert Haberle, *Scientific American Special, Exploring Space,* 1990

"Mars Global Surveyor," *Science,* March 13, 1998

"Opportunity at Meridiani Planum," *Science,* December 3, 2004

"In Search of Martian Seas," Jim Bell, *Sky & Telescope,* March 2005

"The Unearthly Landscapes of Mars," Arden Albee, *Scientific American Special Edition,* 2003

"Mars," James B. Pollack, *Scientific American,* September 1975

Asteroids

Asteroids III, Richard P. Binzel et al. (eds), Univ. of Arizona Press, 2002

Asteroids: A History, Curtis Peebles, Smithsonian, 2000

Impact!: The Threat of Comets and Asteroids, Gerrit L. Verschuur, Oxford Univ. Press, 1997

Rogue Asteroids and Doomsday Comets, Duncan Steel, John Wiley & Sons, 1995

Asteroid Rendezvous: NEAR Shoemaker's Adventures at Eros, Jim Bell and Jacqueline Mitton (eds), Cambridge Univ. Press, 2002

"The Small Planets," Erik Asphaug, *Scientific American Special Edition*, 2003

"The Smaller Bodies of the Solar System," William K. Hartmann, *Scientific American*, September 1975

Jupiter

Jupiter Odyssey: The Story of NASA's Galileo Mission, David M. Harland, Springer–Praxis, 2000

Mission Jupiter: The Spectacular Voyage of the Galileo Spacecraft, Daniel Fischer, Copernicus, 2001

Pioneer: First to Jupiter, Saturn, and Beyond, Richard Fimmel, James van Allen, and Eric Burgess (NASA SP-446), 1980

Voyager Tales: Personal Views of the Grand Tour, David W. Swift, AIAA, 1997

"The Galileo Mission to Jupiter and its Moons," Torrence Johnson, *Scientific American Special Edition*, 2003

"The Hidden Ocean of Europa," Robert Pappalardo, James Head, and Ronald Greeley, *Scientific American Special Edition*, 2003

"Jupiter and Saturn," Andrew Ingersoll, *Scientific American Special, Exploring Space*, 1990

"Jupiter," John H. Wolfe, *Scientific American*, September 1975

Saturn

Passage to a Ringed World, Linda Spilker (ed.) (NASA SP-533), 1997

Mission to Saturn: Cassini and the Huygens Probe, David M. Harland, Springer–Praxis, 2002

Lifting Titan's Veil: Exploring the Giant Moon of Saturn, Ralph Lorenz and Jacqueline Mitton, Cambridge Univ. Press, 2002

"Cassini Reveals Titan," *Science*, May 13, 2005

Pioneer: First to Jupiter, Saturn, and Beyond, Richard Fimmel, James van Allen, and Eric Burgess (NASA SP-446), 1980

Voyager 1 and 2, Atlas of Saturnian Satellites, Raymond Batson (ed.) (NASA SP-474), 1984

Voyager Tales: Personal Views of the Grand Tour, David W. Swift, AIAA, 1997

"Bejewelled Worlds," Joseph Burns, Douglas Hamilton, and Mark Showalter, *Scientific American Special Edition*, 2003

"The Moons of Saturn," Laurence Soderblom and Torrence Johnson, *Scientific American Special, Exploring Space*, 1990

"Jupiter and Saturn," Andrew Ingersoll, *Scientific American Special, Exploring Space*, 1990

"The Outer Planets," Donald Hunten, *Scientific American*, September 1975

Uranus

Planets Beyond: Discovering the Outer Solar System, Mark Littmann, John Wiley, revised edition, 1990.

Voyages to Saturn, David Morrison (NASA SP-451), 1982

Atlas of Uranus, Garry E. Hunt and Patrick Moore, Cambridge Univ. Press, 1989

"Bejewelled Worlds," Joseph Burns, Douglas Hamilton, and Mark Showalter, *Scientific American Special Edition*, 2003

"Engineering Voyager 2's Encounter With Uranus," Richard Laeser, William McLaughlin, and Donna Wolff, *Scientific American Special, Exploring Space*, 1990

"Uranus," Andrew Ingersoll, *Scientific American*, 1987

"Uranus After Voyager," Nigel Henbest, *New Scientist*, July 31, 1986

"The Outer Planets," Donald Hunten, *Scientific American*, September 1975

Voyager Tales: Personal Views of the Grand Tour, David W. Swift, AIAA, 1997

The Planet Uranus: A History of Observation, Theory and Discovery, A.F. O'D. Alexander, Faber & Faber, 1965

"The Outer Planets," Donald Hunten, *Scientific American*, September 1975

Neptune

Neptune: The Planet, Rings and Satellites, Ellis Miner and Randii R. Wessen, Springer–Praxis, 2002

Voyager Tales: Personal Views of the Grand Tour, David W. Swift, AIAA, 1997

The Planet Neptune: An Historical Survey Before Voyager (2nd edition), Patrick Moore, Wiley–Praxis Publishing, 1996

Planets Beyond: Discovering the Outer Solar System, Mark Littmann, John Wiley, revised edition 1990

Atlas of Neptune, Garry E. Hunt and Patrick Moore, Cambridge Univ. Press, 1994

"Bejewelled Worlds," Joseph Burns, Douglas Hamilton, and Mark Showalter, *Scientific American Special Edition*, 2003

"Neptune," June Kinoshita, *Scientific American Special, Exploring Space*, 1990

"The Outer Planets," Donald Hunten, *Scientific American*, September 1975

Pluto and Kuiper Belt

Pluto & Charon, Alan Stern and Jacqueline Mitton, John Wiley & Sons, 1999

"Journey to the Farthest Planet," S. Alan Stern, *Scientific American Special Edition*, 2003

Beyond Pluto: Exploring the Outer Limits of the Solar System, John Davies, Cambridge Univ. Press, 2001

Clyde Tombaugh and the Search for Planet, Margaret K. Wetterer, Carolrhoda Books, 1996

Clyde Tombaugh: Discoverer of Planet Pluto, David H. Levy, Univ. of Arizona Press, 1992

Out of the Darkness: the Planet Pluto, Clyde Tombaugh and Patrick Moore, New American Library, 1981

Exploring the Trans-Neptunian Solar System, Committee on Planetary and Lunar Exploration, National Research Council, 1998

"The Kuiper Belt," Jane Luu and David Jewitt, *Scientific American*, May 1996.

"The Outer Planets," Donald Hunten, *Scientific American*, September 1975

"The 3rd Zone: Exploring the Kuiper Belt," *Sky & Telescope*, November 2003.

Comets

"The Booming Science of Sungrazing Comets," Tony Hoffman and Brian Marsden, *Sky & Telescope*, August 2005

Impact!: The Threat of Comets and Asteroids, Gerrit L. Verschuur, Oxford Univ. Press, 1997

Rogue Asteroids and Doomsday Comets, Duncan Steel, John Wiley & Sons, 1995

Giotto to the Comets, Nigel Calder, Presswork, 1992

Comet, Carl Sagan and Ann Druyan, Guild, 1985

"The Oort Cloud," Paul Weissman, *Scientific American Special Edition*, 2003

Appendix 8: Selected Websites

General

Views of the Solar System (Calvin Hamilton):
http://www.solarviews.com/eng/homepage.htm
The Nine Planets (Bill Arnett):
http://www.nineplanets.org/jupiter.html
NASA Planetary Data System (Geosciences Node):
http://pds-geosciences.wustl.edu/index.htm
Planetary missions (National Space Science Data Center):
http://nssdc.gsfc.nasa.gov/planetary/projects.html
NASA Planetary Photojournal:
http://photojournal.jpl.nasa.gov/index.html
Welcome to the Planets:
http://pds.jpl.nasa.gov/planets/
USGS Gazetteer of Planetary Nomenclature:
http://planetarynames.wr.usgs.gov/
Space Telescope Science Institute:
http://oposite.stsci.edu/
NASA Science Mission Directorate:
http://science.hq.nasa.gov/solar_system/index.html
NASA Discovery programme:
http://discovery.nasa.gov/index.html
Lunar and Planetary Laboratory:
http://www.lpl.arizona.edu/
Russian Space Research Institute (IKI):
http://www.iki.rssi.ru/eng/
Planetary radar observations from Goldstone:
http://wireless.jpl.nasa.gov/RADAR/
Planetary rings:
http://ringmaster.arc/nasa.gov/
Formation of Planets:
http://cfa-www.harvard.edu/COMPLETE/learn/planets/planets.html
New Frontiers in the Solar System: An Integrated Exploration Strategy, Solar System Exploration Survey, National Research Council 2002:
http://www.nap.edu/books/0309084954/html/

Mercury

MESSENGER mission:
http://messenger.jhuapl.edu/
BepiColombo mission:
http://sci.esa.int/home/bepicolombo/index.cfm
Mercury Mariner 10 Image Project:
http://cps.earth.northwestern.edu/merc.html

Venus

Pioneer Venus:
http://nssdc.gsfc.nasa.gov/planetary/pioneer_venus.html
Magellan mission:
http://www2.jpl.nasa.gov/magellan/
Venus Express mission:
http://sci.esa.int/venusexpress

Moon

Digital Lunar Orbiter Photographic Atlas of the Moon, Paul D. Spudis, Lunar and Planetary Institute.
http://www.lpi.usra.edu/research/lunar_orbiter/
Lessons Learned from the Clementine Mission, Committee on Planetary and Lunar Exploration, National Research Council (1997):
http://www.nap.edu/books/0309058392/html/index.html
Clementine (Lawrence Livermore National Laboratory):
http://www-phys.llnl.gov/clementine/
Clementine (Naval Research Laboratory):
http://www.cmf.nrl.navy.mil/clementine/
Lunar Prospector:
http://lunar.arc.nasa.gov/index.htm
Lunar Prospector Spectrometers:
http://lunar.lanl.gov/
Smart-1 (ESA):
http://sci.esa.int/smart-1

Lunar Reconnaissance Orbiter (NASA):
http://lunar.gsfc.nasa.gov/missions.html
Apollo Lunar Surface Journal:
http://www.hq.nasa.gov/alsj/
The Project Apollo Archive:
http://www.apolloarchive.com/

Mars

Mars, by Percival Lowell:
http://www.wanderer.org/references/lowell/Mars/
NASA Center for Mars Exploration:
http://cmex.arc.nasa.gov/
The Quarantine and Certification of Martian Samples
Committee on Planetary and Lunar Exploration,
National Research Council, 2002:
http://www.nap.edu/catalog/10138.html
NASA Mars Exploration:
http://mars.jpl.nasa.gov/
Mars Exploration Rover 2003 mission:
http://mars.jpl.nasa.gov/mer/ and http://athena.cornell.edu/
Mars Express mission:
http://mars.esa.int/
Mars Pathfinder mission:
http://marsprogram.jpl.nasa.gov/MPF
MarsDaily:
http://www.marsdaily.com/
Explore Mars:
http://www.astrodigital.org/mars
Malin Space Science Systems:
http://www.msss.com/

Asteroids

Minor Planet Centre:
http://cfa-www.harvard.edu/iau/mpc.html
Exploration of Near Earth Objects, Committee on Planetary
and Lunar Exploration, National Research Council
(1998)
http://www.nap.edu/catalog/6106.html
NEAR–Shoemaker mission:
http://near.jhuapl.edu/

Dawn mission:
http://www-ssc.igpp.ucla.edu/dawn/
US Naval Observatory Ephemerides of the Largest
Asteroids:
*http://aa.usno.navy.mil/ephemerides/asteroid/astr_alm/
asteroid_ephemerides*

Jupiter

Voyager mission:
http://voyager.jpl.nasa.gov
Voyager 25th anniversary:
http://www.planetary.org/voyager25/index.html
Galileo:
http://www2.jpl.nasa.gov/galileo/
Galileo Imaging Team:
http://www2.jpl.nasa.gov/galileo/sepo/
Satellites:
http://www.ifa.hawaii.edu/~sheppard/satellites/
Ulysses mission:
http://ulysses.jpl.nasa.gov/

Saturn

Voyager mission:
http://voyager.jpl.nasa.gov
Voyager 25th anniversary:
http://www.planetary.org/voyager25/index.html
Cassini mission:
http://saturn.jpl.nasa.gov
Huygens mission:
http://sci.esa.int/huygens
Saturn's rings:
http://science.nasa.gov/headlines/y2002/12feb_rings.htm
History of Saturn's rings:
http://es.rice.edu/ES/humsoc/Galileo/Things/saturn.html
Saturn ring plane crossings:
http://oposite.stsci.edu/pubinfo/SaturnRPC.html
Satellites:
http://www.ifa.hawaii.edu/~sheppard/satellites/

Systema Saturnium, Christiaan Huygens, Smithsonian Institution Libraries:
http://www.sil.si.edu/DigitalCollections/HST/Huygens/huygens.htm

Uranus

Voyager mission:
http://voyager.jpl.nasa.gov
Voyager 25th anniversary:
http://www.planetary.org/voyager25/index.html
Satellites:
http://www.ifa.hawaii.edu/~sheppard/satellites/
Irregular Satellites (Brett Gladman):
http://www.astro.ubc.ca/people/gladman/urhome.html

Neptune

Voyager mission:
http://voyager.jpl.nasa.gov
Voyager 25th anniversary:
http://www.planetary.org/voyager25/index.html
Satellites:
http://www.ifa.hawaii.edu/~sheppard/satellites/

Pluto and Kuiper Belt

The Pluto Portal:
http://www.plutoportal.net/
New Horizons mission:
http://pluto.jhuapl.edu
Distant EKOs—The Kuiper Belt Electronic Newsletter:
http://www.boulder.swri.edu/ekonews/

2003 UB313, the 10th Planet:
http://www.gps.caltech.edu/~mbrown/planetlila/index.html
Great information about Pluto (Lowell Observatory):
http://www.lowell.edu/users/buie/pluto/pluto.html
List of Trans-Neptunian Objects:
http://cfa-www.harvard.edu/cfa/ps/lists/TNOs.html

Comets

Comets (D. Jewitt):
http://www.ifa.hawaii.edu/faculty/jewitt/comet.html
Deep Space 1 mission:
http://nmp.jpl.nasa.gov/ds1/
Deep Impact mission:
http://deepimpact.umd.edu/home/index.html
Rosetta mission:
http://sci.esa.int/rosetta
Stardust mission:
http://stardust.jpl.nasa.gov
Deep Impact mission:
http://deepimpact.jpl.nasa.gov/
Giotto mission:
http://sci.esa.int/giotto
Sedna:
http://www.gps.caltech.edu/%7Embrown/sedna/
SOHO's comets:
http://sohowww.estec.esa.nl/
JPL comet gallery:
http://encke.jpl.nasa.gov/
Comet Shoemaker–Levy 9:
http://www2.jpl.nasa.gov/sl9

INDEX